Cosmic Roots
The Conflict Between Science and Religion and How it Led to the Secular Age

Cosmic Roots
The Conflict Between Science and Religion and How it Led to the Secular Age

Ira Mark Egdall

 World Scientific

NEW JERSEY · LONDON · SINGAPORE · BEIJING · SHANGHAI · HONG KONG · TAIPEI · CHENNAI · TOKYO

Published by

World Scientific Publishing Co. Pte. Ltd.
5 Toh Tuck Link, Singapore 596224
USA office: 27 Warren Street, Suite 401-402, Hackensack, NJ 07601
UK office: 57 Shelton Street, Covent Garden, London WC2H 9HE

British Library Cataloguing-in-Publication Data
A catalogue record for this book is available from the British Library.

COSMIC ROOTS
The Conflict Between Science and Religion and How it Led to the Secular Age

Copyright © 2023 by Ira Mark Egdall

All rights reserved.

ISBN 978-981-125-138-2 (hardcover)
ISBN 978-981-125-247-1 (paperback)
ISBN 978-981-125-139-9 (ebook for institutions)
ISBN 978-981-125-140-5 (ebook for individuals)

For any available supplementary material, please visit
https://www.worldscientific.com/worldscibooks/10.1142/12699#t=suppl

Desk Editor: Rhaimie Wahap

Typeset by Stallion Press
Email: enquiries@stallionpress.com

To Pat, my wife and sweetheart

CONTENTS

Author's Note	xi
Introduction	xv

Part I	**Gods and the One God**	**2**
Chapter 1	The Sag-giga	3
Chapter 2	From Priests to Kings	21
Chapter 3	In the Beginning	29
Chapter 4	One Nation Under God	49

Part II	**Lovers of Wisdom**	**62**
Chapter 5	The Greeks	63
Chapter 6	The Great Natural Philosopher	81
Chapter 7	The City of the Mind	91
Chapter 8	The Great Astronomer	101

Part III	**Fall and Rise**	**112**
Chapter 9	Zeus Versus *YHWH*	113
Chapter 10	Found and Lost	131

Chapter 11	He Has Risen	139
Chapter 12	Jerusalem in Flames	159
Chapter 13	The Thirteenth Apostle	177
Chapter 14	The Blossoming	193
Chapter 15	The Age of Martyrs	207
Chapter 16	By This Conquer	219

Part IV Rebirth — **236**

Chapter 17	The Prophet	237
Chapter 18	The Medieval Universe	251
Chapter 19	The Reluctant Revolutionary	261
Chapter 20	The Observer	285
Chapter 21	The Visionary	299
Chapter 22	The Michelangelo of Science	321
Chapter 23	The Two Chief World Systems	335

Part V Enlightenment — **352**

Chapter 24	The Great One	353
Chapter 25	The Heavens and the Earth	371
Chapter 26	Reason	393
Chapter 27	Two Giants	403

Part VI Life — **424**

Chapter 28	Reluctant Revolutionary Redux	425
Chapter 29	The Origin of Species	445
Chapter 30	The Law of Evolution	463
Chapter 31	The Secular Universe	479

Acknowledgements	487
About the Author	489
Also by Ira Mark Egdall	491
Supplemental Material Online	493

AUTHOR'S NOTE

When I was twelve years old, I wanted to be a Rabbi. I took my religious education very seriously and was Bar Mitzvah'ed the following year. But I am, by nature, a doubting Thomas.

My Hebrew School teacher told the Bible story of how Jonah was swallowed by a whale. "A whale? Really?" I asked. "And Jonah lived?"

"Sit down and be quiet," she said.

That began my personal conflict with religion.

My loving mother was an Ashkenazi Jew — Conservative in belief but lukewarm in practice. She told me that Conservatives were not too strict like the Orthodox and not too liberal like the Reform. Kind of like the Goldilocks of Judaism, I thought.

Her religious feelings were associated with family, home, and her strong sense of being Jewish. This was seared into her psych by childhood experiences in intensely anti-Semitic early 20th century Poland. Nothing makes one feel their ethnicity more than bigotry from an oppressive majority. She left at age sixteen with her family and emigrated to the United States in 1927 — missing the NAZI invasion and Holocaust by some twelve years.

My gentle Dad was also an Ashkenazi Jew. He was American-born and an avowed agnostic like his father before him. He tried to teach me to respect all religious points of view. He felt it was beyond the ability of humans to know for sure whether God exists or not. He refused to practice any religious customs. My mother, dear sister, and I would eat Matzo on Passover. My Dad ate a ham sandwich.

In Hebrew school, I learned of the long and bitter history of persecution of Jews by Christians. I was shocked when I learned that Jesus was Jewish. This made no sense to me.

We were the only Jewish family in an Irish/Italian neighborhood in a working-class suburb of Boston. While I attended Hebrew school, my friends learned about Jesus in CCD classes at the Immaculate Conception Parish Church.

One hot summer night hanging out on the street corner, we discussed our different beliefs. My friends had been taught that only Catholics go to heaven. I was taught that only Jews were God's chosen people.

This "us versus them" aspect of organized religion turned me off.

At around age 15, I had an *epiphany*. I realized I believed in my religion because of my upbringing. My friends believed in their religion because that is how they were brought up. There was no proof either was true.

I began to drift away from the religion of my ancestors towards my father's agnosticism. God became, for me, more of a question than an answer.

I turned to Science, not just as a subject of study and way to make a living, but to find "the Truth" of the universe; its fundamental laws, its creation and formation, and how it came to be what it is today. I loved that, unlike religion, science welcomes skepticism. It insists on it.

As I struggled to understand physics (and still do), the great mystery of uncertainty in quantum mechanics and concept of space and time as a single entity in relativity held me spellbound. It seemed to me that nature possessed a mysticism beyond religion.

At the urging of a work colleague, a devout Christian, I read the four Gospels of the New Testament. I was deeply moved by the compassionate and inspirational words of Jesus. I was shocked by the anti-Semitism.

In later years I have found myself thinking more and more about man's spiritual nature. I have come to appreciate the common thread of ethical monotheism at the heart of Judaism, Christianity, and Islam. For me, their shared vision of one universal God of Righteousness binds the three beliefs into a single religious concept.

If you are a true believer, I congratulate you. For you have the gift of faith. Even so, maybe there are times when you are alone and think about the horrible things that happen in this world. Perhaps you find yourself questioning the existence of God. If you are an affirmed atheist, possibly there are times of great fear when you find you cannot help but whisper a plea to God for help. If you are an agnostic, maybe you find times when you swing one way or the other.

As for me, lately I seem to be in two states at once — like a quantum wave/particle. I alternate between atheist and believer.

This book is a reflection of that internal clash between science and religion which goes on in my own mind.

INTRODUCTION

The sky calls to us.

— Carl Sagan

Over luminous days and crystalline nights, ancient humans watched the Sun, Moon, stars and planets wheel around the sky. The heavenly bodies were gods. Their gift was time. Telling mortals when to sleep, to wake, to sow, to reap.

Thus began a quest to predict the motions of celestial objects — one which would come to throw man out from his vaulted place at the center of the universe. It was an epic journey, covering some 150 generations and nearly 5000 years — punctuated by brilliant leaps of insight, at times against great resistance by religious authorities. From this emerged scientific achievements which stand as testimony to the power and imagination of the human mind.

This book is about that quest. About science. About religion. About their entwined history of conflict and insights which has come to define and shape our modern worldview.

Along the way, we trace the rise and fall of empires, of civilizations, cycles of war and peace, unification and division — and the evolution of religious belief from many gods to one God to doubt to no God (for some) and secularization.

My goal is a book which is comprehensive, entertaining, and enlightening — one which presents the findings of scientists and historical scholars in every-day language.

The emphasis is on science, particularly cosmology and its sister science astronomy. For it was discoveries in these great disciplines which first led to the conflict between science and religion.

Our story focuses on Western civilization. "Science is now international," physicist Steven Weinberg points out. However, the clash between science and religion has mostly occurred "in what may be loosely called the West."[1]

The conflict is an ancient one. In the fifth century BC, Greek philosopher Anaxagoras "was forced to flee Athens for teaching that the Sun is not a god but a hot stone larger than the Peloponnesus."[2]

Religion, it is said, is man's interpretation of God's will. Thus, it is inherently imperfect. We approach religion from an evidence-based perspective. Historical facts and archeological discoveries are emphasized, as well as what we do not know. Results diverge markedly from traditional religious beliefs.

I try to present religious figures as human, with desires, doubts, hopes and fears. To me, seeing them in this light makes their accomplishments all the more remarkable. In the process, we see the insidious influence of power. It corrupts all but the few — religious and irreligious alike.

The story is told in six parts:

Part I — Gods and the One God

In Part I, we watch the stars atop the great Ziggurat of Ur. We uncover the archeology-based history of the ancient Hebrews. We explore their flat-Earth cosmology and fifteen-century passage from paganism to ethical monotheism.

Part II — Lovers of Wisdom

In Part II, we contemplate the virtuous life with Socrates, ponder the Divine Craftsman with Plato, and explore Aristotle's great interlocking crystalline spheres — the first integrated geometric model of the universe.

We visit ancient Alexandria, renowned "city of the mind." Here the great astronomer Ptolemy constructed his circles within circles model of the heavens — its accuracy unchallenged for over a millennium.

Part III — Fall and Rise

In Part III, we ride with rebel commander Judas the Maccabee (the Hammer) into the mountain passes of Judea. We read from the apocalyptic Book of Daniel and the coming of the enigmatic Son of Man.

We hear John the Baptist preach repentance in the Judean wilderness. We ride with Jesus on a donkey colt into Jerusalem and into history. We hear the compassion and transformative power of His words — and examine Gospels writers' motives for anti-Semitism and the sugar-coating of Rome's brutality.

We walk with ecumenical apostle Paul on the fabled road to Damascus and trace the remarkable rise of Christianity across the Roman Empire. We ponder the conversion of Constantine the Great and militarization of Christianity.

Part IV — Rebirth

Part IV begins with the fall of Rome and rise of Islam. We enter Mecca with victorious Prophet Muhammad, and see the deep links between Islam, Judaism and Christianity. We explore the science of the Golden Age of Islam and the great astronomers of the Marāgha School.

We ponder Copernicus as he develops his radical Sun-centered model of the solar system. We learn of its Islamic roots and later conflicts with the Catholic Church.

We read letters exchanged between the devout Johannes Kepler and Galileo Galilei. We kneel with Galileo at the Inquisition as he recants his support of Copernicanism.

Part V — Enlightenment

Part V begins with troubled farm-boy Isaac Newton and his work on calculus and gravity during the plague years. We trace his path to the laws of motion and law of universal gravitation. Was the deeply religious Newton a secret anti-Trinitarian?

We hear Enlightenment thinkers Locke, Hume, Rousseau, the provocative Voltaire, Kant, and Wollstonecraft challenge the political and religious mores of their day.

We meet Pierre-Simon Laplace and his seminal work on the stability of the solar system. His mathematics implied no need for God — centuries before similar claims by physicists Stephen Hawking and Lawrence Krauss.

Part VI — Life

We sail with young Charles Darwin on his voyage around the world in Part VI. We learn how he came to develop his theory of evolution. We hear Darwin explain his radical theory in his own words.

In the last chapter, we present a brief overview of quantum mechanics and Einstein's theories of relativity — and how they further advanced the tenets of modern secular culture.

For me, writing this book has been a joy. I have learned so much in the process. My hope is that you will come to feel the same way.

We begin with what historians generally consider the first human civilization.

PART I
Gods and the One God

Chapter 1

THE SAG-GIGA

> O sweet spontaneous
> earth how often have
> the
> doting
>
>   fingers of
> prurient philosophers pinched
> and
> poked
>
> thee,
> has the naughty thumb
> of science prodded
> thy
>
>   beauty   how
> often have religions taken
> thee upon their scraggy knees
> squeezing and
>
> buffeting thee that thou mightiest conceive
> gods
>
>   (but . . .
>
>   thou answerest
>
> them only with
>
> spring)
>
> — E. E. Cummings[1]

To the nearly rainless land of scorching summers they came. To till the soil and plant the crops. And pray to the gods for life-giving water.

Will the spring floods come? Will the rivers rise and turn the arid soil and marshlands green? Will the harvest sustain us through the searing summer? Will fall rains renew the land? Will there be feast or famine, life or death?

At the dawn of the Bronze Age they came — to the alluvial valley along the southern banks of the Tigris and Euphrates rivers. Their land was called *Ki-en-gir* or "place of the noble lords." Here they settled in c. 4000 BC.[2]

They spoke a strange language. No relation to any other human language has been found. A sample of their beautiful-sounding language is given below[3]:

en-líl-li ì-du nin-líl in-uš
nu-nam-nir ì-du ki-sikil mu-un- ...

English translation:
Enlil walked, Ninlil followed.
Nunamnir walked, the maid followed . . .

— Excerpt from Plate XI, *"Enlil and Ninlil: The Begetting of Nanna."*[4]

Where did these people come from? Some say from Eastern Arabia or Northern Mesopotamia. Others suggest the Caucasus or the Iranian highlands. Or perhaps India or somewhere else in Central Asia. Nobody knows for sure.[5]

They called themselves the *Sag-giga* or "black-headed people." We call them Sumerians.

The First Civilization

The Sumerians settled in the flood plain north of the Persian Gulf in Mesopotamia — now modern-day southern Iraq and Kuwait. Here they founded what is widely recognized as the earliest human civilization.* (The word "Mesopotamia" is Greek for "land between the rivers.")

---

*Anthropologists generally define a "civilization" as a culture with a written language. Urban settlements, social hierarchy, government, and highly developed architecture are also considered typical traits. Source: "Civilizations" nationalgeographic.org. Mar. 4, 2019. Retrieved Aug 13, 2020.

Fed by snow melt from the Taurus mountains of eastern Turkey, the Tigris and Euphrates rivers frequently flooded in late Spring. Torrents of silt-laden water spilled over shallow banks onto the surrounding plain — depositing a thick layer of nutrient-rich soil over miles of parched Mesopotamian land.[6]

The Sumerians built drainage and irrigation canals, dikes, and ditches to control the wild river floods and distribute the life-giving waters. Over time, they developed a massive irrigation system. It transformed the land of sand banks and swamps into a fertile valley ideal for planting crops.[7]

This general abundance allowed people to settle in one place. It fed an ever-growing population. Surplus food freed at least the upper classes from working in the fields and other labor-intensive activities — giving them time to think, imagine, invent, and write.[8]

With that "free time," the Sumerians achieved extraordinary things.

They developed what is believed to be humankind's oldest writing system. They created the first city-states, built monumental architecture, and produced striking works of art.

The "black-headed people" generated the "first known codified legal and administrative systems — including courts, jails, and government records." They built the first formal schools and aquariums. They invented the wheel, the plow, sail boats, glass, carpenter tools, and more.

These ingenious people also developed leather; bronze metallurgy; the first multi-story buildings; and the column, arch, and vault. They established sophisticated business accounting methods and pioneered advances in mathematics.[9]

The first major trading networks and earliest international treaties arose from the Sumerians. Their comprehensive irrigation systems "ushered in the age of intensive agriculture." They practiced the cultivation of "emmer, wheat, barley, sheep (starting as mouflon)" and raised "cattle (starting as aurochs)" for the first time on a grand scale.[10]

The Land

> *O Sumer, great land, of the lands of the universe,*
> *Filled with steadfast brightness, the people from sunrise*
> *to sunset obedient to the divine decrees ...*
> *Thy ... is like heaven, untouchable.*
>
> — Sumerian Tablet: *Enki Decrees the Fate of Sumer*[11]

6   *Cosmic Roots*

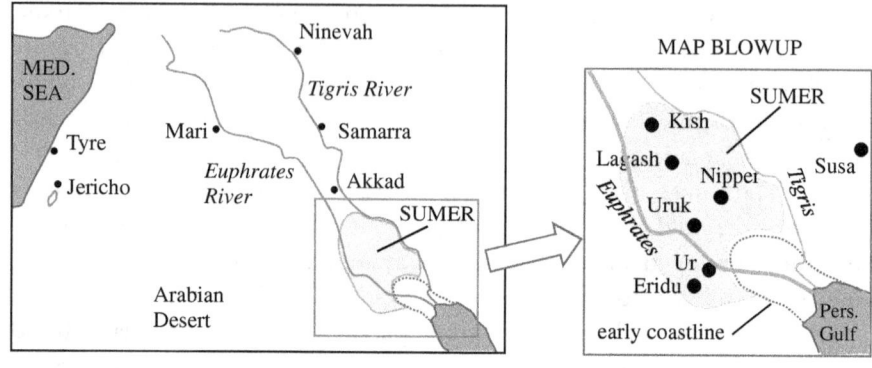

Sumer and the ancient Near East.   The City-states of Sumer.

Figure 1.1. Sumer. The ancient Near East is shown on the left. Major city-states of Sumer are shown on the right blowup. The coastline of the Persian Gulf (dashed lines) has receded since the days of Sumer.

The land of "Sumer," as it is called, extended from near present-day Baghdad in the north to the Persian Gulf in the south. It was bounded to the east by the Arabian desert and the Iranian Plateau to the west. (See Fig. 1.1.)

Early towns and villages grew to include some twenty city-states. Chief among them were Eridu, Larsa, Uruk, Kish, Ur, Lagash, Umma, Adab, Isin, and the holy city of Nippur. All Sumerian city-states resided near the Tigris or Euphrates rivers or one of their tributaries.[12]

Of the Sumerians' many inventions, arguably their greatest was their writing system:

The Written Word

With blunt reeds, Sumerian scribes marked picture symbols on wet clay tablets. They then set them aside to dry and harden in the hot Mesopotamian Sun. Over time, these picture symbols became wedge-shaped signs — each representing a syllable.[13] (See Fig. 1.2.)

It seems once they began to write, they could not stop. Sumerians inscribed their "cuneiform" script (from the Latin *cuneus* for wedge) on "stone and clay tablets, bricks, door sockets, bowls and vases, mortars and mace heads, plaques and monuments, statues and statuettes."[14] (See Fig. 1.3.)

Scribes trained to read and write recorded the deeds and thoughts of the Sumerians. There were "junior scribes, high scribes, and temple scribes." Highly specialized administrative scribes kept business and legal

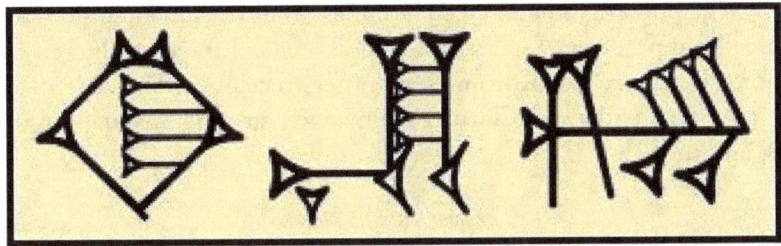

Figure 1.2. The Word Ki-en-gi(r) in Sumerian Script. Believed to be the Akkadian word for Sumer, the three symbols represent three syllables which mean "place, lords, noble" or "place of the noble kings."

Figure 1.3. Sumerian Tablet. Inscription in monumental archaic style, c. 26th century BC.

records. "Royal scribes documented correspondence between kings and their officials."[15]

Writings ranged from the sublime to the mundane. It began with the recording of commercial transactions. Scribes recorded business partnerships and dealings, debt/payment certificates, messages between merchants and trade agents abroad, and "inventories of all types."

To identify ownership of goods stored or shipped, the Sumerians invented "cylinder seals." These small stone cylinders were little works of art, often engraved with religious or mythological designs. Sumerians rolled them onto wet clay to record their signature and mark business transactions.[16]

The Sumerians generated law codes. They documented subpoenas, lawsuits, legal precedents, marriage and divorce contracts, wills, and court decisions. They recorded promissory notes, receipts, and deeds of sale. They also wrote the first farmer's almanac, farming manuals, musical scales, and more.[17]

The poor have no power.

— Sumerian proverb.[18]

Their rich literary works included moral maxims, precepts, proverbs, and sayings. The Sumerians generated epic stories of gods and heroes. They wrote poems, prayers, letters, and songs of love. They composed "essays long and short," as well as disputations, lamentations, elegies and "eloquent funeral chants."[19]

Also recorded in cuneiform script were incantations, divine hymns, votive inscriptions, cosmological myths, songs of victory, and lists of kings. They wrote military summaries, treaty documents, records on social reforms, textbooks, dictionaries, and mathematical treatises.[20]

The inscribed clay tablets of the Sumerians left a permanent record of human thought recorded for the first time — to be read a moment later, a day later, a season later — or five thousand years later by modern archeologists who marvel at how contemporary-sounding are the words of man's first known civilization.

Beneath the Sands

Serendipity — a discovery unlooked for.

Evidence of Sumer lay hidden beneath the shifting sands and shapeless mounds of Mesopotamia for some 5000 years. "Nothing was known about the Sumerians, not even the fact that they existed" until the mid-nineteenth century AD, renowned scholar Samuel Noah Kramer tells us.[21]

In the 1840s, French, German, and English archeologists searched in northern Mesopotamia for the remnants of Akkadian, Babylonian, and Assyrian civilizations. Among the ruins were the remains of unknown buildings, strange pottery, and baked clay tablets with mysterious signs — puzzling artifacts which could not be identified.[22]

At around the same time, British archeologist Austen Henry Layard was examining the Kuyunjik mound just outside present-day Mosul in northern Iraq. Here he unearthed the remains of ancient Nineveh — capital of Assyria and largest city in the world in c. 700 BC.

A few years later, his assistant at Kuyunjik, Iraqi archeologist Hormuzd Rassam excavated the great palace complex of Assyrian King Ashurbanipal (668–627 BC). Rassam and his team uncovered thousands of clay tablets in the famed palace library. One in particular contained "alphabets, grammars, and vocabularies" in *two languages*: Assyrian and some yet to be unidentified ancient tongue.[23]

Irish Assyriologist and clergyman Edward Hincks studied this curious second language. He found consonant and vowel characteristics which were not present in Assyrian or other Semitic languages. He could not trace most "syllabic values of signs to Semitic words." He began to suspect this odd tongue was not Semitic.[24]

British major-general and orientalist Henry Rawlinson also puzzled over this unrecognized language. He too concluded it was not Semitic. He declared in 1853 that it was in fact "an entirely new and hitherto unknown language."[25]

The work of Hincks and Rawlinson was the highlight of their professional lives. Here was the discovery of a new written language — one apparently more ancient than all others in recorded history. The question remained: who were the people who spoke it and where did they come from?

Some 16 years later, French-Jewish Assyriologist Julius Oppert translated a common Akkadian inscription found on a number of ancient tablets. It read: "King of Sumer and Akkad." From this he proposed "these (hitherto unknown) people and their language should be called *Sumerian*."[26]

The search for the Sumerians was on. From 1877 to 1900, French and Americans archeologists excavated the ruins of ancient Lagash and Nippur in southern Mesopotamia. They unearthed statues, steles, and well over 30,000 tablets and tablet fragments.

What had been suspected was confirmed. An ancient civilization — the oldest yet discovered — had thrived in the southern valley of the Tigris and Euphrates rivers.

The 20th century gave even stronger confirmation. Excavated sites revealed a treasure trove of Sumerian culture. They included walled villages and cities; monumental sculptures and soaring temples; glorious royal palaces; ancient tombs; and statues of gods and kings.

Also unearthed were bronze tools; precious objects of gold and silver, headdresses of gold, lapis lazuli and carnelian; as well as exquisitely carved ivories. Joyous archeologists dusted off thousands of years of sand from alabaster vases; stone, metal, and terra cotta artifacts; and "fantastically worked heads of bulls, harps, and lyres, sledges and chariots." (See Fig. 1.4.)[27]

In 1927, famed British archeologist Leonard Woolley excavated sixteen royal graves at the ancient Sumerian city of Ur. The site dating to 3500 BC revealed the skeletal remains of "six to eighty bodies entombed with each king."

No evidence of violence was found — only gold or clay cups next to the bodies. It appears those who accompanied the king died willingly of mass poisoning.[28] By the time of king *Ur-Nammu* c. 2050 BC, this custom was no longer practiced.[29]

Figure 1.4. Sumerian Artifacts. From left to right: TOP ROW — Mask of Warka 3200–3000 BC; Enthroned King of Ur with Attendants c. 2600 BC; Bull's head of Queen's lyre from Pu-abi's grave; MIDDLE ROW — Naked priest offering libations to Sumerian temple, Ur, 2500 BC; Diorite statue of Gudea, prince of Lagash; Pottery jar, late Ubiad period. BOTTOM ROW — Queen Pu-abi's jewelry (recent reconstruction); Master of animals' motif in a panel of harp soundboard; Foundation figure of Ur-Nammu holding basket.

Thousands upon thousands of Sumerian cuneiform tablets have since been unearthed — the great majority broken or fragmentary. Fortunately, ancient scribes commonly produced more than one copy. And they generated a number of texts in both Sumerian and Akkadian; a godsend to translators.[30]

From ruins and artifacts, translations of cuneiform tablets, and scientific dating methods, archeologists and scholars have reconstructed the ancient history, literature, religion, and cosmology of Sumer.[31]

It has been a difficult, painstaking task — beset by limited data in some cases, disturbed artifacts in others, and blurred by "the obscuring haze of Millennia," as American author and rabbi Chaim Potok put it.[32]

Let's take a look at the Sumerian creation story, the earliest for which we have written records.

Creation

> *After An had carried off heaven,*
> *After Enlil had carried off earth,*
> *After Ereshkigal had been carried off into*
> *Kur (the netherworld) as its prize ...*
>
> — The Epic of Gilgamesh[33]

In the beginning there was the sea-goddess *Nammu* and the boundless Primordial Sea.

The mother of all gave birth to *An*, the god of heaven and *Ki* the goddess of earth united.

An and *Ki* in turn begat the air-god *Enlil*, who separated *An* and *Ki*, the heavens from the earth.

An carried off the heavens. *Enlil* himself carried off his mother *Ki*, the earth — and let water fall from the sky.[34]

Enlil then begat the moon-god *Nanna* to give light to the darkness.

Nanna's wife *Ningal*, the "Great Lady," spawned the sun-god *Utu*. It is *Utu* "who lights the world with rays issuing from his shoulders."[35]

Earth-goddess *Ki* then gave birth to the planets.[36]

Then the great air-god *Enlil* "brought forth seed from the earth" — life both vegetable and animal.

Finally, *Enki*, god of wisdom, and his half-sister *Ninki* put a god to death, mixed his body with clay, and created man — to "bear the labor of the gods."³⁷

> *In the clay, god and man*
> *Shall be bound,*
> *To a unity brought together;*

— Sumerian Tablet at Nippur³⁸

Enlil "took pity on humans" and established the *Hegel* — abundance and prosperity in the land. He then organized the earth and established law and order.³⁹

So goes the creation story of the Sumerians.* Now let's take a look at the first recorded cosmology in human history:

<u>Sumerian Cosmology</u>

The dark-headed people believed the universe is a great closed dome — surrounded by *Nammu*, the saltwater Primeval Sea. (See Fig. 1.5.)⁴⁰

Contained within this dome are the heavens *An* and the earth *Ki*: The earth is stationary and flat — a square of finite thickness with four corners. It floats on *Nammu*, the Primeval Sea and forms the base of the closed dome. The dome or firmament in turn rests on the edges of the earth.⁴¹

Above the earth is *Enlil*, the air — filled with *lil* or atmosphere.

Above the air are three heavens:

The first heaven contains the moon *Nanna*, Sun *Utu*, and the stars — which are also made of *lil* but luminescent.

The heavenly abode of the gods lies in the second heaven.

The third heaven is highest of all — the residence of the heaven-god *An*.⁴²

In the first heaven, the Sun-god *Utu* arises each morning from behind the mountains of the East (the Zagros mountains of Iran). He rides the Sun in a chariot across the sky and sets West under the earth in the evening.

---

*There is no extant tablet which tells the Sumerian creation story in its entirety. The brief chronicle here is culled from translations and interpretations of a number of Sumerian myths, epic tales, and hymns. Primary source: Kramer, *History Begins at Sumer*, Chapter 13.

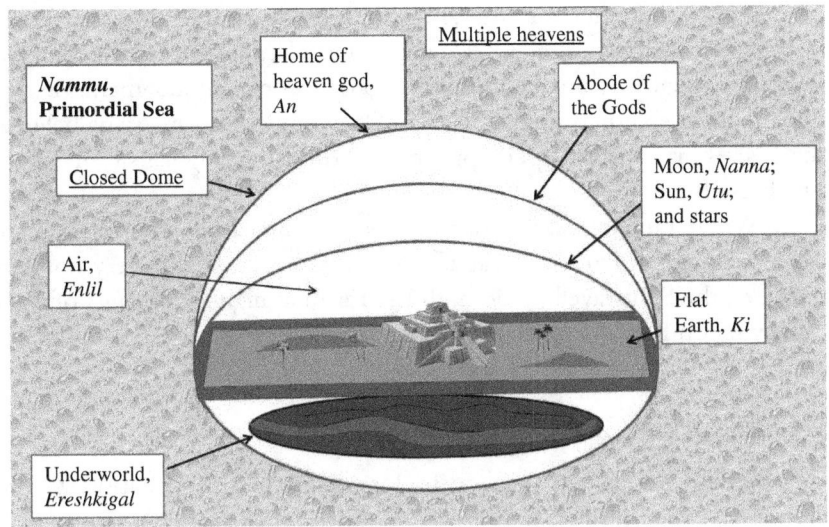

Figure 1.5.   The Sumerian Cosmos.

Here the moon God *Nanna* "sails the moon from east to west across the sky at night" in a *gufa* (a round wicker boat). He brings "light to the pitch-black lapis lazuli sky."[43]

Beneath the earth is the foreboding Netherworld.

The Great Below

O Father Enlil, my pukku fell into the nether world,
My mikku fell into the nether world;
I sent Enkidu to bring them up to me, the nether world has seized him*

— *Epic of Gilgamesh*, Sumerian Tablet[44]

Just beneath the earth lies the abyss ruled by wisdom and water-god *Enki*. Also here is the subterranean sea — the *Apsû*. Sumerians believed the *Apsû* provided "the fresh waters of the Tigris and Euphrates."[45]

Beneath the *Apsû* is the Land of No Return — the dim Netherworld or *Kur*. It is ruled by *Queen Ereshkigal*, goddess of death. She and her husband *King Nergal* reside here in their temple of lapis lazuli.[46]

---

*Pukku and *mikku*. Translation uncertain, "perhaps a drum and drumstick . . . or ball and stick." Source: Webster, Michael. "The *Huluppu*-Tree / Gilgamesh, Enkidu, and the Netherworld" World Mythology, Grand Valley State University, gvsu.edu. http://faculty.gvsu.edu/webstern/Huluppu2.htm. Retrieved Oct 7, 2018.

Kur is where humans are said to go after death. The dead are ferried by a boatman across the river *Hubur*. On the other side waits the Sun-god *Utu* to pass judgment on their souls. Though a positive judgment yields a happy life in the Netherworld, it is an unlikely outcome. Sumerians believed humans are born sinful — perhaps a precursor to the original sin of Eve in the Hebrew Bible.[47]

Where humans dwell is dark, dusty, and unpleasant. People are made to go naked with only dust to eat and no water to drink. It is very hot here due to *Utu*, the Sun travelling beneath the Earth at night. Offerings of food by the living on Earth can ameliorate the situation.[48]

Cosmological Considerations

The Sumerians concocted their cosmology with the innate curiosity, vivid imagination, and a spiritual yearning endemic to our species. Their construct was based on their environment.

The vast plains of Southern Mesopotamia did suggest a flat earth. The lack of perceived motion did imply a stationary earth (more in this in later chapters). Water did bubble up when one dug in the soil of the Tigris/Euphrates River valley — hence the waters of the *Apsû* below.

The Sun does appear to rise in the east, travel across the sky, set in the west, and return the next morning. The night sky does appear like a great dome with all stars the same distance from earth. (The separation between our eyes is way too small to perceive depth at star distances.)

And what of our modern cosmological beliefs? Are they not based on our "environment" — extended many light-years in distance by telescopes and other modern instruments? Are not our beliefs founded on observation coupled with imagination?

Modern science conflicts with literal religious views. The age of the universe is but one example. This dichotomy did not exist in ancient Sumer.

To Sumerians, religion and science — their gods and their cosmology if you will — were the same thing. *An* was both the visible heavens and the invisible heaven-god within it, as Chaim Potok points out. *Ki* was both the earth and the invisible god within. And so on for all the gods.[49]

There was no conflict.

Goddesses and Gods Galore

May you find favor before your god ...

— Excerpt from Sumerian essay[50]

There were many hundreds of deities in the Sumerian pantheon.⁵¹ They ranged from divine beings for love and war to deities for ditches and sheep. My favorite is *Ninkasi*, goddess of beer — the "lady who fills the mouth."⁵²

There were also demi-gods and heroes, spirits and demons, and legendary beasts. These included scorpion-man, as well as a monster with a lion's face and dragon's teeth — and scariest of all to a writer, a female demon known as "she who erases."⁵³ (See Fig. 1.6.)

Every Sumerian had a personal god — a kind of guardian angel who served as their "representative in the assembly of gods." ⁵⁴

And there was the judge-god *Sataran*. He was in charge of settling complaints. Please contact him if you have issues with this book.⁵⁵

What were the gods like? They were anthropomorphic — they looked and acted like us. They "ate, drank, married, made love, slept, and had children." They "quarreled, felt jealousy, hatred, and rage," and fought amongst themselves.⁵⁶

Unlike humans, the gods were immortal. If "wounded and killed, they somehow returned to life. If ill to the point of death, they always recovered." Where do I sign up to be a god?⁵⁷

Figure 1.6.   Enkidu and the epic hero and demi-god Gilgamesh. The goddess of creation formed Enkidu "to rid Gilgamesh of his arrogance."⁵⁴

> *Enlil, whose command is far-reaching, whose word is holy ...*
> *Who perfects the decrees of power, lordship, and princeship,*
> *The earth-gods bow down in fear before him,*
> *the heaven-gods humble themselves before him . . .*
>
> Sumerian hymn[58]

Enlil was the king of the gods, king of the universe. "For some unknown reason, *Enlil* took over from *An* as supreme ruler," Samuel Noah Kramer tells us.

The Me — It was *Enlil* who created the *"me,"* a set of cosmic laws which govern every aspect of existence.[59] He assigned a *me* to each goddess and god. These instructions told each deity how to guide and maintain her or his portion of the universe.[60]

The *me* included directives for the sun, moon, planets, and stars. It had decrees for the "sea and air, wind and storm, rivers, mountains, and plains." It included rules for "city-states, dykes and ditches, and fields and farms." Everything, including all human institutions — whether political, religious, or social — were governed by the *me*.[61]

Perhaps the Sumerians, lovers of law and order, came up with the *me* to establish stability in a world of unpredictable river floods.

City Gods

Enlil was the patron god of Nippur, the holy city of Sumer. Every city-state had its own god or goddess — selected from high up in the Sumerian pantheon.

If the city had a goddess, a high priest ruled. If the city had a god, a high priestess ruled. Citizens were expected to do what the gods asked of them, as directed by the high priest or priestess.[62] If I can't be a god, I want to be a high priest. What power!

Where did this invisible city god or goddess live? At times in the city temple and at other times in heaven — thought to be quite near in the sky.[63]

Stairway to the Gods

The first city temples were small mud brick shrines. They were set on raised mud brick platforms to protect from floods. They also provided a stage for priestly ceremonies.

These shrines would evolve into great temple complexes — the heart of the city's economic, political, and spiritual life. At their center were immense temple *ziggurats* — stepped pyramid-like towers which soared up to 100 feet (30 meters) into the sky.[64]

First appearing in c. 3500 BC, these magnificent structures of Sun-baked mud brick displayed a series of rising square platforms, each smaller than the one below — like a tiered wedding cake. "Every major Sumerian city had at least one."[65]

A ziggurat's outer walls were covered with fired brick. Some also incorporated glazed tiles with different colors — decorated with geometric mosaics and frescos of plants and animals. Outer areas of terraces were often planted with trees and other flora in containers.[66]

On the uppermost platform or terrace was the temple or shrine to the city god or goddess — placed as close to heaven as possible to ease the travel of the divine being between heaven and Earth.

Some suggest priest-astronomers used the tops of ziggurats for astronomical observations.

The Ziggurat of Ur

The great Ziggurat of Ur, c. 3100 BC, soared an estimated 75 feet (23 meters) into the Mesopotamian sky — the height of a modern seven-story building (Fig. 1.7). Its base was 200 feet wide and 159 feet deep (61 by 48 meters), covering well over two-thirds of an acre.[67]

The great Ziggurat stood on a raised terrace, augmenting its great height. This highest point by far in the city was visible for many miles.[68]

The mighty edifice was named the *Etemenniguru* or "the house whose foundations create terror."[69] The sheer enormity of it was meant to inspire awe and fear — a demonstration of the power and wealth of Ur.

Three monumental stairways of a hundred steps each stood at the front face of the Ziggurat (again see Fig. 1.7). They rose from the ground and met at the gateway to the second terrace. A single flight of stairs then led upward to the third and top terrace and the door to the shrine of the moon god *Nanna*, protector of Ur.[70]

At the top, adorned with gleaming blue enamel brick, was the shrine of *Nanna* — the *Egishirgal*, "house of the great light." Here *Nanna* resided when he chose to come down from heaven to earth. Only priests and priestesses were allowed to enter this sacred place.[71]

18   *Cosmic Roots*

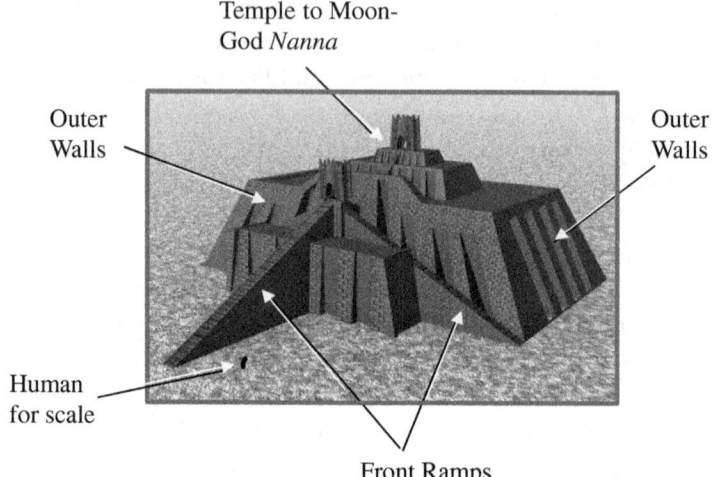

Figure 1.7.   *Etemenniguru* — **the Ziggurat of Ur.** c. 3100 BC.

Priests held daily sacrifices of food, wine, and beer to the gods here. Special feasts took place on the day of the new moon, and 7th, 15th, and last day of the month. Sumerians observed the New Year on the first full moon after the Spring Equinox with twelve days of raucous celebration. They knew how to party!

A double-walled compound surrounded the great Ziggurat. Within the vast compound were shrines, a courthouse, workshops, courtyards, and treasure chambers. Here temple personnel lived, "children went to school, farmers and traders stored their goods," and the poor were fed from granaries.[72]

Across from the compound and separated by a paved street was the *Giparu* — a large temple to *Ningal*, goddess of reeds and wife of *Nanna*.[73]

Ziggurat Construction

A masterpiece of ancient architecture, the ziggurat or "holy mountain" required a staggering amount of manual labor to build. Its interior was a solid mass of mud bricks — each brick some 11.5 by 11.5 by 2.75 inches (about 29 by 29 by 7 cm.).[74]

Mud brick degenerates over time. Ziggurats required constant restoration. Later ziggurats were often built on the ruins of earlier ones — rising higher and higher over the Mesopotamian plain. Despite millennia of

Figure 1.8. Restored Ruins of the Great Ziggurat of Ur. Only the lowest of its three levels remains. Its façade was restored under Saddam Hussein in the 1980's. For scale, please note people on top.

wear, the remains of some 33 ziggurats have managed to survive in some form to this day — including the (restored) Great Ziggurat of Ur (Fig. 1.8).

Unprecedented Change

Humankind's first city-states would morph from rule by priests to reign by hereditary kings. With royal ambition came conquest, subjugation, rebirth, and collapse — a cycle which would repeat in various forms for millennia across our planet.

The political structure, cultural advances, religious views, and technical achievements of Sumer would spill across the ancient Near East and into Asia and Europe. It would plant the seeds for the birth of astronomy and religious conflict which resonate to this day.

This is the subject of the following chapters.

Chapter 2

FROM PRIESTS TO KINGS

He who possesses many things is constantly on guard.

— Sumerian proverb[1]

Major Sumerian city-states began to form around c. 3500 BC. They consisted of "walled cities surrounded by villages and hamlets." Over time, a bureaucracy developed within the city-state to administer its political and economic life — led by a high priest or priestess in the name of the city god.[2]

Under the auspices of the high priest or priestess, "secular officials, workers, and slaves" built and maintained the city-state's irrigation systems. They also worked temple farmland outside city walls and managed temple buildings.[3]

Democratic assemblies initially ran the secular operations of city-state governments. In these early days, "free citizens met in a two-branch assembly, an upper house of 'elders' and lower house of 'men'," Kramer tells us — much like the British parliamentary system of today.[4]

By around 3000 BC, intercity conflicts began to break out. Threats from neighboring city-states pressured city assemblies to appoint a *Lu-gal* or "Big man" to organize defenses.[5]

Thus began the evolution to rule by hereditary kings.

The Kings of Sumer

As independent kingdoms, the opulent city-states of Sumer formed the economic backbone of the ancient Near East — especially Ur, Lagash, and Uruk (where Iraq gets its name). "Industries from as far away as Jericho near the Mediterranean Sea and Catal Huyuk in Asia Minor competed for trade" with the black-headed people. This in turn spread Sumerian technology, culture, and beliefs throughout the region and beyond.[6]

Populations expanded. The city-state of Lagash grew to an estimated 100,000 people. Uruk was of similar size. At its peak, the sacred city-state of Nippur contained some 40,000 people with several hundred surrounding villages.[7]

The Sumerians had a good thing going — irrigation works which (mostly) tamed the floods, bountiful food supplies, a rich culture, and great wealth. It was not enough.

Led by warrior-kings, Sumerian city-states soon fought each other over fertile land, water rights, and ego.

The flat plains of southern Mesopotamia did not help. They offered no natural barriers to thwart or slow what was to become perpetual combat between neighbors. Neither a common language nor a shared culture would stop the carnage.[8]

For over seven bitter centuries they fought: Kish against Uruk; Uruk against Aratta near the Caspian Sea; a tri-city clash for supremacy between Kish, Ur, and Uruk; Kish and Akshak against Lagash; Lagash versus Isin; and Lagash opposite Ur. A "hundred years war" was fought between Umma and Lagash over the fertile plain between them, known as the *Guedena*. Blood of the first civilization spilled again and again onto the sands and irrigated soil of Sumer.[9]

At times, a city-state king would conquer a vast area of Sumer and beyond. A few examples:

If the "Sumerian King List" is to be believed, *Etana* of Kish was first to control all of Sumer in c. 2800 BC. *Gilgamesh* of Uruk is the king for whom the famous *Epic of Gilgamesh* is loosely based. He ruled over Sumer in c. 2670 BC. The famed Stele of the Vultures unearthed by French archeologists in the 1880s commemorates the victories of King *Eannatum* of Lagash. His empire was said to extend from all of Sumer to parts of Elam in the east in c. 2400 BC. (See Fig. 2.1.)[10]

Note to Reader: The "Sumerian King List" is a text which lists Sumerian kings and their reigns. It declares that kingship "descended from heaven." Its earliest parts are legendary — assigning reigns of 3600 years for monarchs.[11]

Figure 2.1. Sumerian King *Eannatum*. Period 2455–2425 BC. Statuette in Alabaster. Anonymous Sculptor. Louvre Museum, Paris.[12]

24   *Cosmic Roots*

Figure 2.2. "War" panel of the Standard of Ur, c. 2600 BC. Shows parading men, onagers, and four-wheeled chariots.

Over time, Sumerian forces evolved into armies of professional soldiers. Troops fought in phalanx formations with eight-man fronts six files deep. High ranking nobles wielded bronze battle axes and lances. Pairs of warriors rode two and four-wheeled chariots into battle, pulled by four onagers (wild Asian donkeys).[13] (See Fig. 2.2.)

Regular infantry wielded short spears, slings, and simple bows. Light infantry brandished battle-axes, daggers, and spears. Copper helmets shielded their heads. Overlapping plate body armor protected their bodies.[14]

Highly trained armies, advanced bronze weapons, and sophisticated tactics meant slaughter on a scale never before seen — one of the "benefits" of human civilization.

The nearly constant warfare weakened Sumer. It was a major factor in its downfall.

In c. 2600 BC, Sumer became a vassal of the Elamite kings of what is now southwest Iran. Some one hundred years later, Sumerian king *Lugal-Annemundu* of Adab overthrew the Elamites. Sumer was free again.

It did not last.[15]

The End

> *That [Sumer's] rivers flow with bitter water ...*
> *That in the banks of the Tigris and Euphrates ...*
> *there grow sickly plants ...*
> *That well-founded cities and hamlets be counted as ruins,*
> *That its teeming black-headed people be put to the mace ...*
>
> — Sumerian Lamentation[16]

He was a foreigner who lived in the Sumerian city of Kish — an Akkadian from the town of Azupiranu on the banks of the Euphrates in northern Mesopotamia. According to the King List, he worked his way up to the prestigious position of cup bearer to King *Ur-Zababa* of Kish.[17]

The story goes that *Ur-Zababa* sent the cup-bearer to negotiate peace with King *Lugal-Zagesi* of Umma. Fearing the growing power of his cup-bearer, *Ur-Zababa* also sent along a secret note asking the king of Umma to kill the Akkadian.

Lugal-Zagesi instead asked the cup-bearer to defect and join him in an invasion of Kish. (Whether he showed the note to the cup bearer is not known.) Together they invaded Kish and crushed *Ur-Zababa's* army.

The cup-bearer in turn overthrew *Lugal-Zagesi*. He then took the name *Sarru-ukin* or "the king is legitimate." We know him as Sargon. He and his army went on to subdue the rest of *Lugal-Zagesi's* empire.[18]

It was not enough.

Sargon then conquered Akkad, his homeland to the north — uniting northern and southern Mesopotamia. His armies invaded the land of the Elamites across the Tigris to the east. They captured Mari on the banks of the Euphrates in present day Syria. They seized lands of the Amorites — Semitic nomads from the Syrian and Arabian deserts.[19]

During his reign (c. 2335 to 2280 BC), Sargon ruled over some sixty-five cities of various cultures and languages — the first truly international empire in history.[20]

"Sargon the Great," as he is known, was a man of overpowering ambition. He was one in a long line of megalomaniacs for whom — indifferent to the deaths of others — the need to conquer is paramount. What is it in our species that draws us to such men? And why do we exalt them as "Great"?

The "Barbarians"

Some eight decades later, in c. 2200 BC Sargon's great grandson *Naram-Sin* ruled. He was beset by Elamites from the east, Amorites from the west, and nomadic barbarians called Gutians from the central Zagros mountains in present day southwest Iran.[21]

> *... the snake and scorpion of the mountains.*
>
> — Sumerian narrative poem on the Gutians.

Around the same time, extreme drought enveloped the region. It would last some two to three centuries. The precious soil of the Tigris-

Euphrates valley would become increasingly salty.[22] Food production fell dramatically. Vast numbers of Sumerians migrated north. The population of Sumer declined by an estimated three-fifths.

In c. 2190 BC, the hated Gutians invaded en masse. They razed the capital city of Agade and ended Akkadian rule. Sumer suffered the ignominy of rule by barbarians. Canals fell into disrepair. Famine and death increased. Cities were abandoned.[23]

A Brief Revival

After over a century of misery and abuse, Sumerian king *Utu-Hegel* of Uruk "routed the Gutian forces" and liberated the land of the civilized kings.[24] Upon his death, his son-in-law *Ur-Nammu*, governor of Ur, seized the reins of power (Fig. 2.3). He proceeded to reconquer much of Sumer — including Lagash, Eridu, and Uruk.[25]

Ur-Nammu initiated vast building projects throughout Sumer. It was he who ordered construction of the Great Ziggurat of Ur. He built magnificent city walls said to be "high as a shining mountain." He also had a number of ziggurats and temples rebuilt in other cities.[26]

Thus began the "Sumerian Renaissance" — a period of unprecedented technology advancement and artistic expression.[27] It too would not last.

Figure 2.3. King Ur-Nammu — Sumerian Head Sculpture. Note the "headdress, heavy-lidded eyes, rounded nose, and angle at which the ears are bent." Photo courtesy of Vladimir Korostyshevskiy.

Fleeing drought, Amorite tribesmen began a mass migration from the west in the 21st century BC. They attacked Sumerian outposts on the border and established settlements within the empire. Within several generations, Amorite communities occupied a significant portion of Sumerian lands.

Seeing their opportunity, Elamites from the east sacked the great city of Ur in c. 2000 BC. They carried off *Ibbi-Sin* — last of the *Ur-Nammu* dynasty.[28] The Amorites in turn drove the Elamites from the land.

Over the next two centuries, Amorites would merge with what was left of the Sumerian population. They would establish the fabled city of Babylon some 20 miles (32 km) northwest of Kish as their capital. The land became known as Babylonia; the populace Babylonians.[29]

This marked the end of the Sumerians as a distinct people.

Legacy

Sumer's impact was initially felt in later Mesopotamian cultures and their neighbors in the ancient Near East. Sumerian beliefs were the "inspiration for much of the subsequent religion, mythology, and astrology in the region." Babylonians and Assyrians in particular "worshipped many of the same Sumerian gods, though under different names."[30]

The "dead" Sumerian language remained in use for scholarship, education, the court, and religion for some two thousand years after its demise. It was still in practice at the time of Christ.[31]

Great River Civilizations — Three other world civilizations would arise along the flood plains of rivers:

(1) The Egyptians along the Nile c. 3000 BC; (2) the Harrapan along the Indus River valley c. 2500 BC (in present day western India and Pakistan); and (3) the Chinese along the Yellow and Yangtze rivers c. 2200 BC.[32]

Like the Sumerians, invasion and assimilation would end ancient Egyptian and Harrapan civilizations. On the other hand, ancient Chinese culture would survive with modification to this day — making it the oldest continuous civilization in the world.[33] Sumerian culture would affect the Egyptians and Harrapan.

Some scholars argue that Egyptian writing, religion, cosmology, architecture, art, law, and literature have roots in Mesopotamia. Much like early Sumerian kings, the pharaohs of Egypt were buried in elaborate tombs along with members of the royal family and attendants.[34] The Egyptian Sun-god *Aten* (and the Greek Sun-god *Apollo*) drove a chariot across the sky, just as the Sumerian Sun-god *Utu* did.

The wheeled vehicle, the potter's wheel, and other Sumerian inventions made their way to the Nile — as well as the Indus valley. Indian (as well as Greek) epic writings show "a striking similarity in form and content" to Sumerian epic literature — including "heroic narrative stories in poetic form." There are indications that "Sumerian sailing vessels reached the valley of the Indus River" as well.[35]

The mark of Sumerian civilization would span three continents. Divinely sanctioned kingship, written law, trade practices, Sumerian inventions, bronze metallurgy, mathematics, and more would make their way into Central Asia, North Africa, and Europe.[36]

The Modern World — Sumerian culture has also reached across some five thousand years of separation to the present era. Sumerian schools, technology, architectural advances, innovations in art and sculpture, moral ideals, proverbs and sayings, love songs and lullabies, aquariums, farmer's almanacs, and libraries — to name a few — are all found in one form or another in modern times.[37]

The Sumerian mathematical system was based on the number sixty.[38] It was they who came up with the sixty-second minute, sixty-minute hour, and twenty-four-hour day, as well as the 360-degree circle. Sumerians also "set a limit on work days" and established the concept of days off for holidays.[39] Ya gotta love 'em for that!

Look up your Horoscope? Perhaps it says you will find good fortune. Or be wary of your pride. Or some such nonsense. Its roots are found in Sumer.

The Sumerians believed in the pseudo-science of astrology. To them, positions of stars and planets were "omens for people on Earth." They also used the stars as "calendars to determine when to plant crops." Their astrological observations were likely significant factors in the development of astronomy.[40]

In science, the Sumerian creation of the me — cosmic rules to govern the universe — can be seen as a precursor to the modern search for laws of physics which govern all phenomena in the cosmos.

As blogger and Sumer lover David Darcey put it; "The real legacy of the Sumerians is civilization itself."[41]

The Next Chapter

The rich mythology of Sumer would come to have a profound effect on the foundations of Western religion. Of all its influences, none is deeper and less well known. This is the subject of the next chapter.

Chapter 3

IN THE BEGINNING

In the beginning, God created the heavens and the earth.

— Genesis 1:1

Do you believe in God? Yes? No? On the fence? When you read this question, did it occur to you to ask, "Why God — why not gods?" Or even better, "why not gods and goddesses?"

The notion of multiple deities likely did not occur to you. This is due to an ancient people whose story began some 4000 years ago. It is they who, over time, established our modern concept of God — particularly for those whose roots are in the West.

The God of Abraham

Barukh atah Adonai elohenu melekh ha-olam
Blessed art Thou, Lord, our God, King of the universe

From the beginning of human civilization in Sumer to the height of the Roman Empire, polytheism was the common practice in Europe, North Africa, West Asia, and beyond. There was an exception — a small group of people living in the highlands of the southern Levant: the Hebrews.

From their holy book, the Torah — the first five books of the Bible — the Hebrews trace their belief in a single God and lineage to the patriarch Abraham. (See Fig. 3.1.)[1]

Figure 3.1. A Torah Scroll. The Pentateuch of the Hebrew Bible: Genesis, Exodus, Leviticus, Numbers, and Deuteronomy. Glockengasse Synagogue, Cologne (reconstruction).

Unlike pagan deities, the "God of Abraham" is a God of morality, of righteousness. Kindness to fellow humans is at the core of what scholars call "ethical monotheism."[2]

God is not always depicted as kind. The book of Joshua in the Hebrew Bible, for example, describes a series of attacks by his army on Canaanites. In city after city, the Bible tells us, the Hebrews "utterly destroyed all that breathed, as the Lord, the God of Israel, commanded." (Joshua 12:24.)[3]

This is the part of the Bible which so disturbed me as a youth. How can one justify the slaughter of men, women, *and children* ordered by God? Whether we like it or not, the divinely sanctioned murder of innocents is part of our heritage.

Abraham

According to the Hebrew Bible, Abraham was from the former Sumerian city of Ur (Genesis 11:31). It appears that Bible writers wanted to depict the "great father of the Hebrew people" as originating from "the very heart of the civilized world — the famed city of Ur."[4]

Religious scholars suggest Abraham may have been born sometime between 1900 and 1700 BC. He is believed to be an Amorite — the semi-nomadic people who put an end to Sumerian civilization.

The Hebrew Bible tells us Abraham traveled with his family to the land of Canaan — today's Israel, West Bank, Gaza, southern Lebanon and Syria. The One God tells Abraham: "And I will give unto thee, and to thy seed after thee, the land of thy sojournings, all the land of Canaan, for an everlasting possession." (Genesis 17: 8.) Thus the religious justification to conquer this land and keep it forever — a conflict that goes on to this day.[5]

People of the Book

The Hebrew Bible has been called "the single most influential literary and spiritual Creation" in human history. Coupled with the New Testament of Christianity, it stands as the world's all-time best seller. An estimated 5 billion copies have been printed to date.[6]

The earliest Bible texts were likely written before 1000 BC; the last in the second century (100's) BC. Unlike the stone tablets of Sumerians, the Hebrew Bible was originally written on papyrus. This paper product has a limited life. Over the years, some scrolls were lost, others dutifully copied by successive generations. Biblical stories and histories were reworked and refined, often "incorporating new influences."[7]

The extraordinary detail in the Bible makes it seem real. It was written over centuries and rewritten after the fact to reflect God's impact on events. It is hard to believe Bible authors made so much up.

The Bible tells the story of the Hebrews — their triumphs and defeats, their evolving relationship with the God of Abraham, their struggles between righteousness and sin. Filled with glorious miracles, the 28-book Hebrew Bible has become the very foundation of Western culture.

Who hasn't heard of Adam and Eve, the first humans; or the story of their firstborn son Cain's murder of his brother Abel? Or tales such as Noah and the Ark; Abraham and the sacrifice of his son Isaac stayed by the hand of God?

There is the destruction of Sodom and Gomorrah; Moses parting the Red Sea; the seduction of Samson by Delilah; Jonah swallowed by a whale; and David's defeat of the Philistine giant Goliath with a slingshot. Later Bible stories tell of Daniel in the Lion's Den and the epic travails of Job.

In reading the Bible as an adult, I discovered a rich, complex narrative with the human struggle for dominance interweaved with the desire to behave righteously. I was stuck by the existential angst of Ecclesiastes; the eroticism of Song of Songs; and the lyrical and haunting "The Lord is My Shepherd."

I realized the "good book" depicts its characters as real people with human flaws. This along with its vivid portrayals and profound humanitarian message is surely a key to why it has stood the test of time. For example, the story of Saul, the first King of Israel, is an unrelenting depiction of jealousy, rage, and the "melancholy of defeated greatness," as Rabbi and author Chaim Potok put it. It is "the first portrait in the history of our species of the *humanity* of a king."[8]

Despite all the colorful characters in the Bible, God is the main protagonist. Throughout the Bible, events experienced by the Hebrews — whether natural or man-made — are caused by God to reward the righteous and punish evil.

Here is a world where God is supreme, where justice rules.

> *I am God Almighty ... And I will establish My covenant with thee and thy seed after thee throughout their generations for an everlasting covenant ...*
>
> — Genesis 17:1-8

A major theme in the Hebrew Bible is the covenant between God and the Hebrews. God tells Abraham that he will be "the father of a multitude of nations" and, as noted, give "all the land of Canaan to thee, and to all thy seed after thee ... and I will be their God." (Genesis 17:5-8.) Here Bible writers use the covenant to bind the Hebrews and their descendants to their God and their land. The Bible also includes covenants between God and Noah, Moses, and David. "No other ancient people had such a thing."[9]

The word covenant is *berit* in Hebrew, which means contract. We see here the roots of the Sumerian emphasis on written legal processes. A number of stories and beliefs in the Hebrew Bible are rooted in the writings of Sumer.

Ancient Roots

It was a great shock. Tablets hidden for millennia showed that a number of major stories and themes in the Bible were not original. They began, as so many things do, with the Sumerians.

> *And it came to pass in the days of Amraphel king of Shinar,*
> *Arioch king of Ellasar, Chedorlaomer king of Elam ... that they made*
> *war with Bera king of Sodom, and with Birsha king of Gomorrah ...*
>
> — Genesis 14:1-2

The name "Shinar" is mentioned eight times in the Hebrew Bible. It refers to the land we call Mesopotamia. Some scholars argue that it is the Hebrew name for Sumer. Others say no, as it encompasses both the southern and northern regions of "the land between the rivers." Whatever its exact meaning, the connection between the bible and Sumer is deep and extensive.[10]

Here are some examples from Samuel Noah Kramer's brilliant book *The Sumerians: Their History, Culture, and Character.*

Like the Sumerians, the Hebrews considered themselves the "chosen people," selected by the gods (or God) amongst all the peoples of the Earth. Armies fought in the name of their gods in Sumer, as did the Hebrews for the God of Abraham. The Sumerian day began at sunset, just as in the Hebrew calendar.[11]

In Sumer, if you commit an offense against the gods, you are punished with calamity and misfortune. The Gutian invasion of Sumer, for example, was blamed on Akkadian king *Naram-Sin's* ransack of the holy city of Nippur. We find this theme throughout the Hebrew Bible, except it is the God of Abraham who does the punishing.[12]

The city temple of Sumer was the god's house. The Jerusalem Temple of the ancient Hebrews was also the house of God. In the mythology of Sumer, the gods voiced their intentions and it was so. In Genesis, "God said let there be light, and there was light." (Genesis 1:3.)

Enlil is the father of the gods and King of the Universe. The Hebrew God is also called "King of the Universe."[13]

The Sumerian Law Codes of *Ur-Nammu* were decrees "descended from the gods." The King was merely the "administrator of those laws." The Laws of Moses — the Torah, the first five books of the Hebrew Bible — were commandments from the One God that even the King must obey.[14]

For Sumerian kings, war was fought for and sanctioned by the city god. The Hebrews invoked the One God to support the slaughter and fought in His name.[15]

And God said: Let us make man in our image, after our likeness ...

— Genesis 1:26[16]

Like the Sumerian gods, the God of Abraham looks like us (see Figure 3.2).

Saturday Night Live comedian Father Sarducci (aka Don Novello) once quipped, "If God made man in his own image, why aren't we, like ... invisible." (I'm certainly going to Hell for repeating that one.)

Figure 3.2. The Creation of Adam. Section of the Sistine Chapel Ceiling by Michelangelo Buonarroti (1512)[17] Being a European artwork, God and Adam are of course depicted as white.

Devout Jewish scholars argue that "in our (God's) image" refers to the spiritual rather than the physical. I respectfully disagree. The Bible passage above says, "after our likeness." It seems to me this emphasizes physical qualities.

We see roots of Sumer in the biblical Garden of Eden story (Genesis 2:8). In Sumerian writings, humans once lived in a "god-created paradise ... a divine garden green with fruit-laden trees and meadows." It is located in *Dilmun*, which is somewhere *east* of Sumer, just as the biblical paradise is "eastward in Eden" (Genesis 2:8).[18]

The Sumerian version of Noah and the Ark is found in the Epic of Gilgamesh. Here the gods bring a Great Flood to destroy mankind. *Enki*, the god of wisdom, takes pity on humans. He tells a reverent king named Ziusudra to "build a great boat to save his life and 'the seed of all living creatures.'" Like Noah, Ziusudra "releases a bird in search of dry land."[19]

Then the Lord God formed man of the dust of the ground ...

— Genesis 2:7

The Book of Genesis tells us that man was made from clay (Genesis 2:7). Again, *Enki* and his half-sister *Ninki* created man out of clay — the dust of the ground in Mesopotamia.[20]

A Sumerian poetic essay contains themes identical to the biblical Book of Job: human suffering and submission to the gods (or God). Its introductory plot is the same. Nonetheless, the brief Sumerian poem "in no way compares with the Bible story in breadth, depth, and beauty," writes Kramer.[21]

Through Canaanite, Hurrian, Hittite, and Akkadian writings, the cosmology of Sumer also found its way into the Hebrew Bible.[22]

Ancient Hebrew Cosmology

In the beginning, God created the heavens and the earth.

— Genesis 1:1

The phrase above begins the Hebrew Bible. Note the word *created*.

"Recent scholarship suggests that the Hebrew word for 'creation' in Genesis 1:1 is *bara*," writes Michael Shermer in *Scientific American*. "More accurately translated, this verb means to 'separate' or 'divide.'" The famous passage in the Bible should read, "In the beginning, God *separated* the heavens from the earth."[23]

Recall the Sumerian creation story: the primordial sea-goddess *Nammu* gave birth to *An* and *Ki*, the heaven and earth *united*. Their son, air-god *Enlil separated* the heavens from the earth. If the above Bible translation is accurate, it appears to have been derived from Sumerian beliefs.

There are a number of other connections between Sumerian cosmology and the Bible:

The Book of Genesis depicts a dome-shaped universe surrounded by cosmic waters. Within the dome are multiple heavens above and a flat Earth and Netherworld (S*he'ol* in Hebrew) below. Sound familiar? It is Sumerian *flat-earth* cosmology.

Both biblical and Sumerian accounts propose that primordial waters existed *before* creation. In Sumerian mythology, there is *Nammu* the sea-goddess and the Primeval Sea. In Genesis, we have "Darkness upon the face of the *deep*." (Genesis 1:2.) (My italics.)[24]

In both stories, a "firmament" divides the waters to separate heaven and earth. The late John McDermott, Professor of Philosophy at Texas A&M, pointed out that the word "firmament" is *raqia* in Hebrew. Its root means "to beat or spread out" as in hammering out a dome.[25]

And gather together the scattered of Judah,
from the four corners of the earth

— Isaiah 11.12

For he looketh to the ends of the earth.

— Job 28.24[26]

The Hebrew Bible is filled with expressions "the ends of the Earth" and "the four corners of the earth." They imply a flat Earth.

Summary

The Hebrew Bible depicts the beliefs of the ancient Near East in the first millennium BC — the cosmology of Sumer. This includes a flat earth. They never told me that in Hebrew school. This is summarized in Table 3.1.

Table 3.1. Sumerian roots of the Hebrew Bible.

Sumerian Mythology	Hebrew Bible	Reference
Enlil divides heaven	In the beginning, God separated the heavens and the earth.	Genesis 1.1
Nammu and the primeval Sea	*The earth was formless and void, and darkness upon the face of the deep.*	Genesis 1.2
The gods voice their intentions and it is so	*God said, "let there be light," and there was light.*	Genesis 1.3
Anthropomorphic gods	*God said, "Let Us make man in Our own image, according to Our likeness ...*	Genesis 1.27
Flat earth	*And gather together the scattered of Judah, from the four corners of the earth."*	Isaiah 11.12
	"For he looketh to the ends of the earth ...	Job 28.24
Earth supported by pillars	*For the pillars of the earth are the Lord's.*	1 Samuel 2.8
Water above heavens	*"Praise Him, ye heavens of heavens, And ye waters that are above the heavens*	Psalm 148.4
Nether-world	*The nether-world from beneath is moved for thee To meet thee at thy coming.*	Isaiah 14.9, 15

The Name of God?

The God of Abraham undergoes a curious name-change in the Bible. In Genesis, He is known as *Ēl* to the Patriarchs — Abraham, his son Isaac, and grandson Jacob. God changes Jacob's name to *Israel*, which translates: "May *Ēl* persevere." (Genesis 35:10.)[27]

> *I appeared unto Abraham, unto Isaac, and unto Jacob,*
> *as Ēl Shaddāi, but my name YHWH I made not known to them.*[28]
>
> — Exodus 6:2-3

In Exodus, the second book of the Bible, God reveals His real name to Moses: *YHWH*. Hebrew consonants are represented here in English. The vowels are lost to history. The name *YHWH* is then used extensively throughout the rest of the Bible. It perhaps means "bring into being" or "creator." It is considered a sacrilege to pronounce God's name out loud.[29]

Ēl is the name of the chief god of the Canaanites, "father of the gods" and "creator of all living things." He was called *Ēl* the Kindly, *Ēl* the Merciful. He "rejoiced in human happiness, mourned for human pain," and forgave those who had atoned for their transgressions.[30]

Ēl and his wife, mother goddess *Asherah* were parents to a number of gods in the Canaanite pantheon. This included *YHWH* — who, the Bible tells us, was a member of *Ēl's* council.[31]

> *Who is like YHWH among the sons of Ēl? In the council of the holy ones, Ēl is greatly feared; he is more awesome than all who surround him.*
>
> — Psalm 89:6-7[32]

YHWH declares in the first commandment "Thou shalt have no other gods before Me." Here He explicitly acknowledges the existence of other gods.

This reflects the ancient Hebrew belief in a number of deities. As we shall see, the path to exclusive worship of *YHWH* was a long and troubled one.

Now let us examine Bible stories and see how they compare to archeological evidence.

Of Villages, Kings and Exile

<u>Note to Readers</u>: *Much of the section below is based on archeologists Israel Finkelstein and Neil Silberman's excellent book,* The Bible Unearthed: Archaeology's New Vision of Ancient Israel and the Origin of Its Sacred Texts.

The historical accuracy of the Bible is a controversial subject, to say the least. The views expressed here are not accepted by all archeologists and Bible scholars. Future discoveries may change our current understanding. With this mealy-mouthed caveat, let us begin.

<u>Did the Patriarchs Really Exist?</u>

The Genesis narrative on the Patriarchs is troubled by a number of anachronisms. Bible stories tell of camel herds, certain Arabian trade items, and Philistines which did not yet exist at the time of Abraham, Isaac, and Jacob.[33]

The Bible mentions other details which are misplaced in time. These include the partition of territory with Arameans, the kingdom of Edom, the Kedarite dynasty, specifics of the Assyrian and Babylonian empires, and certain geographical details. All these things actually occurred in the ninth to the sixth centuries BC — over a thousand years *after* the patriarchs are said to have walked the Earth.

We know that Amorites migrated from Mesopotamia to Canaan in the period attributed to the Bible story (c. 1800 BC). Yet according to archeologist William G. Dever, he and his colleagues have all but "given up hope of recovering any context that would make Abraham and Isaac and Jacob credible historic figures."[34]

A number of Bible stories are also questioned by secular scholars.

No Exodus?

> *And the children of Israel journeyed from Rameses to Succoth, about six hundred thousand men on foot, besides children.*
>
> — Exodus 12:37

Did the epic escape of Hebrew slaves from Egypt in the time of pharaoh Ramesses II (c. 1200 BC) really occur? How did some six hundred thousand people avoid the extensive series of manned Egyptian forts guarding its eastern border? Why is there not a single mention of this massive flight of escaped slaves in Egyptian records?[35]

Archeological surveys add to the mystery. Surely the travels of such a large group over 40 years through the Sinai wilderness to the border of Canaan would show some trace of evidence. Yet there is nothing. Not a campsite, temporary shelter, building, or any sign of occupation "from the time of Ramesses II, his immediate predecessors or successors."[36]

The Exodus story tells how Moses slew an Egyptian for beating a Hebrew slave. He then "fled from the face of Pharaoh and dwelt in the land of Midian." (Exodus 2:11-14.) Here he stayed with a nomadic tribe for 40 years. He then heard the voice of *YHWH* inside a burning bush and led the Hebrews out of Egypt to the promised land. (Fig. 3.3). Is there a kernel of truth to this story?

One theory suggests a small group of Canaanite slaves did escape from Egypt. On their way to Canaan they passed through Midian — now

40 *Cosmic Roots*

Figure 3.3. Moses and the Burning Bush. Fresco from the Dura Europos in Syria c. 244 AD, the world's oldest preserved Jewish synagogue.

southern Jordan and Northern Saudi-Arabia. Here they met Semitic-speaking nomads called the *Shasu*.[37]

Egyptian texts on the period mention a place where the *Shasu* lived called *YHW*. Scholars tell us this was "likely the name of their patron god." Once they met up with the great mass of Hebrews who were already in Canaan, the fleeing Canaanites told them of the *YHWH* Moses story. The theory remains controversial.

No Invasion of Canaan?

The Bible tells us that after the death of Moses, his lieutenant Joshua leads the Hebrews on a lighting campaign of conquest in the promised land of Canaan. Nearly all ancient Canaanite cities mentioned in the attacks have now been excavated by archeologists. They include Jericho, Ai, Gibeon, Lachish, and Hazor. Findings contradict the Bible story.[38]

Did the biblical trumpets of rams' horns really blow the walls of Jericho down (Joshua 6:1-27)? Archeological evidence shows that at the time of the story, according to the biblical chronology (13th century BC), there were no walls surrounding Jericho or any Canaanite city. And there are no signs of destruction of Jericho, Ai, or other kingdoms of Canaan and Transjordan.[39]

Egyptian records show Canaan was a vassal of Egypt during this period (Middle Kingdom). Egyptian garrisons were spread throughout the region. Archeological digs confirm this. Yet there is no mention of Egyptians in the biblical invasion story.

It is possible Israelites may have destroyed a few cities, such as "Hazor in the upper Galilee or a site or two in the south." The general consensus is that the invasion of Canaan as described in the Bible did not happen.[40]

Who Are the Israelis?

> *The Canaan has been plundered into every sort of woe ...*
> *Israel is laid waste and his seed is not ...*
>
> — Victory Stele of King Merneptah
> — c. 1207 BC[41]

A hieroglyphic inscription of Egyptian king Merneptah in 1207 BC tells of a people called Israel who lived in Canaan. This is the first known document to mention such a group.[42]

Archeological excavations in present day Israel have uncovered early Israelite communities in the central hill country of Canaan dating to 13th and 12th centuries BC (1200s to 1100s). Over three hundred little villages were unearthed from the hills of Samaria to the north and the Judean hills to the south. Each village typically extended over about an acre of land — populated by some 50 adults and 50 children.[43]

Unlike Canaanite cities and villages, Israelite sites contained no palaces, no temples, and no public buildings. Amenities were simple; with rough pots, no weapons, and little jewelry. No sign of pork was found.

It appears the roots of ethical monotheism are in tiny villages of the peasant poor.

These nomadic people gradually developed a pastoral life of herding and farming. They later evolved into large cities, market centers, and small villages. By the eighth century BC (700s), two kingdoms — Israel in the north and Judah in the south — were established with some 500 sites and some 160,000 people.[44]

Kings David and Solomon

> *(I killed Je ho)ram son of (Ahab) king of Israel, and (I) killed (Ahaz)iahu son of (Joharma kin)g of the house of David ...*
>
> — Inscription of King Hazeal of Damascus[45]

This fragmentary part of a broken black basalt monument was discovered in 1993 at Tel Dan in northern Israel. It is dated to 849 BC — within one hundred and fifty years of the biblical lifetime of King David (c. 1000 BC). It is the first clear evidence that the "house of David" did exist. That is if the translation from Aramaic, the language of ancient Aramean Syria, is correct. More recent findings also hint at David's existence but are less clear.[46]

The Bible tells of a Golden Age where King David united Israel and Judah. It tells of Solomon the wise, the son of David, who ruled in an era of great prosperity. The first Temple to *YHWH* in Jerusalem is said to have been built by Solomon.

Extensive digs in Jerusalem find no trace of Solomon's magnificent temple or his great palace. Archeological evidence of the period reveals not a city but a hill country village.[47]

Up to and past the presumed time of David and Solomon, Judah was an undeveloped rocky region of "dense scrub and forest," Finkelstein and Silberman tell us. It contained not a single "major urban center." If a United Kingdom of Israel and Judah existed, it apparently was on a very small scale.[48]

North versus South

According to the Bible, the sons of the patriarch Jacob became the progenitors of the twelve Hebrew tribes. Ten of the tribes — Asher, Dan, Ephraim, Gad, Issachar, Manasseh, Naphtali, Reuben, Simeon, and Zebulun — formed the Kingdom of Israel in the north in c. 930 BC. The tribes of Judah and Benjamin established the Kingdom of Judah in the south. (See Fig 3.4.)[49]

The Bible tells of a long history of enmity between ancient Judah and Israel, its Hebrew sister to the north. Israel at times even allied with Syria and Aram against Judah.[50]

Israel enjoyed a land of fertile valleys, rich agriculture, and a generally hospitable climate. It possessed gentle eastern slopes, the bucolic Jezreel

In the Beginning 43

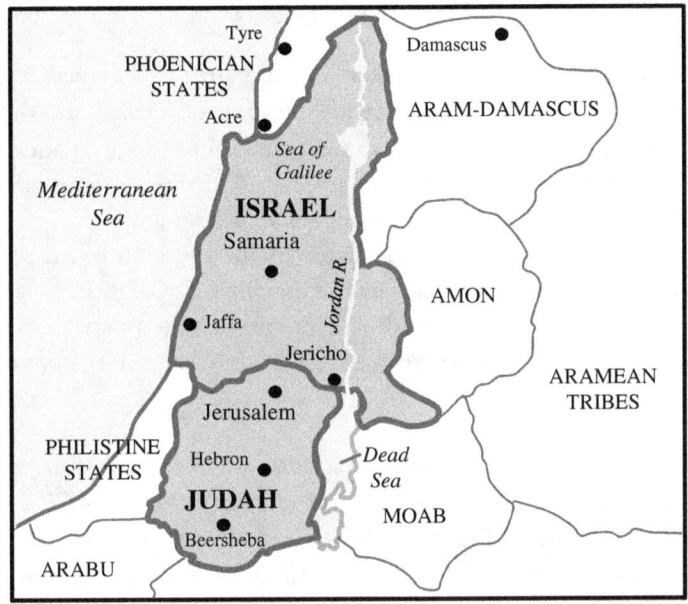

Figure 3.4. Map of Ancient Kingdoms of Israel and Judah.

valley to the north, and the less rugged western desert. This gave it easy access to the Canaanite coast, Mesopotamia, and the Jordan valley. Trade flourished.[51]

In the ninth century BC, the mighty Omride dynasty ruled the northern kingdom. Samaria, Israel's capital, was an impressive metropolis some 35 miles north of Jerusalem. Here archeologists have uncovered a sophisticated city of some 20 acres.

The city was surrounded by outer and inner walls constructed of "ashlar blocks of the highest quality." Contained within were an elaborate palace, royal tombs, and public buildings of limestone pilasters and proto-ionic capitals. Unearthed artifacts revealed "over five hundred ivory plaques and fragments ... in glass, enamel, and lapis lazuli inlays," as well as others with gold leaf. There were figures from nature: plants, flowers, trees, and wild animals.[52]

In contrast, Judah at the time was a poor, sparsely populated, rocky region.[53] The southerners of Judah also spoke with a different accent. They used expressions like *"Good ol' boy"* and *"He ain't got the good sense God gave a rock."* (My lame attempt at humor, y'all.)

The Lost Tribes

Prosperous Israel was trapped between Mesopotamia to the north and Egypt to the south. Through alliances, playing one power against the other, and good fortune, the Israelis managed to maintain a precarious balance.

This ended in 724 BC when Israeli king Hoshea attempted to make a deal with the Egyptians. He would be Israel's last king.

Enraged by Hoshea's negotiation with his arch-enemy, Assyrian king Shalmaneser V marched on Israel. His vast army of "chariots, cavalry, archers, slingmen, spearmen, shield-bearers, and battering rams" crushed city after Israeli city. After a three-year siege, his son Sargon II conquered its capital, Samaria.[54]

> *The inhabitants of Samaria, who ... plotted with a king hostile to me ... I fought against them with the power of the great gods, my lords. I counted as spoil 27,280 people, together with their chariots and gods, in which they trusted ... I settled the rest of them in the midst of Assyria. I repopulated Samaria ... (with) people from countries conquered by my hands ...*
>
> — Chronicle of Sargon II[55]

Tens of thousands of Israelis — including "officials, scribes, army personnel, and craftsmen" — were taken away and dispersed to "scattered regions of the Assyrian empire." (See Fig. 3.5.)[56]

Figure 3.5. Assyrian Relief — Israelis Deported into Exile after Capture of Lachish. *"And the king of Assyria carried Israel away into Assyria ..."* (2 Kings 18:11.)

Sargon II resettled Israel "with refugees from other Assyrian conquests." The ten tribes — the so-called Lost Tribes of Israel — vanished from history as an integral people. The mixing of remaining Israelites with foreign refugees came to make up a people called the "Samaritans" — the source of the Good Samaritan story of Jesus' time (more on this later). They still exist in tiny numbers today and continue to worship the Hebrew God on Mount Gerizim in Israel.[57]

> *Woe to them that are at ease ...*
> *in the mountains of Samaria ...*
> *That lie upon beds of ivory,*
> *And stretch themselves upon their couches ...*
> *That drink wine in bowls ...*
>
> — Amos 6:4-6

The fall of Israel left Judah to write the history of Samaria. Now they would have their revenge on their uppity cousins to the north. The Bible depicts the Omride kings of Israel as evil men who worshipped idols and promoted social injustice (1 and 2 Kings).[58]

In turn, Judahite writers romanticized and embellished the kingdom of Judah — especially the reigns of David and Solomon. They assigned God as the agent of all calamities and good fortune in both kingdoms. This depended on how faithful to *YHWH* and his commandments they were. The north, of course, fared far worse than the south in this Godly judgment.[59]

Then it was Judah's turn to face Assyria.

Jerusalem on the Precipice

Assyrian king Sargon II died in 705 BC. His death set off a series of revolts throughout the empire. Judah's king Hezekiah joined the rebellion.[60]

The new king of Assyria — Sargon's son Sennacherib — marched out of the glorious Assyrian capital of Nineveh to quell the uprisings. At his disposal was an army of 100,000 troops or more. He subdued the ever-restive sacred city of Babylon. Sennacherib then stormed through Syria and Phoenicia. He reconquered Sidon and Tyre without a fight, and routed Egypt.

Then he turned towards Judah.

"All of Hezekiah's immediate neighbors" submitted to Sennacherib.

In preparation for an attack, Hezekiah strengthened the fortifications of Judahite cities — including the walls of Jerusalem. He had a tunnel built in the capital city to provide water when under siege. This engineering marvel ran from the Achon Spring within city walls some 1750 feet (530 meters) to the Pool of Siloam outside.[61]

Sennacherib's army advanced through Judah. With "ramps, battering rams, mines, breeches and siege engines," they "laid siege to 46 walled cities." They were victorious in all and captured some 200,000 prisoners.[62] This according to Assyrian records.*

Jerusalem was next. The frightened Hezekiah gathered all the treasures from the Temple and palace he could scrape up and sent them to the new Assyrian king. He is even said to have "cut off the gold from the doors of the temple of the Lord ..." (2 Kings 18:16.)

Unimpressed, Sennacherib sent "a great army unto Jerusalem." (2 Kings 18:17.) They surrounded the Holy City and began a "parley with Judean officials" on its walls. All in the city held their breathe.

Then a miracle happened. Or so says the Hebrew Bible. In the dark of night *YHWH* sent a plague to assail the Assyrian army. Multitudes perished. Those troops who survived departed the following morning. Jerusalem was spared.

> *And it came to pass that night, that the angel of the Lord went forth, and smote in the camp of the Assyrians a hundred four score and five thousand; And when men arose early in the morning, behold, they were all dead corpses.*
>
> — 2 Kings 19:35

Bible scholar Philip Stern offers a more prosaic explanation. "Word reached Sennacherib that Babylon had once again risen in revolt." He abandoned the siege of Jerusalem and rushed his army to Babylon to quell the uprising.[63]

Before his army departed, Sennacherib extracted heavy tribute from Hezekiah. How do we know this? The annals of Sennacherib give a detailed list of these "gifts" delivered over the years to Nineveh by Judah. No mention of this is made in the Bible.[64]

What if Sennacherib had destroyed Jerusalem and dispersed survivors throughout his empire — as his predecessor Sargon II had done in Samaria? Judah would most likely have been another lost tribe. The God of

---

* Archeological remains verify the invasion of Sennacherib. The number of captives may be an exaggeration. Source: F&S, pp. 252, 259.

Abraham would have been forgotten. There would have been "no Judaism, hence no Christianity or Islam," Stern points out.

The Judeans had been saved. The long road to arguably the world's first monotheistic religion had begun. This is the subject of the next chapter.

Chapter 4

ONE NATION UNDER GOD

I am the Lord thy God ...
Thou shalt have no other gods before Me.

— Exodus 20:2-3

In the morning of their deliverance, Jerusalemites stared out from the walls of the city on a hill — their eyes filled with tears of joy. They looked to the Mount of Olives in the east, towards Hebron in the south, over the valley of Hinnom in the west, and north towards Samaria. The Assyrians had vanished.

Were they dreaming? Was this a cruel trick? As time went by, the dread and foreboding passed — drifting up into the heavens like harmless wisps of smoke. Their hearts filled with love for their pious king, Hezekiah. They fell prostrate in thanks to *YHWH*, the God of their ancestors, who had saved them.

This singular episode would set off a series of events — played out over a century and a half — that would give the world the God of Abraham and the Hebrew Bible as we know them.

The Roots of Monotheism

We turn back to the Assyrian invasion of Samaria under Sargon II in 723 BC. At the time, an increasingly vocal religious faction had arisen in Judah. It was perhaps influenced by pious priests and prophets who had escaped from the north.

At the heart of this belief was the adoration for only *YHWH*. And the condemnation of widespread idol[*] worship in Judah. Historian Morton Smith dubbed this the "*YHWH* alone movement."[1]

Hezekiah became king of Judah in 715 BC. He would take up the banner of the "*YHWH* alone" movement and institute a series of religious reforms.

According to the Bible, he ordered the exclusive worship of *YHWH*. He declared the Temple in Jerusalem the only legitimate place to worship Him. He commanded pagan idols and religious sites be destroyed throughout Judah.[2]

> *All ... that were present went out to the cities of Judah, and broke in pieces the pillars, and hewed down the Asherim [female idols], and broke down the high places and the alters out of Judah and Benjamin, in Ephraim also and Manasseh, until they had destroyed them all.*
>
> — 2 Chronicles 31

---

[*] Archeological findings of the 8th century appear to confirm that idolatry was ubiquitous in both Judah and Israel at the time.

Hezekiah's intent was to unite Jerusalem and the countryside into one nation under *YHWH*. It would not be easy. It clashed with the "time-honored" tradition of the independence of the countryside.[3]

Did these religious reforms really occur or were they embellishments by later Bible writers? Perhaps. Evidence for Hezekiah's reforms is scanty and questionable. Archeological evidence does appear to confirm the birth of the *YHWH* alone movement and the religious and social evolution it stimulated.[4]

Whatever the extent of Hezekiah's reforms, the movement saw the Assyrian invasion of Judah under Sennacherib in 701 BC as *YWYH's* punishment for idol worship. Why was Jerusalem spared? Because of the piety of Hezekiah.[5]

Outside the capital city, things were not so sanguine. The invasion had left large parts of Judah devastated. And "a significant number of Judahites had been deported to Assyria."[6]

Some outside Jerusalem felt abandoned by *YHWH*. Farmers in particular held on to their pagan fertility deities to assure good harvests. This religious conflict between the elites of the capital city and the common people of the countryside would rage throughout the 7th century BC (600s).

After the death of Hezekiah in 698 BC, King Manasseh reigned. He was followed by his son Amon in 641. The latter two kings gave their support to pagan cult worship to gain the co-operation of the countryside. Later Bible writers would denounce them for doing "evil in the sight of the Lord." (2 Kings 21:20.)[7]

Amon was assassinated in a coup d'état in 640 BC — likely because of his allegiance to the hated Assyrians. The elite of Jerusalem in turn killed the conspirators and placed Amon's eight-year-old son Josiah on the throne.

Josiah would prove a very pious king. Beginning in about 621, he enacted the "most intense puritan reforms in the history of Judah." His first target was idol worship within Jerusalem and the Temple itself.[8]

And the king commanded Hilkiah the high priest ... to bring forth out of the temple of the Lord all the vessels that were made for Baal, and for the Asherah, and for all the host of heaven; and he burned them ... †

— 2 Kings 23:4

King Josiah then ordered pagan shrines and alters in Bethel and the cities of Samaria be destroyed. (2 Kings 23:19-20) Josiah was seen by many as a true messiah who would "restore the fallen glory of the house of David."[9]

Figure 4.1. Josiah hearing the book of the law. From: *The Story of the Bible from Genesis to Revelation*. Author Unknown (1873).

<u>The Fifth Book</u>

In 622 BC, a strange and marvelous thing happened. Deuteronomy — the last book of the five books of the Torah — was "discovered." The ancient scroll was found by Josiah's high priest during restoration of the Temple (Figure 4.1). At least this is what the Bible tells us.

> *And Hilkiah the high priest said unto Shaphan the scribe:*
> *'I have found the book of The Law in the house of the Lord.'*

— 2 Kings 22:8

A number of secular scholars dispute the "discovery of Deuteronomy" story told in the Bible. They generally believe the fifth book of the Torah was written during or just before the reign of Josiah. Deuteronomy is "strikingly similar in form to early 7th century BC Assyrian vassal treaties,"

Finkelstein and Silberman point out, "thus placing it as a work *contemporary with the post-Sennacherib period*." [My italics.]

Deuteronomy represents the heart of ethical monotheism. It contains "ethical laws and provisions for social welfare that have no parallel anywhere else in the Bible," Finkelstein and Silberman tell us. Here the exclusive worship of the One God in the Jerusalem Temple is espoused. Here the call is given for a national observance of the major Hebrew festivals, Passover and Tabernacle. Here is bequeathed "a range of legislation on social welfare, justice, and personal morality" found nowhere else in the ancient Near East.[10] (More on this later)

The writing of Deuteronomy was accompanied by extensive compilations and editing of the first four books of the Torah.[11] Why did Josiah stage this "discovery" of the Deuteronomy scroll? Perhaps out of deep piety or as a clever political ploy to unite his people, or maybe both.

Specific accomplishments of Josiah's reign have been difficult to verify. Archeological evidence does appear to show that, despite the reforms of King Josiah, polytheism continued to be practiced throughout Judah. It didn't help that the four kings that followed ended government suppression of idol worship.[12]

In the last decade of the 7th century (600s) BC, a new power would arise in Mesopotamia. It would provoke the greatest tragedy yet for the people of Judah. Ironically, this devastation would prove key to the establishment of monotheism in the land.

Exile

The closing decades of the 7th century BC saw the decline of Assyria as a world power. Disintegration accelerated with the death in c. 627 BC of its great and cruel king Ashurbanipal — grandson of Sennacherib.[13]

In 612 BC, Chaldeans and allied forces sacked the Assyrian capital, the beautiful Nineveh. The great Assyrian Empire was no more — replaced by the "Neo-Babylonian" Empire.

Under its king, Nebuchadnezzar II, Babylon became the greatest city in the world. Within this metropolis of over a hundred thousand people were a thousand temples, eight magnificent gates, and, as legend has it, the glorious hanging gardens of Babylon.[14]

Judah soon found itself in the crossroads between warring Babylon and Egypt. Judean king Jehoiakim decided to align with Egypt. Wrong choice again. Babylon defeated Egypt at the Battle of Carchemish in 605 BC. Nebuchadnezzar then carried out a series of invasions of the "Holy Land."

> *In the third year of the reign of Jehoiakim king of Judah came Nebuchadnezzar king of Babylon into Jerusalem and besieged it.*
>
> — Daniel 1:1

In the invasion of 605 BC, Judah's King Jehoiakim was forced to submit to save Jerusalem. The Assyrian king extracted tribute from the city treasury as well as temple artifacts. He also deported some members of the royal family and nobility as hostages to Babylon.[15]

Nebuchadnezzar attempted an invasion of Egypt in 601 BC. He suffered heavy losses. A number of states in the Levant revolted. Judah's new king — Jehoiachin, son of Jehoiakim — joined them.

(Watch out. The two kings' names are almost identical. This drove me crazy during my research.) The Babylonian king invaded Judah again in 597 BC. This time he took King Jehoaichin, the rest of Jerusalem's aristocracy, and its priesthood to Babylon — some 10,000 people according to the Bible.

> *... And he (Nebuchadnezzar) carried away all Jerusalem, and all the princes, and all the mighty men of valour, even ten thousand captives, and all the craftsmen and the smiths; none remained, save the poorest sort of the people of the land.*
>
> — 2 Kings 24:10-14

Nebuchadnezzar appointed Zedekiah as puppet-king in Judah. He too revolted and aligned with Egypt. Talk about continuing to poke the tiger in the eye. The enraged Nebuchadnezzar invaded Judah for a third and final time. Archeological evidence verifies that one by one the cities of Judah fell.[16]

On 587 BC, Babylonian troops surrounded Jerusalem and let its inhabitants slowly starve. Then, on the infamous 9th of Av in the Hebrew year 3174, the Babylonian army entered the Holy City. They tore apart the Temple of *YHWH* and reduced the city to rubble. Survivors were sent off to Babylon, except peasants and the poor.[17]

As for Zedekiah, he was forced to watch the murder of his two sons. Then they poked his eyes out, bound him, and took him to Babylon where he died in prison (2 Kings 25:1-7).

An estimated fifteen to twenty thousand Judahites were deported in the Babylonian Exile — the *Galut Bavel* (Figure 4.2). Some settling in the city of Babylon and others on the Chebar river near Ur in an area called Tel-Aviv.

One Nation Under God 55

Figure 4.2. The Babylonian Exile. "The Flight of the Prisoners" by James Tissot. c. 1896–1902, gouache on board.

Judah's royal family was ensconced in the southern palace of the Babylonian kings.[18]

> *By the waters of Babylon,*
> *There we sat down, yea, we wept,*
> *When we remembered Zion ...*
> *If I forget thee, O Jerusalem,*
> *Let my right hand forget her cunning.*
> *Let my tongue cleave to the roof of my mouth,*
> *If I remember thee not;*
> *If I set not Jerusalem*
> *Above my chiefest joy.*

— Psalms 137:1, 4-6

Among the exiled aristocracy and priesthood in Babylon were devout followers of the *YHWH* alone movement. They had lost their homes, their temple, their holy city and the land they believed *YHWH* had promised them for eternity. Some perhaps began to forsake the God who it seemed had abandoned them.[19]

A devout few held to their faith and tried to make sense of the great tragedy. Surely this was God's punishment for the continued worship of pagan gods. Out of what must have been a soul-searching time of doubt, the few rose up to find what God called for them to do.

Slowly at first, they drew out the sacred scrolls they had taken with them. They began to read them, edit and rewrite them, and compose new scrolls. They consolidated their works with the vision of a single God, a jealous God, a God of mercy and righteousness.[20]

In what seemed to be endless years of exile, they told and retold the stories of the Hebrew people — with emphasis on the exclusive worship of YHWH. Here in the land of pagans they forged the core of the great masterpiece, the Hebrew Bible, essentially into its final form.[21]

Note to reader: Bible writings during and after exile show influence from Zoroastrianism, a quasi-monotheistic religion of Persia. Biblical concepts such as free will, independent angels, Satan, and in part the later heaven and hell are believed to have come from this Persian religion. Please see endnote for more details.[22]

Then, as the Judahites believed, a miracle happened.[23]

Second Exodus

In 539 BC, Babylonia fell to the Persians. To ease acceptance of his rule, Persian King Cyrus encouraged reconstruction of ancient temples throughout his empire. He issued an edict that same year — some forty-six years after the Exile had begun. It permitted the exiles from Judah to return to Jerusalem and rebuild their temple.[24]

> *In the first year of Cyrus the king, Cyrus the king made a decree;*
> *Concerning the house of God at Jerusalem, let the house be builded ...*
>
> — Ezra 6:3

Bible writers saw this as an act of God. The Book of Isaiah tells us that YHWH spoke "to his messiah, to Cyrus, whom I [YHWH] took by his right hand to subdue nations before him ..." (Isaiah 45:1).

Here "messiah" means the "anointed one." Many kings, prophets and high priests were "anointed in the Hebrew Bible, and are referred to as the *messiah* or *anointed one*," writes Bible scholar Lisbeth Fried. Cyrus is the only instance of a foreigner called out in this way.[25]

Under Cyrus, Judah or *Yuhuda* in Hebrew was now a Persian province. The Hebrews of Judah were now known as *Yuhudim*. This eventually became "Jews" in English. This terminology will be used from now on.[26]

Led by Sheshbazzar, son of King Jehoiachin, the first wave of Jews journeyed back to the Holy Land. Though many remained in the comforts of Babylon, more would come — an estimated thirty to fifty thousand in all.[27]

They found post-exile Judah nearly devoid of people. Archeological surveys suggest a reduction in population by as much as 80 percent. This made it easier for the returnees to establish *YHWH*-only worship in the region.[28]

The returning Jews began construction of a new Temple in Jerusalem. Samaritans from the north volunteered to help. They were turned away. The second Temple to *YHWH* was completed in 516 BC on the site where the first had been destroyed seventy years earlier.[29]

> *And all the people shouted with a great shout, when they praised the Lord, because the foundation of the house of the Lord was laid.*
>
> — Ezra 3:11

In the post-exile era — the Second Temple period as it is called — a governor appointed by Persia directed the secular affairs of Judah. Jewish priests held religious authority, led by the high priest at the Jerusalem Temple.[30]

Priest scribes continued to edit and add to bible scrolls. They completed final redaction of the first five books, the Torah. The Hebrew Bible ends with the release of Jehoiachin — the last king of the lineage of David — from prison in Babylon.[31]

Monotheism Achieved

Blessed may he be by YHWH and his Asherah

— 8th century BC Hebrew inscription at cemetery west of Hebron.[32]

Modern excavations and archeological surveys have uncovered a plethora of pagan figurines in the soil of ancient Judah. In particular, thousands of nude female terracotta idols were unearthed "in tombs, in households, everywhere," archeologist William Dever tells us. These artifacts date from the 10th to the 6th century. Dever suggests they are representations of the goddess *Asherah* (Fig. 4.3).

Figure 4.3. Asherah Figurines. Mother goddess and Consort of *El* and then *YHWH*.

Recall that in the Canaanite pantheon of gods, *Asherah* was the consort of *El* and mother of *YHWH*. It appears she became associated as the consort of *YHWH* in Judah.

> *Thus sayeth the Lord God; Return ye,*
> *and turn yourselves from your idols;*
> *and turn away your faces from all your abominations*

— Ezekiel 14:6

In the period *after* the return of the Jews from Persia, "not a single cultic figurine [was] found," archeologist Ephraim Stein points out. It appears that exile in Persia was the key. Those who came home to Judah fulfilled the dream of the *YHWH* alone movement. The fifteen centuries long passage to ethical monotheism was at last complete.[33]

Unheralded and unknown to the rest of the world, this seminal achievement set the foundations of ethical monotheism for all humanity.

Now let us take a brief look at the commandments of the Hebrew Bible and their unique religious and moral perspective.

The Law

In one of the most famous passages of the Hebrew Bible, God instructs Moses, "*I will send thee unto Pharaoh, that thou mayest bring forth my people the children of Israel out of Egypt."* The Bible tells of Moses leading his people out of slavery in Egypt to the borders of the "promised land."

The story's climax is the appearance of God on Mount Sinai.

> *And He gave unto Moses ... upon Mount Sinai ...*
> *two tablets of stone written with the finger of God.*
>
> — Exodus 31:18

Here Moses receives the *Decalogue* — the Ten Commandments — from God. (See Figure 4.4.)[34] Like the Code of Babylon's King Hammurabi, The Law of the Hebrews is quite harsh.[35]

Figure 4.4. The Ten Commandments. Parchment by Jekutheil Sofer (1768). Emulates Ten Commandments at Amsterdam Esnoga synagogue in 1675.

Commandments six through ten are briefly:

(6) *Thou shalt not murder; (7) Thou shalt not commit adultery; (8) Thou shalt not steal; (9) Thou shalt not bear false witness; and (10) Thou shalt not covet.* (Exodus 20:13-14.)

These directives are "common to other ancient Near Eastern civilizations." The Egyptian Book of the Dead, for instance, reads: "I have not cursed my local god ... I have not stolen ... I have not killed men ... I have not been covetous ..."[36]

The first five commandments are unique to the Hebrew Bible. They are, in brief:

(1) thou shalt have no other gods before Me; (2) Thou shalt not make unto thee a graven image; (3) Thou shalt not take the name of the Lord thy God in vain; (4) Remember the Sabbath day ... the seventh day ... thou shalt not do any work; and (5) Honor thy father and mother. (Exodus 20:3-12.)

The last one is my favorite, though I must admit I didn't always obey it in my teenage years.

As Biblical scholar Joshua Berman points out, the commandments of the Hebrew Bible are directed towards the *individual*. The Covenant is between God and the people, not just the king.[37]

The Ten Commandments are followed by an extensive law code. Here the sanctity of human life is raised to a new level. Unlike Mesopotamian codes, "the Covenant Code of the Hebrews does not punish any crime against property by death ..." Chaim Potok points out, "[nor does it] allow any act of murder be paid for in property."[38]

Deuteronomy

As noted, promulgated by the *YWYH* alone movement, the Book of Deuteronomy in the Hebrew Bible adds key tenets to the Laws of Moses.

They include a balance of power in government, an independent judiciary, and education of the masses. In addition, the king is not a god. He is subject to The Law like everyone else (Deuteronomy 17:14-20).[39]

> *Thou shalt not judge unfairly; thou shalt show no partiality; thou shalt not take bribes ... Justice, justice shalt thou pursue.*
>
> — Deuteronomy 16:18-20

Legal judgments are based on written law, not arbitrary decisions of a king or judge. The laws make it easier for the average man to own land, to remain solvent under tax laws, and to get a loan. They also call for charity: Every three years, the tithe is to be distributed to the poor.[40]

Debts are released every seven years. A woman rejected by her husband has inheritance rights. Resident aliens are protected from discrimination. Slaves are freed after six years of servitude (Deuteronomy 14, 15, 26).[41]

Global Impact

Unlike the Sumerians, ancient Jews did not produce ground-breaking inventions, build monumental architecture or create great works of art. Nor did they establish mighty empires. The Jews' gift to the world was and is the Hebrew Bible.

Through Christianity, Islam, and related religions of the God of Abraham, ethical monotheism and its moral underpinnings now constitute the religious beliefs of over 50% of people on this planet.

The new religious paradigm begun by the Jews would come to conflict with the science of cosmology. This would begin with the discoveries of the greatest natural philosophers in the ancient world: the Greeks. This is the subject of the next chapter.

PART II
Lovers of Wisdom

Chapter 5

THE GREEKS

... the unexamined life is not worth living

— Plato, The Trial and Death of Socrates[1]

The land was rocky. There was little arable terrain in this mountainous peninsula. There were no river floods to turn the arid soil green. Yet they produced one of the greatest civilizations known to man. What was their secret?[2]

The sea.

The Greek mainland and islands were never more than 50 miles from the Aegean Sea, Ionian Sea, or Sea of Crete. There was an abundance of good harbors. Overseas trade became the main source of survival.[3]

Here at the crossroads of the Eastern Mediterranean, they built the first great seafaring society. Greek ships were laden with pottery, perfume, wine, and their chief export — olive oil. Through trade, they made contact with and were influenced by great ancient civilizations — especially Mesopotamia and Egypt.

Ah, the glories of ancient Greece: timeless philosophical treatises, profound works of drama, grand epic poetry, elegant temples, exquisite sculptures, and the world's first democracy.

Their mythology was rich with "gods and goddesses, demigods and spirits around every corner." To the Greeks, "everything in nature was imbued with a divine essence."[4]

Small city-states soon formed. And almost as soon, they were at war with each other. But land areas were generally isolated from each other — preventing centralization of power.

Its harsh, isolated geography fostered strong individualism. This characteristic of independence would drive their politics and science and come to be a hallmark of Western civilization.[5]

By the early 5th century BC (400s), Greeks had spread throughout the Mediterranean region. There were Greek colonies in southern Italy, Sicily, and the Iberian Peninsula to the west; along the shores of the Black Sea to the north; in Asia Minor and Cyprus to the east; and North Africa in the south. (See Fig. 5.1.)

Greek civilization would produce the greatest scientists in the ancient world — though they were called *natural philosophers* as in philosophers of nature. The term "scientist" did not come into vogue until the 19th century AD, as we shall see. From the Greek homeland and its colonies would come unparalleled breakthroughs in both mathematics and natural philosophy — especially astronomy.

Cosmic Roots

Building on Babylonian, as well as Persian, Indian, and Egyptian astronomy, the Greeks used "elaborate geometric constructions" along with

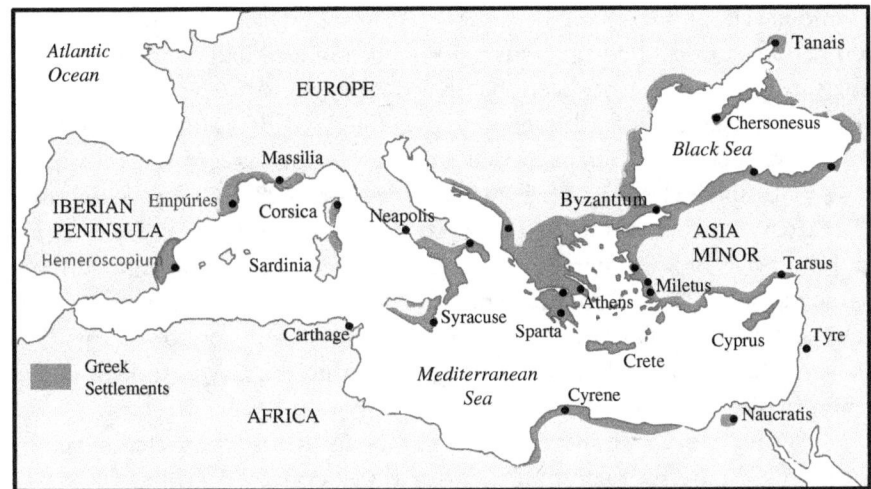

Figure 5.1. Spread of Ancient Greeks — c. 800 to 480 BC.

"long-term systematic observations" to revolutionize our understanding of the cosmos.[6]

Thales of Miletus (c. 624–c. 547 BC) is said to be the first natural philosopher, based on his attempts to "give rational explanations for physical phenomena." Unfortunately, none of his writings have survived to this day.[7]

Thales was allegedly the first person to successfully forecast an eclipse of the Sun. According to ancient Greek historian Herodotus, it occurred on the date predicted by Thales in 585 BC. This claim is doubted by modern historians.

Do you remember the Pythagorean theorem from high school geometry? It is said to have come from Pythagoras of Samos (c. 569 BC–c. 475 BC), the great Greek mathematician and philosopher.* He founded

---

*Where or when exactly the so-called Pythagorean theorem was first discovered is not clear. Egyptian writings in 2500 BC list the "Pythagorean" triplet, e.g., 3, 4, 5, and 5, 12, 13, but no hard evidence of the theorem. Babylon c. 1800 BC shows a "systematic understanding of producing solutions to that equation." The first statement of the theorem is found in India c. 800 BC in the Shuba Sutra of Baudhayan. The first known proof of the theorem is found in China in school textbooks tracing back to between 1046 and 356 BC. Source: Rajendran, P. "Did India discover Pythagoras theorem? A top mathematician answers" rediff.com. Jan 9, 2015. https://www.rediff.com/news/special/did-india-discover-pythagoras-theorem-a-top-mathematician-answers/20150109.htm#:~:text=That%20first%20occurs%20about%20 800,the%20Shuba%20Sutra%20of%20Baudhayan.%22&text=%22The%20Shuba%20 Sutras%20do%20contain,%2D%2D%20after%20the%20Shuba%20Sutra.%22. Retrieved Oct 12, 2020.

a famous school, the secret *Semicircle of Pythagoras*, which was "half-religious and half-scientific." Pythagoreans held that numbers "had divine meaning." They are also said to have been the first to suggest *Earth is a sphere*.[8]

Anaxagoras of Clazomenae (c. 500–c. 428 BC) and Heraclides of Pontus (c. 390–c. 315 BC) proposed that the *Moon reflects light from the Sun*.[9]

Empedocles of Agrigentum (c. 500–c. 428 BC) proposed that all matter is made of the elements earth, air, fire and water. Democritus (c. 460–c. 370 BC) was particularly prescient. He declared that Empedocles' elements are made of indestructible, invisible "atoms." Democritus also proposed that the Milky Way is made up of "thousands of unresolved stars."[10]

Heraclides of Pontus also suggested that the Earth, though at the center of the universe, *rotates*.[11]

These declarations may seem obvious to us, even elemental. In ancient times, they were nothing short of revolutionary.

Sidebar: Egypt and the Roots of Geometry

We are flying from Paris to Cairo. We cross the Mediterranean into north Africa. Below is the vast reddish expanse of the Sahara Desert. A haze of rising sand obscures the horizon. Leaning against the window, I spot it. A long, thin, gently winding dark blue line cutting south to north. It is the Nile — the life-giving Nile.

I smile. The scene reminds me of my awkward attempts to draw the great river weaving through the Egyptian desert with colored crayons in grammar school. This is where my love of ancient history began.

For this is the fabled land of mummies, magnificent stylized art, timeless architecture, mighty pyramids, and the enigmatic Sphinx. Famed 5th century BC Greek historian Herodotus wrote of "wonders more in number than any other land."[12]

The Foundations of Geometry

Along with the Babylonians, the Harrapan, and the Chinese, the ancient Egyptians were among the earliest known practitioners of the geometric arts. From here, it is said, the fundamentals of geometry first made their way to the West.[13]

Herodotus wrote:

> *They say that that king [Sesostris]* distributed the land among all of the Egyptians, each one having an equal lot in a square shape... If the river*

---

*King Sesostris I, reigned 1908–1875 BC. Source: "Sesostris I: King of Egypt" *Encyclopedia Britannica* https://www.britannica.com/biography/Sesostris-1. Retrieved Mar 1, 2021.

bore away a part, the owner announced the loss, and officials were sent to observe the extent to which the plot had been diminished for the purpose of adjusting the tribute payment. It is my feeling that this indicates the invention of geometry here, prior to its passing to Greece.

— Herodotus[14]

Thales of Miletus and Democritus are reportedly said to have studied geometry in Egypt. "Plato spent thirteen years living with Egyptian priests, reading texts of mathematics and theology" and studying their works in astronomy.[15] At the time of these Greek philosophers, Egypt was a venerated ancient culture already over 2500 years old.[16]

Hieroglyphic writing on surviving papyruses and ancient stone carvings indicate that Egyptian geometry was rudimentary. Unlike the later Greeks, the practical Egyptians were not interested in abstract mathematical reasoning. To them, a close approximation was good enough.[17]

Among other things, they used geometry to estimate areas and volumes, including for the pyramids. In mathematics, they employed simple addition and subtraction — and a cumbersome method of multiplication and division which "relied on doubling or halving." They used unit fractions only, such as $\frac{1}{8}$, $\frac{1}{3}$, $\frac{1}{4}$, $\frac{1}{2}$ in their calculations."[18]

Egyptian surveyors using geometry and knotted ropes to "reestablish property boundaries" after the receding of annual Nile flood waters. The later Greeks "called them *arpedonapti* or 'those who knot ropes.'"[19]

The Pyramids

From about 2700 BC, pharaohs were buried in massive pyramids.[20] Egyptian engineers showed "extremely accurate measuring and geometric skills in their construction."

The 146-meter-high Great Pyramid of Giza (Fig. 5.2) is a striking example. This tomb of Pharaoh Khufu outside present-day Cairo is thought to have been built between c. 2590 and 2500 BC. A veritable army of "tens of thousands of skilled workers" (not slaves) built the Great Pyramid.[21]

It was constructed from some two and a half million blocks of stone. They averaged about 2.5 tons each.[22] Workers extracted fine-grained white Tura limestone from nearby quarries for its outer casing. They ferried some of the largest granite stones along the Nile from Aswan over 500 miles (800 km) away — the biggest weighing "up to 80 tons."[23]

To build this architectural wonder, the Egyptians are said to have used wooden wedges, inclined planes, long ramps or perhaps switchbacks, ropes, log

Figure 5.2. The Great Pyramid of Giza Today. Note people in foreground for scale. The Great Pyramid has weathered in the span of some 450 centuries. Looted, shot at, its outer casing of highly polished white limestone gone, it still stands tall. It is "a monument to human labor, human engineering, and geometry."[24] Photo by Nina Aldin Thune, CC BY 2.5

rollers, copper tools, wooden sledges, water-filled troughs to determine level, and extraordinary ingenuity.[25]

The precision is remarkable. Each side of the square pyramid's base is over 230 meters (750 feet) long. The four base lengths match to within 4.4 cm (1 ¾ inches). The base is level "to within 2.1 cm (less than an inch)."[26]

Within the Pyramid's interior, the Egyptians constructed an "intricate geometric arrangement of carefully calculated internal passageways and chambers, corridors, and air drafts." Here resides the stunning Grand gallery and red granite King's chambers. "The king's sarcophagus, also carved from red granite, sits at the exact central axis of the pyramid."[27] Construction was completed in some 22 years.

Oldest of the seven wonders of the ancient world, the Great Pyramid of Giza remains the largest manmade stone structure on Earth.[28]

Geometry and the Universe

Geometry would go on to become the fundamental language of the cosmos. As we shall see, ancient Greek philosophers and astronomers would use geometry to model the motion of stars and planets. Later Islamic astronomers would extend Greek astronomical models with geometric devices of their own.

> Renaissance European astronomers and natural philosophers would pioneer new geometry-based models of our universe which would clash with religious dogma. Isaac Newton would later use a *geometric version* of calculus to produce his seminal law of universal gravitation.
>
> In modern times, Albert Einstein would use *curved surface geometry* to model "spacetime curvature" — the warping of space and time by massive objects such as the Earth and Sun. This would form the basis of his landmark gravitational theory of general relativity.[29]

In what historians call the Greek Classical Period (500–323 BC), three great philosophers established ways of thinking which resonate to this day. Their ideas would plant the seeds for the next advances in astronomy and cosmology in the West.

Lovers of Wisdom

It was the Age of Democracy. It was the era of Socrates, Pericles, and Sophocles, the home of Plato's *Akademia* and Aristotle's *Lyceum*.

> *Our constitution is called a democracy because power is in the hands not of the minority but of the whole people ... everyone is equal in the eyes of the law... in positions of public responsibility, what counts is not membership of a particular class, but the actual ability which the man possesses.*
>
> — Pericles, 431 BC[30]

Here in Athens, democratic rule began in c. 508 BC — though restricted to adult males who were not slaves. Eighteen years later, 10,000 Athenians and Plataeans (city south of Thebes) vanquished a Persian army under Darius I two to three times its size at Marathon. Some six thousand Persians lost their lives. Reportedly, there were only 200 Greek casualties.[31]

Thus began the Greco-Persian wars. For almost a half a century (492–449 BC), Persia attempted invasion after invasion of Greece. Again and again, they were repulsed by a "collective defense mounted by the Greeks."[32]

Despite Persian incursions, Athenian democracy continued almost without interruption for 180 years. Here humankind took the first tentative steps towards a world view where reason as well as religion played a role in the creation story of the universe.[33]

It began with the "gadfly" of Athens.

Socrates

Western philosophy is said to have begun with the illustrious Socrates (c. 470 BC–399 BC) — a radical non-conformist who questioned everything. Unlike earlier Greek philosophers, he emphasized morality and ethics over physical science. (The word philosophy comes from the Greek *philo* and *sophia* or "lover of wisdom" in English.)[34]

Note to Reader: No works of Socrates have survived, if indeed there were any. Almost all of what we know of him is from the writings of Plato — a student of Socrates. We do not know to what extent Plato's portrayal of Socrates may have been influenced by his own views or by his admiration for his beloved teacher.[35]

In his early years, Socrates was a sculptor like his father. He served in the Athenian army and fought at the Battle of Potidaea in 432 BC, a precursor to the Peloponnesian War against Sparta. Here Socrates saved the life of General Alcibiades — a deed which would come back to haunt him, as we shall see.[36]

Later in life, Socrates devoted himself to philosophy and education. It is said that he refused to take any money for his teaching. Rather than give lectures, he engaged his student in dialogue. He listened to their views. Then he asked a series of probing questions to guide and educate. My gentle father used this "Socratic method" to help me learn how to think on my own.[37]

At some point, it got into Socrates' head to walk around Athens and challenge the ruling elite. Barefoot and clad in old clothes, he hung out in the city square with the money changers. There he badgered prominent Athenians — reproaching them and demanding they practice the highest moral principles in government, business, and daily life.[38]

Since I was a child, a divine voice has spoken to me, has given me signs, telling me to me to prod and question other men in order to put them on the road toward true wisdom.

— Socrates[39]

Socrates' ethics were based on "individual conscience" rather than loyalty to the state — a radical notion in ancient Greece. His relentless, high-minded public attacks created enemies. It also attracted a number of young idealistic followers.

"No man on earth who conscientiously opposes either you or any other organized democracy and flatly prevents a great many wrongs and illegalities from taking place in the state to which he belongs can possibly escape with his life," Plato has Socrates say in *Apology*. "The true champion of justice, if he intends to survive even for a short time, must necessarily confine himself to a private life and leave politics alone."[40]

According to Plato, Socrates scorned the idea that the uneducated could vote wisely. He was also associated with other anti-democratic leaders.

Alcibiades, whose life he had saved in battle, had become his student and close friend — and for a time, his lover. The Athenian general subsequently defected to Sparta, then Persia. In 411 BC, he plotted to overthrow the democratic government in Athens.[41]

Socrates was also a friend and teacher to Critias — one of the hated "Thirty Tyrants." The group had seized the government of Athens in 404 BC after its defeat by Sparta in the Peloponnesian Wars. The Tyrants had sentenced hundreds of Athenians to death and forced a number into exile. Athens was returned to democracy a year later.[42]

Socrates' friendship with Alcibiades and Critias put him under suspicion by the democratic government of Athens — then and now the largest city in Greece.

It came to a head in 399 BC — a mere four years since the terrors of the Thirty Tyrants. The now seventy-years-old Socrates was arrested and put-on trial by the Athenian state. He was charged with "denying the gods recognized by the state, introducing new divinities, and corrupting the young."[43]

The trial was held in the courthouse "at the foot of the Acropolis." Socrates — a "strongly built squat man with a white beard, bald pate, and pug nose" — stood up.[44]

With the gallery looking on, he addressed the jury. Here in part is speech according to Plato (who was ill and did not attend the trial)[45]:

> *Athenians, I am not going to argue for my own sake ... but for yours, that you may not sin against the gods by condemning me, who are a gift to you. For if you kill me you will not easily find a successor to me, who... am a sort of gadfly, given to the state by the gods ...*

Socrates was then questioned by the prosecutor, Miletus. Below are excerpts from their dialogue, again according to Plato[46]:

> <u>Socrates</u>: *... why would I want to corrupt people when I know that doing so will make them want to harm me ...? As for religion, do you accuse me of teaching different gods or of being an atheist?*

> Miletus: *You are an atheist.*
>
> Socrates: *But you say I teach spiritual concepts and believe in strange divine beings. How curious it is that I believe in gods and not believe in gods at the same time. The fact is, it appears that you and others here are going to condemn me simply because I have the courage to tell the truth.*

It was apparent Socrates had no intensions of stopping his public attacks. Nor did he take responsibility for the effects of these actions.[47]

A jury of five hundred and one citizens selected at random sat on wooden benches. They had a difficult decision to ponder — a choice between democracy and free speech. One by one they placed their votes into an urn. The majority, some 280 jurors, voted guilty.

By Athenian law the accusers and the accused would then both recommend a penalty. The jury was to choose between the two.

Socrates' accusers called for death.[48]

Apparently surprised by all this, Socrates joked that for punishment he should be honored with "free meals in the Prytaneum." This was "a place reserved for benefactors of Athens and heroes of Olympic games." Then he offered to pay a small fine of 100 drachmae, one-fifth of his meager property holdings.[49]

After this second act of defiance, an even greater majority voted for death.[50]

Socrates was executed by the standard method. He was given a beverage to drink which contained poison hemlock. It is said he "drank the contents as though it were a draught of wine."[51]

There are many questions. Why didn't Socrates request exile as a penalty, as his friends had urged? This may have spared him. Did he feel himself too old for exile? Did he wish to accept the death penalty rather than renege on his principles?[52]

One friend had come up with an escape plan. Why did Socrates refuse this as well? According to Plato, Socrates felt it was his civic duty to accept the judgment of the court.[53]

Whatever the motives and justifications, the execution of Socrates *for his words* made him a martyr for the ages — one who stands, at least in some eyes, as a symbol of courage, integrity, and free speech.

The Immortal Soul

> *True philosophers make dying their profession*
> *and that to them of all men death is least alarming.*
>
> — Socrates[54]

Awaiting the cup of poison Hemlock, Socrates (as told by Plato) explained to his friends why he did not fear death: "When I have drunk the poison, I shall remain with you no longer, but depart to a state of heavenly happiness... It is only my body that you are burying."[55]

In life, the soul is trapped. "When using the sense of sight and hearing or some other sense," Socrates said, "the soul is dragged by the body into the realm of the changeable and wanders and is confused."[56] This soul of which Socrates speaks is immortal. It "existed prior to this life" and continues to exist after death.[57]

For the true philosopher, at death the soul "passes into the realm of the pure and everlasting and immortal and changeless ... to a place that is, like itself, glorious and invisible — the true Hades or unseen world."[58] According to Socrates, only lovers of wisdom will go to this glorious and pure "unseen world" when they die.

What about the rest of us? "As punishment for their bad conduct," the wicked will become "shadowy apparitions" for a time. They will then pass into "perverse animals like donkeys for the gluttons, selfish, or drunkards" and "wolves and hawks and kites" for the "irresponsible lawlessness and violent."

Those who are "good citizens" will "probably" pass into "some other kind of social and disciplined creature like bees, wasps, and ants, or even back into the human race again."[59] "Of course, no reasonable man ought to insist that the facts are exactly as I have described them," Socrates said. "But that either this or something very like it is a true account of our souls."[60]

We see hints of Hindu and Buddhist beliefs here. The passing of the soul to animals and "back to the human race" is suggestive of reincarnation. The notion that desire is the root of our troubled minds has similarities to the Buddhism tenet that desire and ignorance are at the root of all suffering. Though Buddhists do not believe in an everlasting soul.

The ancient Greek concept of the immortal soul was also foreign to Jewish beliefs. Nonetheless, it would be later adopted by Christianity, as we shall see.

After his death, followers of Socrates founded a number of schools of philosophy in Athens. The first and longest lasting was that of his most celebrated student, Plato.[61]

Plato

Such was the end of our friend, a man, I think, who was the wisest and justest, and the best man I have ever known.

— Plato on the death of Socrates.[62]

Plato (428 – 348 BC) is one of the most influential philosophers in history. Like his mentor Socrates, the Athenian aristocrat was "more concerned with human affairs than nature." Along with his works on the life of his teacher, Plato wrote seminal works on political philosophy, aesthetics, epistemology, theology, and the philosophy of language.[63]

Plato's views were often clothed in symbolism and allegory. Interpretations remain controversial to this day.[64]

Though more philosophical than physical, Plato's vision of the world was steeped in *geometry*. Here was mystery. Here was magic. Something beyond human imperfection and the capriciousness of the gods. Something permanent, unchanging, pure, divine.

Plato argued that the shapes of heavenly bodies are spherical, their motions circular. The elements of the world are made of *geometric* atoms — earth of cubes, air of octahedrons, fire of tetrahedrons, water of icosahedrons, and the cosmos of the dodecahedron. These five "Platonic Solids," as they came to be called, are the only convex, regular polyhedrons. (See Fig. 5.3.)[65]

This view of the elements would remain universally accepted in the West "until the birth of modern chemistry in the 18th century."[66]

The Divine Craftsman

This cosmos is beautiful, and its Craftsman good.

— Plato, *Timaeus*[67]

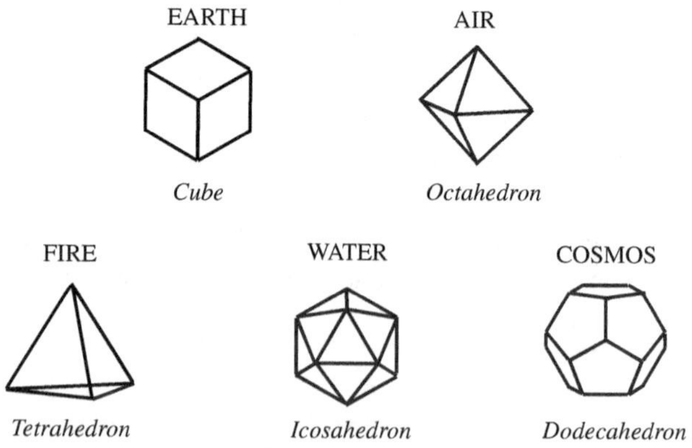

EARTH

Cube

AIR

Octahedron

FIRE

Tetrahedron

WATER

Icosahedron

COSMOS

Dodecahedron

Figure 5.3. The Elements and the Five Platonic Solids. According to Plato, earth atoms are associated with the cube; air atoms with the octahedron; fire atoms with the tetrahedron, water atoms with the icosahedron, and the cosmos with the dodecahedron.

Plato presents his version of creation in *Timaeus*. It begins with the primordial universe, which is in a state of chaos. The *Demiurge* or Divine Craftsman orders the cosmos out of this chaos.[68]

The Craftsman then "brings forth into being the Sun, the Moon, and five other stars (the visible planets)." He then delegates the creation of humans to "the lesser, created gods."[69]

Two Worlds

For Plato, there are two levels of reality: the transient "everyday world of the senses" and "the underlying concepts, rules and laws which are eternal." The logic inherent in mathematics and geometry speaks to Plato of this higher, hidden reality.[70]

The physical world in which we live is an appearance only. It is changeable. It "comes to be and passes away, but never really is."[71]

The second, underlying world is the "World of Forms," of Ideas. It is the *real* world. According to Plato, the true objects in this hidden world, "are apprehended by reason and intelligence."[72]

Plato was not dogmatic. He called his construct "likely," one that provides "the best possible account."[73]

I am the wisest man alive, for I know one thing, and that is that I know nothing.

— Plato, *The Republic*

The great thinker was concerned about the new Greek science. He worried that it could lead to atheism. He felt that "a religious life was a necessary foundation to morality and law."[74]

In his *Laws*, Plato proposed that "anyone who denies the gods are real and [that they] intervened in human affairs would be condemned to five years of solitary confinement." If they then did not repent, they should be executed.[75]

This was a somewhat popular view. At the time "open expressions of atheism were dangerous," writes physicist Steven Weinberg. Two examples: Anaxagoras taught that "the Sun was not a god but a hard stone." After contemplating the nature of the gods, Protagoras concluded "I cannot know that they exist, nor yet that they do not exist." Both were forced to flee Athens for their impiety.[76]

This religious intolerance would come to affect Plato and his students, as we shall see.

Plato's Academy

In c. 387, some twelve years after the death of Socrates, Plato established the famed *Academia* or Academy in the suburbs of Athens. Here among olive and plane trees, temples and statues, and tombs of illustrious men the brilliant moralist held court.[77]

The Academy drew scholars from all over the Mediterranean. Among his esteemed pupils were the inventor Archytas from Tarantum in southern Italy and the mathematician Eudoxus from Cnidus in Asia Minor. Students also included astronomers Heraclides from Pontus on the southern coast of the Black Sea and Aristarchus from Samos, a Greek island in the eastern Aegean.[78]

Here Plato taught his "First Principle" approach to deductive reasoning — a method common to Greek philosophers of the day. It stated that

(1) a first principle is one that is obviously true, thus
(2) anything derived from it must also be true.[79]

Plato's First Principle of cosmology was that *the heavens are perfect*. What was the basis for this declaration? The apparent unchanging appearance of heavenly objects. The Sun, Moon, planets, and stars appeared unchanged in form during centuries of observation.

From his First Principle of cosmology, Plato argued that the only perfect shape (in two dimensions) is a *circle*. Why? For one thing, it has perfect symmetry. No matter how it is rotated, no matter how it is flipped, it remains unchanged. The circle is the only 2-D geometric figure with this property (see Fig. 5.4).

In three-dimensions, a sphere was considered perfect, as it is perfectly symmetrical. Rotate it, spin it, flip it — it too always looks the same.

Plato contended that the only motion which is perfect is circular motion — more precisely, *constant* circular motion. This is motion around a circle which transverses an equal angle in an equal amount of time. (More on this later as well.)[80]

Plato's famous dialogue *Timaeus* asserts that the universe is perfect and unchanging, celestial objects eternal and divine. From this First Principle, he declares therefore that all heavenly motion is circular.

Plato was deeply influenced by the Pythagoreans. A century earlier, they had professed that the planets are eternal and divine, and thus their motion must be uniform and circular.[81]

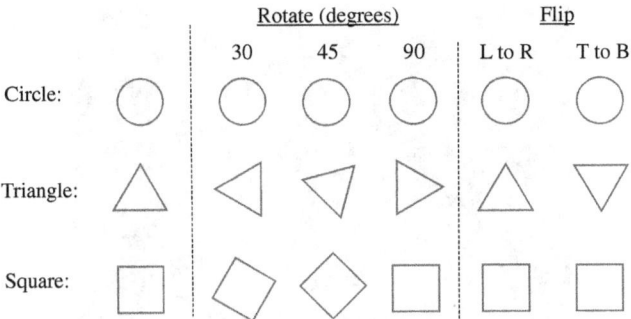

Figure 5.4. Symmetry in Two Dimensions. The circle is the only 2-D geometric form which remains unchanged through any rotation angle or flip axis.

This religious belief in perfect circles and constant circular motion would lead to no end of troubles in astronomy. It would not be fully resolved until the work of Johannes Kepler in the seventeenth century AD.[82]

The Wandering Gods

Looking up at the night sky, the ancients saw, as we do, a great carousel of stars slowly wheeling counterclockwise around the Earth.[83] The stars remain in the same place relative to each other — in the same configuration from night to night. Like the Babylonians, the Greeks labeled certain star groupings "constellations" and named them after their gods.

There were anomalies — five wandering stars called "planets"* which *move* against the background of "fixed" stars. The Greeks named them Hermes, Aphrodite, Ares, Zeus, and Cronos. We know them from the Roman gods as Mercury, Venus, Mars, Jupiter, and Saturn, respectively.[84]

If you watch a planet night after night, you will see it gradually move eastward across the sky — most of the time. Every so often, it does something odd. As seen from Earth, the planet turns and moves in the *opposite direction* or westward. Astronomers call this "retrograde motion."

Take Mars for example. It moves eastward for 780 days. During retrograde, it turns and moves westward for 83 days.[85] And, like all planets,

---

*They are called planets based on the ancient Greek word for wanderer (*planētoi*). The Sumerians called then wild sheep. They named them *Enki, Inanna, Gigalanna, Enlil*, and *Ninurta* after their gods. Source: "Missing Planet" May 18, 2012 http://carolynsmissingplanet.blogspot.com/2012_05_01_archive.html. Retrieved Jun 14, 2018.

78 *Cosmic Roots*

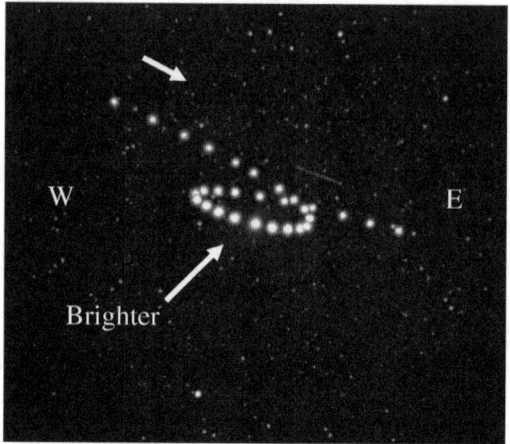

Figure 5.5. Modern Image of Retrograde Loop Motion of Mars as Seen from Earth. Composite image of superimposed images taken on twenty-nine different dates. (Faint moving dots show motion of Uranus.)

it appears *brighter* in the sky during retrograde — as though it is closer to the Earth. This is shown in Fig. 5.5, a series of modern photographs of Mars taken over a number of nights.

This strange motion was a problem for science and religion.

Plato's Challenge

"Many people in Athens did not like the idea of philosophers thinking about the irregular motion of planets," writes Richard Lindgren, Professor of Physics at the University of Virginia. All celestial deities are supposed to move in perfect circles around the Earth — including planets.[86]

Plato refers to this in his last and longest dialogue, the *Laws*: "It is impious to use the term 'planet' of the gods in the heavens," an Athenian says, "as if they and the sun and the moon never kept to one uniform motion course but wandered hither and thither."[87]

Plato knew full well that Socrates had been executed for "denying the gods." Plato was fearful for his students, especially those studying the motion of the planets (as well as, perhaps himself).[88]

Plato challenged his students to find the "real" movements of the planets. That is the perfect circles and uniform circular motion behind the "apparent irregular wanderings of the planets."

In his own words, he asked: "What are the uniform and ordered movements, by the assumption of which the *apparent* [not real] movements of the planets can be accounted for." (My italics.)[89]

Plato hoped that the discovery of the true, hidden motions of the planets would "make safer the work of his students and those that followed."[90]

Later writers said that Plato made his famous challenge to "save the phenomenon" — that is to produce a geometric model which *matched observations*. It appears Plato's true motive may have been to save his students.[91]

Summary

An early reviewer of this chapter asked: "How could Plato maintain that the planets travelled in perfect circles when [as viewed from Earth] they obviously do not?"[92]

In modern science, observation is what counts. We see or measure how nature behaves. Then we attempt to come up with the mathematics, e.g., geometry, to best fit the data.

This is not how Plato and a number of other Greek philosophers thought. Again, to Plato, what we observe is not necessarily "real." There is a hidden world behind what we observe. This hidden world is the "real" world.

He instructed his students to come up with a model which used combinations of *circles* in *uniform circular motion* (the hidden real world) to match the observed (not real) world of retrograde motion of planets.

Where did Plato (and the earlier Pythagoreans) come up with this circle idea? Again, the planets are gods. Thus, they must travel in perfect "divine" circles. Or combinations thereof.

This is an example of how religious beliefs can lead to irrational thinking.

Plato's challenge to his students to match observations with perfect circles and uniform circular motion would set off a two-millennium quest to understand the behavior of the planets. This would come to clash with established religion in the West — and ultimately result in the separation of science and religion, as we shall see.

Next Chapter

In 367 BC, a budding seventeen-year-old scholar from the small city of Stagira in Thrace (now mostly Bulgaria) would be sent to study at Plato's Academy. He would take up Plato's challenge and produce the first comprehensive geometric model of the universe. His name was Aristotle.[93]

Chapter 6

THE GREAT NATURAL PHILOSOPHER

... there is a science higher than
natural science...
It is the principles and causes of the things...
the science of it is First Philosophy — and such
a science is universal just because it is first.

— Aristotle

Aristotle
(c. 384–322 BC)

Aristotle (c. 384–322 BC) is regarded as the greatest natural philosopher of the ancient world. The philosophy of his mentor Plato was founded on the "reality of abstract ideas." Aristotle was more interested in the material world.[1] Plato asked "Why does the world exist? Aristotle asked "How does the world work?"

Scholars estimate Aristotle wrote hundreds of books in his lifetime. Only thirty-one have survived to this day. They are, for the most part, unfinished works — lecture notes, drafts, and the like. Parts were written up by his assistants under his guidance.[2]

They cover an extraordinary number of subjects. They include physics, astronomy, empirical biology, zoology, medicine, dreams, metaphysics, psychology, ethics, aesthetics, poetry, theater, music, rhetoric, linguistics, law, politics and government. Aristotle is also credited with "the first formalized system of logic and inference."[3]

Aristotle's mostly unfinished works were brilliant in scope and revolutionary in content. They would become the principal source of knowledge and basis for nearly all academic studies in Medieval Europe, as we shall see.[4]

The master of them that know.

— Dante on Aristotle,
13th century AD.[5]

One book in particular stands out as the chief source of Medieval thinking on cosmology — Aristotle's famed *De Caelo* or *On the Heavens*. It has been called "the most influential treatise of its kind in history."[6]

On the Heavens

"Nature leaves nothing to chance."

— Aristotle, *On the Heavens*

In *On the Heavens*, Aristotle engages without polemics. His tone is gentle, thoughtful, without rancor or overconfidence. He presents reasons for his position, lists the views of those who disagree, and is open to further investigation — an admirable precedent for future scientific inquiry.

Aristotle takes Plato's challenge one step further. He searches not just for mathematical models to match celestial observations, but for the physics behind them. He develops his cosmology from his theories on motion and the nature of the elements. His arguments are based on belief (Plato's and his own) and observation.[7]

Like Socrates and Plato before him, Aristotle's universe is teleological. That is, all "things are what they are because of the *purpose* they serve." Nonetheless, there is no divine Craftsman. To Aristotle, it is nature itself which acts purposefully.[8]

Aristotle's core tenet is that all things *move* according to their nature. Let's look at an example.

Aristotle argues that the Earth is spherical. If so, why don't we fall off? Because "the natural place of the element earth is downward." Thus solid bodies fall towards the center of the Earth. "Sparks fly upward because the natural place of fire is in the heavens."[9]

He contends that the Earth and the heavens are governed by *different* physical laws.

Earth is subject to decay and change. Unless disturbed, Earthly objects move in straight lines.[10] Earth is made of the elements earth, air, fire, and water.

The heavens are incorruptible and eternal. Heavenly bodies — the stars, planets, Moon, and Sun — move in perfect circles or combinations thereof, ala Plato. They are made of a fifth element: imperishable *aether*.[11]

Like Plato, Aristotle argues that it is observation which confirms that the heavens are eternal. "For in the whole range of time past, so far as our inherited records reach," he writes, "no change appears to have taken place either in the whole scheme of the outermost heaven or in any of its proper parts." Of course, we now know that the stars do move with respect to each other, but their distances from us are so great we generally cannot detect these motions with the naked eye over many human lifetimes.[12]

> *The ancients gave to the gods the heavens*
> *or upper place, as being alone immortal.*
>
> — Aristotle[13]

Why are heavenly bodies permanent and unchangeable? Because "whatever is divine, whatever is primary and supreme," Aristotle writes, "is necessarily unchangeable."[14]

And why do heavenly bodies move in perfect circles? Because circular motion has no beginning or end. It is eternal. "The activity of [the gods] is immortality," Aristotle writes, "... therefore the movement of that which is divine must be eternal."

> *... it is the earth which is at rest at the center.*
>
> — Aristotle[15]

Aristotle proposes that Earth is at rest at the center of the universe. In this anthropocentric world view, all heavenly objects revolve around us. A stationary Earth "is required," Aristotle writes, "because eternal movement in one body (the heavens) necessitates eternal rest in another."[16]

Not all concurred. "There is no general agreement," Aristotle tells us, as to the position of the Earth or whether it is at rest or in motion. Pythagoreans, for instance, believed all bodies including the Sun, the Earth and a counter-Earth revolve around a "central fire invisible to human eyes." They also believed, as did Socrates, that Earth is flat.[17]

Aristotle gives the common argument for a stationary Earth. If it moves, an object "thrown straight upward would be left behind (and) fall to a place different from where it was thrown."[18]

He also contends that we'd see an effect called "stellar parallax." Stars would appear to change their positions with respect to each other due to changes in Earth's location. "Yet," Aristotle notes, "no such thing is observed."[19]

To get an idea of what parallax is, place your index finger pointed upward in front of and close to your nose. Now open only the left eye, and then only the right eye. Repeat. Do you see the finger appearing to move left and right? This is parallax. The location of the finger appears to move against the background scene because you are looking at it from two different positions; the left eye position and right eye position.

Now stretch your arm out to put that same index finger far away from your face. Repeat the opening of only the left eye, then only the right. The finger still appears to move, but a much smaller amount. You see that the *farther away* an object is, the *less the parallax* effect.

Aristotle had no idea how very far away the stars are. Our nearest star outside the Sun is Proxima Centauri. Modern measurements put it some 4.2 light years or 25 trillion miles (40 trillion kilometers) from Earth. At that distance, stellar parallax is way too small to see with the naked eye. It was not detected by telescope until 1838 AD.[20]

There were also ongoing disputes about the shape of the Earth as well. "Some think it is spherical," wrote Aristotle, "others that it is flat and drum-shaped." As we saw, a flat Earth has roots in Sumerian cosmology.

The drum-shaped rather than rectangular perimeter most likely comes from Egyptian cosmology.[21]

"Others say the earth rests upon water," Aristotle wrote. "This, indeed, is the wildest theory that has been preserved, and is attributed to Thales of Miletus."[22] Recall the "water" idea has roots in Sumerian cosmology as well.

A spherical Earth

Aristotle backs up his argument that Earth is spherical with observations:

The shadow of the Earth on the Moon is curved during a lunar eclipse — "In an eclipse the outline is always curved," Aristotle writes, "and since it is the interposition of the earth that makes the eclipse, the form of this line will be caused by the form of earth's surface, which is therefore spherical."[23]

We see new stars when moving north or south — When you travel south at night, new stars appear to rise over the horizon. The reverse is seen when traveling north. "Indeed there are some stars seen in Egypt and in the neighborhood of Cyprus which are not seen on the northerly regions," Aristotle writes, "and stars, which in the north are never beyond the range of observation, in those (former) regions rise and set."[24]

Ships appear to sink over the horizon — As Greek navigators knew, a ship at sea appears to sink as it moves over the horizon.

Sun elevation varies with latitude — Other evidence known to the Greeks: the further south one travels, the higher the elevation of the Sun at noon.[25]

This is summarized in Fig. 6.1.

Given a spherical Earth, Aristotle proposes that objects fall towards its *center*. "That the center of the earth is the goal of their movement," he writes, "is indicated by the fact that heavy bodies moving towards the earth do not [fall] parallel but so as to make equal angles, and thus to a single center, that of the earth."[26]

Earth's "shape must necessarily be spherical," Aristotle continues. "For every portion of earth has weight until it reaches the center, and the jostling of parts greater and smaller would bring about ... compression and convergence of part and part until the center is reached ... If the earth was generated (created rather than always existing), then, it must have been formed in this way, and so clearly its generation was spherical."[27]

Wow! This description of how Earth formed presaged Isaac Newton's law of gravity some two millennia later. It is remarkably close to our modern understanding.[28]

86 *Cosmic Roots*

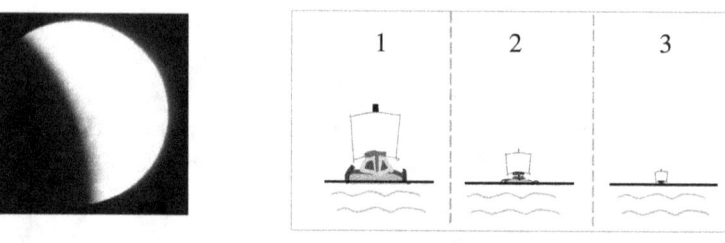

(A) - Curved shadow of Earth on Moon during eclipse

(B) - Ship appears to sink over horizon

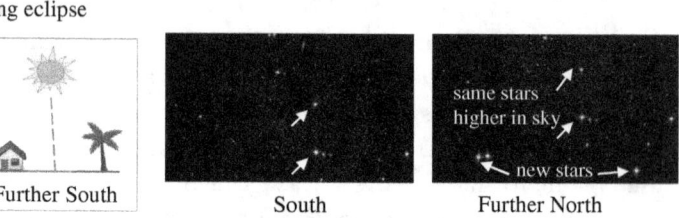

(C) - Sun elevation varies with latitude

(D) - See new stars when moving north or south

Figure 6.1. Aristotle's observations which support a spherical Earth.

Aristotle also ponders questions which haunt the thoughts of theoretical physicists to this day. Are there other universes, he asks? He argues for only one. Is the universe finite or infinite? He contends the universe is eternal in time — but finite in space.[29]

This eternal, no beginning of the universe contradicts Genesis in the Hebrew Bible. It would be the belief of physicists until the Big Bang Theory would come to prominence in the mid-20th century.

Does time exist beyond the universe? No, the master maintains. "It is clear that there is neither place, nor void, nor time, outside the heaven."[30]

There was a more "practical" application for Aristotle's brilliance. Recall that Plato had challenged his students to come up with a *geometric* model of the universe — one that would replicate the observed motion of the heavenly bodies using perfect circles and spheres. Aristotle would take up the challenge and produce a cosmic model which would stand for some 1800 years.[31]

Aristotle's Model

We imagine Aristotle studying the Athenian sky night after night, attempting to come up with a geometric scheme to match what he sees.[32] What is his construct? A universe of great concentric, interlocking *crystalline spheres*, all rotating around stationary Earth at the center.[33]

Aristotle's construct is a modification of a model developed by Greek astronomer Eudoxus of Cnidus (c. 390–340 BC) and his student, Calippus of Cyzicus (c. 370–c. 300 BC). It proposes that the stars, the planets, the Sun, and the Moon are "imbedded in the rims" of invisible celestial sphere.[34]

This conception is yet another which appears to have originated with the Pythagoreans. In about 530 BC, they imagined that the stars and planets sit on concentric spheres.[35]

In Aristotle's model, the individual stars, planets, Sun, and Moon are themselves at rest — or as Aristotle puts it, "not self-moved." It is the great spheres that do the moving, carrying the heavenly bodies around with them.[36]

One sphere holds all the "fixed" stars. Each of the five known planets has its own celestial sphere. Another sphere holds the Sun and the innermost sphere holds the Moon. Each sphere slowly rotates at a different but constant angular velocity — from "one month for the Moon to thirty years for Saturn."[37]

What *causes* Aristotle's nested spheres to rotate? The great natural philosopher imagines each celestial sphere is moved by the "Prime Mover," an outermost sphere whose rotation makes all the inner spheres rotate.[38]

> *Since everything that is in motion must be moved by something ...*
> *there must be some first mover.*
>
> — Aristotle, *Physics*[39]

The axis of each sphere is attached to the sphere which surrounds it, so they are all connected. Thus, when the Prime Mover rotates, all other spheres do so as well.

A simplified diagram of Aristotle's overall structure is shown in Fig. 6.2. As you can see, it shows the spheres (shown as circles in the 2-D figure) holding the stars, planets, Sun, and Moon.

Is there anything we left out? What about comets?

Fiery messengers

> *... the heavens themselves blaze forth the death of princes.*
>
> — Shakespeare, *Julius Caesar*[40]

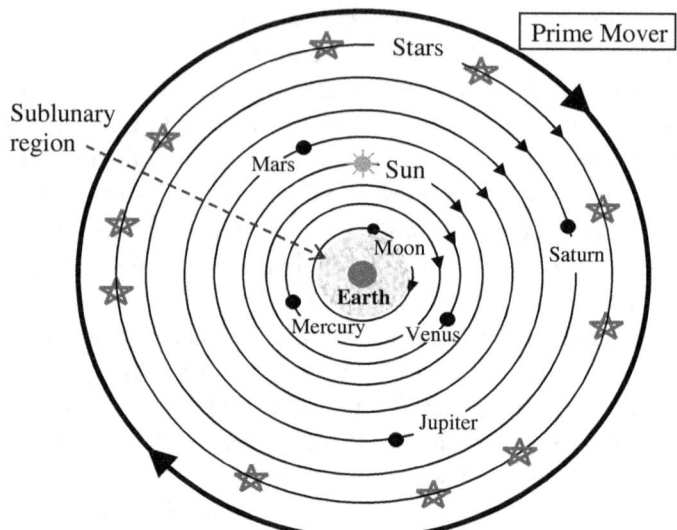

Figure 6.2. Aristotle's Earth-centered Universe (simplified) — All celestial objects (stars, planets, Sun, and Moon) are embedded in a series of concentric spheres which rotate in uniform motion (at different rates) around the Earth. Rotation of the outermost sphere, the "Prime Mover" causes all other rotations. (Not to scale.)

Aristotle declares that all objects in the heavens are eternal and unchanging. Not comets. They suddenly appear, blaze across the sky, and disappear.

How does Aristotle deal with this conundrum? He labels the region *below* the Moon the "sublunary sphere." Contained within it are the space between the Earth and Moon, Earth's atmosphere, and Earth itself. Inside this realm, all is corrupt and changeable. All things from the Moon and above are eternal and unchanging.

Comets, Aristotle contends, exist *only* within this sublunary sphere — that is below the orbit of the Moon. (Again, please refer to Fig. 6.2.)

Now let's look at the detailed structure of Aristotle's model.

The Eudoxus model

Recall Plato's challenge: to find perfect circles behind the movement of the planets. How does one hold to Plato's rules and replicate the irregular retrograde motion of the planets with a bunch of spheres?

It was Plato's student Eudoxus of Cnidus who first came up with an attempted solution. He represented the motion of each planet with a combination of *four interlocking concentric spheres* rotating on various tilted axes. The idea here was to produce *compound* motion.[41]

How does it work? A planet is attached to the rim of the innermost sphere. As it rotates, the other three spheres rotate at different rates, on different axes, and in different directions. These parameters are adjusted to reproduce the retrograde motion of the attached planet.

This is certainly not circular motion. It is a *combination* of circular motions. Plato's perfect circles remain sacrosanct. How tenaciously we humans cling to our beliefs.

To account for the apparent drift of the Moon and Sun seen in the sky over the months and years, Eudoxus included three interlocked spheres for each body. Only the "dome" of fixed stars remained with a single sphere.[42] All in all, the Eudoxian model totaled 27 spheres. His student Calippus added 7 more for improved accuracy.[43]

Science historian Michael Crowe calls Eudoxus' model "among the most impressive product of Greek genius."[44]

Aristotle's Construct

Aristotle adopted the Eudoxus/Calippus model and made key modifications of his own. The Eudoxus/Calippus model had a *separate* set of spheres for each heavenly body, e.g., a set of spheres for the Sun, another separate set for the Moon, a set for Venus, etc. These individual models were not linked together into an overall system.

It was Aristotle who *connected all the spheres*. He then inserted a series of spheres as counter-turners. They were required so the motions of outer sphere sets were not passed on to inner ones. All in all, his model contained 56 spheres.

Aristotle's connected model with the Prime Mover and counter-turners was the first integrated geometric model of the universe in human history.

Eudoxus considered his construct only a mathematical model. Aristotle proposed that his model was *"physical."* To him, the crystalline spheres really existed.[45]

How closely did Aristotle's integrated model match observations? Not too well. It agreed to some degree with the observed retrograde motions of Jupiter and Saturn. It did poorly for Venus and Mars. It failed for Mercury.[46]

There was another problem. In Aristotle's model, a planet is attached to a rotating sphere — thus always the same distance from Earth. Thus his model failed to account for the observed increase in brightness of a planet when in retrograde.

There was yet another issue. Ancient astronomers knew Venus always appears close to the Sun in the sky. It may be seen in the west in the early evening near to where the Sun has set. Or it can appear in the east in late evening close to where the Sun will rise.

Mercury also appears close to our home star. As seen from Earth, it is never more than about 28 degrees from the Sun — Venus never more than about 45 degrees. Aristotle's model failed to account for this so-called "bounded elongation" as well.[47]

For more torture, I mean for a more comprehensive description of the Eudoxus/Calippus/Aristotle model, please see Appendix A.

Aristotle died in 322 BC — the same year democracy came to an end in Athens. Despite its limitations, his geometric model of the universe would take its place alongside his physics to dominate natural philosophy in Europe and beyond.*

A New City

Ancient Macedon was a mountainous country on the northeast border of Greece. In the mid-fourth century BC, its king, Philip II, transformed it into a great military power. Through a series of wars and clever political machinations, he came to dominate most of the Balkan peninsula. By 338 BC, he was ruler or hegemon of all city-states of mainland Greece except Sparta.[48]

Sixteen years later, Athens revolted and was crushed in the Battle of Crannon. Democratic rule was replaced by an oligarchy of wealthy men. The city of Socrates, Plato, and Aristotle began its long decline.[49]

A new city founded by Philip's son would come to replace Athens as the center of learning and science in the ancient West. Here a Greek astronomer would come to produce an accurate model of the heavens that was the wonder of the classical world.

This is the subject of the next two chapters.

---

*About the same time as Aristotle, astronomers in China began making groundbreaking observations of the heavens. They were "the most persistent and accurate observers of celestial phenomena in the world," writes British biochemist and historian Joseph Needham. "So far as we know, the first people to map the positions of stars were the Chinese astronomers Shi Shen, Gan De and Wu Xian in the third and fourth century BC." During the summer of 365 BC. Gan De apparently saw "Jupiter's 3rd satellite Ganymede [or 4th satellite Callisto] with his naked eye." This was nearly 2,000 years before Galileo's discovery of the moons of Jupiter with his telescope (see Chapter 22). The Chinese were especially zealous about *time* e.g., the timing of lunar eclipses, etc. A striking example is fifth century AD astronomer and polymath Zu Chongzhi (429-500). "Using self-designed instruments, he determined that the year is 365.24281481 days long." This agrees with modern measurement to within less than a minute. Sources: Needham, p. 498.; Shuttleworth, "Ancient Chinese Astronomy" explorable.com.; "Early Star Maps and Astrology" history.aip.org.; Kronzek, Rochelle. email to author, Mar 20, 2018.

Chapter 7

THE CITY OF THE MIND

Euclid alone hath looked at Beauty bare.

— Edna St. Vincent Millay[1]

From his mother, Olympias, he was descended from Achilles, hero of the Trojan War. From his father, King Philip II, he was descended from Heracles (Hercules to the Romans), mortal son of Zeus, king of the gods. Or so legend has it.[2]

He led troops into battle at age 16 and again at age 18, victorious both times. He became king of Macedon at age 20. His name was Alexander III Philipou Makedonon. We know him as Alexander the Great.[3]

The God-King

Alexander (Fig. 7.1) became king when his father, Philip II, was assassinated in 336 BC. The young monarch's first actions were to execute those deemed complicit in his father's murder. He then had all possible rivals to the throne killed. His mother, Olympias in turn killed Philip's other wife and her infant daughter — common practices in the treacherous Macedonian royal court.

Through conquest, diplomacy, and bribes, Philip II had subjugated most of the Greek city-states. This included Athens, Thebes, and Corinth, but not Sparta. Shortly before his untimely death, he had announced his plan to attack Persia.

Upon his rise to the throne, Alexander quickly subdued Greek revolts. He then took up his father's dream.[4] With a mere 40,000 troops — primarily Macedonians as well as Greeks of the Hellenic League — Alexander

Figure 7.1. Bust of Alexander the Great. Called Hermes Azara, it is a copy of 1st or 2nd century AD bronze sculpture made by Lysippos. Found in Tivoli, East of Rome, Italy.

set out in 334 BC to vanquish the world's greatest power, the Achaemenid Persian Empire.[5]

Alexander was a fierce warrior, brilliant strategist, and brutal conqueror. He honed the army he had inherited from his father into a fighting force the likes of which the world had never seen. His marches were a "marvel of planning, speed, and endurance; his supply strategies and implementations miracles of military logistics," historians Thomas Martin and Christopher Blackwell point out.[6]

City after city opened their gates in fear. Those that resisted — such as Thebes and Gaza — were crushed, their buildings leveled, "every male defender killed, every woman and child sold into slavery." He outmaneuvered and out-fought the forces of Persian king Darius II in two epic battles — defeating an army two to three times his size.[7]

Within ten years Alexander was "King of Macedonia, hegemon of Greece, Pharaoh of Egypt, and King of Asia." His empire (Fig. 7.2) extended from the Adriatic to the fringes of western India — an area of some 2 million square miles (5.2 million km².)[8]

Alexander encouraged the belief that he was divine — much to the chagrin of his Macedonian and Greek compatriots. He saw the advantages of self-deification and its power to hold a diverse empire together. His

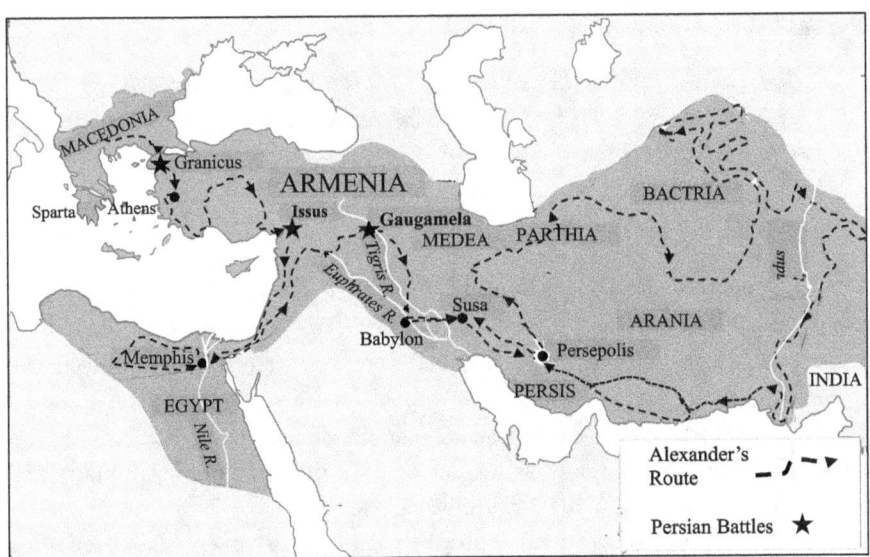

Figure 7.2. The Empire of Alexander the Great. He defeated Persian King Darius II in two epic battles, Issus and Gaugamela (see black stars).

campaign was effective. At the river Kabul, for example, rajas (Indian kings and princes) told Alexander they had come to see the visiting god.[9]

Immortality

Alexander died on June 10, 323 BC under mysterious circumstances in Babylon — the ancient home of the Amorites and now capital of his empire. He succumbed after a ten-day fever. Was it from natural causes? Or was it from a slow acting poison, as some scholars have suggested? He was 32 years old.[10]

Like his supposed "ancestor" and personal hero, Hercules, Alexander became a god after his death. Athenians built an altar dedicated to Alexander, "the invincible god." Over time, a number of people over the known world came to worship the great conqueror.[11]

Plutarch added to the legend in the 1st century AD. According to the famed Greek biographer, Alexander's mother Olympias told him before he departed for Asia that she had been impregnated by a bolt of lightning from Zeus, king of the gods. He, Alexander, was born from this miraculous conception.[12]

Dissolution

> *And a mighty king shall stand up, that shall rule with great dominion ... his kingdom shall be broken, and shall be divided towards the four winds of heaven ...*
>
> — Daniel 11:3-4.

After his death, Alexander's empire fell apart like sand castles against a rising tide. None of his generals — "Successors" as they were called — possessed his strength of will, political acumen, or charisma to hold the empire together.

Alexander's war of conquest had produced massive casualties on both sides. Perhaps as many as a quarter of a million people, military and civilian, were killed. Now his empire was gone.[13]

What did Alexander accomplish other than mass slaughter? For one thing, he ended nearly two centuries of war with Persia. They would never again threaten Greece and the West. And his campaign of conquests produced a new culture.

The Hellenistic Age

Alexander's empire opened vast areas of the East to Greek civilization. Greco-Macedonian language, art, literature, philosophy, science, and populations filled the lands of the former Persian empire and beyond. These included the Levant, Egypt, Mesopotamia, Iran, the lands of central Asia, and India.[14]

A hybrid Greco-Asian culture developed, especially amongst the upper classes. Modern scholars dubbed it "Hellenistic." (Greeks trace their origins to some forty or so Hellenic tribal nations.)[15] The death of Alexander in 323 BC marked the beginning of the so-called "Hellenistic Age."[16]

The core of Hellenistic culture "was essentially Athenian," argues historian Peter Green. "Town planning, education, local government, and art . . . were all based on Classical Greek ideals."[17]

The process of Hellenization fostered a major increase in trade between east and west.[18] For example, land routes linked Mediterranean ports in Syria and Palestine to Bactria (modern-day Afghanistan, Uzbekistan, and Tajikistan). Bactria in turn had access to the markets of India. As a result, we find Greek-like realistic statues and monumental stone sculpture in India for the first time.[19]

Greek science made its way across Eurasia as well. Archeologists have uncovered "Greek astronomical instruments dating to the 3rd century BC in the Greco-Bactrian city of Ai Khanoum in modern-day Afghanistan."[20]

The Hellenistic Empires

Several decades of civil war followed the death of Alexander. Three victors — all Macedonian generals under Alexander — came to rule the Hellenistic world. Seleucus reigned from Asia Minor and Syria to Mesopotamia and Bactria. Antigonus and his son Demetrius held Macedonia, Thrace, and parts of northern Asia Minor. Ptolemy governed in Egypt.[21]

Alexander had founded some twenty cities in his name.[22] The first and most famous was Alexandria in Egypt in 332 BC. Here in this hybrid Greek-Egyptian city, the next great advances in mathematics and astronomy would occur.

The City of the Mind

After the demise of Alexander's empire, his general and childhood friend, Ptolemy Lagides (c. 367 BC–283 BC), seized Egypt and declared himself Pharaoh. He called himself Ptolemy I Soter or "Savior."[23]

He made Alexandria his capital and created a metropolis that would become the largest city in the ancient world.[24]

The great city stood on the banks of the Mediterranean at the western edge of the Nile delta. Various scholars describe the city: Its bustling, crowded streets ran by exquisite stone works, magnificent royal palaces, "temples, theatres, administrative buildings, a coin mint, and a zoo." Markets featured expensive foods, fine wines, pungent spices, and "gorgeous silks and fabrics from the East."[25]

Streets were laid out in a rectangular grid system — "its two principal streets said to be over 100 feet wide and paved with gold."[26]

In its harbor stood the first lighthouse ever built. Situated on the island of Pharos, it was one of seven wonders of the ancient world. It stood some 400 feet (125 meters) high. Reportedly, ships at sea could see it's great light at night from up to 100 miles away.[27]

Situated on the city's acropolis was the Serapeum. It has been called "one of the grandest monuments of Pagan civilization." Inside this temple of gleaming marble stood the massive statue of Greek-Egyptian god Serapis — a "dazzling colossus of multiple precious stones and metal," writes classical scholar Dustan Lowe.[28]

The Kingdom of the Ptolemies lasted three centuries and came to encompass "southern Syria in the east to Cyrene in present day Libya in the west, and the frontier of Nubia to the south."[29]

Seat of the Muses

When Alexander was in his early teens, Aristotle was his tutor. From the great natural philosopher, Alexander developed a lifelong commitment to the advancement of science. He brought with him on his mission of conquest "explorers, engineers, architects, and natural philosophers."[30]

Ptolemy sought to emulate his deceased friend and former leader. He wanted to dispel the common Greek notion that Macedonians were semi-barbarians. Under his patronage, Alexandria would become the premier center of learning and science in the known world.[31]

Adjacent to the royal palace, Ptolemy and his son built the Alexandrian Museum — the Seat of the Muses, goddesses of knowledge and culture. This complex of buildings housed researchers, laboratories, observatories, botanical gardens, and "richly decorated lecture and banquet halls."[32]

Scholars from all over the Greek world came here and wrote works in "science, history, applied science, mathematics, optics, psychology, applied medicine, botany, hydraulics, engineering, and mechanics."[33]

Within this state-funded complex was the fabled Library of Alexandria — its mission "to acquire a copy of every book in the world."[34]

At its peak, the library is thought to have contained over 500,000 volumes. They included Greek, Roman, and Oriental works, as well as "ancient Egyptian texts, Hebrew scriptures, and writings of the Persian prophet Zoroaster."[35]

Among the scholars who worked at the Library were three outstanding Greek mathematicians.

Archimedes (c. 287–211 BC) from Syracuse studied in Alexandria. He is considered to be the greatest mathematician of the ancient world. The "Great Geometer" Apollonius (c. 240–c. 190 BC) from Perga wrote his celebrated thesis on conics here.

Euclid of Alexandria (c. 365–c. 275 BC) collected the mathematics of the ancients into a single work: the *Stoicheia* (*Elements* in English). This "comprehensive thirteen volume masterpiece" has been called "the most enduring mathematical work of all time."[36]

Three astronomers also studied in Alexandria. With observations and mathematics, especially geometry, they made great leaps in our understanding of the cosmos.

Aristarchus

Aristarchus (c. 310–c. 230 BC) from Samos estimated "the sizes of the Sun and Moon and their distances from Earth" (in terms of the diameter of the Earth). His mathematics were impeccable, but his estimates were way off due to naked eye observational errors.[37]

In about 280 BC, Aristarchus made a radical proposal: Earth not only rotates but *orbits the Sun* along with the other planets. This stimulated numerous arguments against such a foolish notion[38]

They included: If the Earth moves, why don't we feel any motion? A perpetual wind would blow over the Earth. An object thrown straight up into the air would land behind the thrower.[39] And my favorite — "why would the center of the universe be moving?" More on this later.[40]

Greek stoic philosopher Cleanthes of Assos urged that Aristarchus be charged with impiety for making such a blasphemous claim — yet another early clash between science and religion.[41]

Eratosthenes

Eratosthenes (c. 276–c. 195 BC) from Cyrene in Egypt used geometry to calculate the *circumference of the Earth*. His answer was between 1% and 16% of the modern value. Why the range of percentages? We are not sure of the exact conversion from his units of length (stadia) to modern units of length (such as miles or kilometers).[42]

Hipparchus

Hipparchus (c. 190–120 BC) was from Nicaea in modern north-western Turkey. He was a pioneering mathematician and geographer. Some credit him with the invention of trigonometry. He was also the first to use latitude and longitude to determine terrestrial locations. It is for his work in astronomy that he is most remembered.[43]

Hipparchus made stellar observations at Bithynia in what is now Turkey; north of the island of Rhodes; and in Alexandria. From these, he determined the positions of stars to the greatest accuracy yet. He listed his findings in a compendium of some 850 stars.[44]

Some thought to catalogue these heavenly gods was also a grave impiety.[45]

Hipparchus calculated the orbits of the Moon and Sun, as well as their sizes and distances from Earth. His values for the size and distance to the Moon were excellent. His estimate for the size and distance to the Sun were way too small.[46]

He also calculated the length of the year — to within a remarkable 6½ minutes. In doing so, he discovered the so-called "Precession of the Equinoxes." We now know this effect is due to the long-term wobble of the spinning Earth's axis. Its period is some 26,000 years.[47]

Despite their seminal accomplishments, the Hellenic astronomers of Alexandria never developed a satisfactory model for planetary motion. This would come from an Alexandrian astronomer in a new era.[48]

The Ascent of Rome

Like her predecessors, Cleopatra VII Philopator (69–30 BC) was Pharaoh of Egypt, a Macedonian, and a descendent of Ptolemy I Soter. She learned to speak Egyptian — a rare act which endeared her to her people. The famous

Queen also had a child with Julius Caesar and later became Marc Antony's lover.[49]

In 31 BC, the combined forces of Cleopatra and Marc Antony were routed by Gaius Octavius aka Octavian in the Battle of Actium on the Ionian Sea. The nephew and adopted son of Julius Caesar then seized power in Rome. He became its first and greatest Roman emperor: Caesar Augustus. Egypt with its vast agricultural riches along the Nile was the crown jewel of his empire.[4848]

A year later Cleopatra VII — the last of the Ptolemies — took her life in Alexandria. Her lover Mark Antony soon followed.

Historians mark the Battle of Actium as the end of the Hellenistic Period and the start of the Roman Era.[50]

Pax Romana

Over some four centuries, the Romans established a veritable military and political machine. With its vaunted legions, fearsome war apparatus, extensive network of roads, and ruthless suppression of revolts, Rome captured *and held* territories on three continents.[51]

Imperial Rome reached the height of its empire in the second century (100s) AD. A single city state now ruled over Italy, Greece, Spain, Gaul, and north Africa. Its empire extended from southern Britain in the west to Judaea, Syria, Mesopotamia, Assyria, and Armenia in the east.[52]

A general peace and stability reigned throughout the sprawling empire. It was a time when "the conditions of the human race were most happy and prosperous," according to famed English historian Edward Gibbons.[53]

Perhaps for Roman citizens. Or at least the wealthy ones. It was certainly not for slaves. Nor for Jews and Christians, as we shall see.

The Greco-Romans

> *Captive Greece held captive her uncouth conqueror (Rome) and brought the arts to the rustic Latin lands.*
>
> — Horace, Roman lyric poet.[54]

Greece was to Romans what Sumer had been to the Amorites — a high culture to be venerated and copied. Romans emulated classical Greece in their dress, literature, art, architecture, and theatre. Roman gods were mostly Greek gods, only with different names. Numerous Roman paintings and poems were based on Greek myths[55]

Upper class Romans generally hired educated Greeks to care for and teach their young. The best Roman libraries had "separate sections for Greek and Latin books."[56]

Romans were generally more practical-oriented than Greeks. Think aqueducts and concrete. Roman science was nowhere as innovative and cosmic as that of the Greeks. Romans were more engineers than natural philosophers.

The Other Ptolemy

Hellenistic Alexandria remained a prosperous center of learning under the Romans. By the second century AD, it was second only to Rome in size and wealth. Here the next great leap in astronomy would be put forward — not by a Roman but by a Greek named Ptolemy (no relation to the governing Ptolemies).

Working at the great Museum in Alexandria, Ptolemy tried to solve the problem of the wandering planets.[57] Recall that Aristotle's crystalline spheres were not very accurate, especially in modeling the retrograde motion of the planets. To better account for the movement of the celestial gods, Ptolemy created a remarkable geometric construct — the world's first accurate model of the heavenly bodies.[58]

This is the subject of the next chapter.

Chapter 8

THE GREAT ASTRONOMER

... when I search out the massed wheeling circles of the stars, my feet no longer touch the Earth, but, side by side with Zeus himself, I take my fill of ambrosia, the food of the gods.

— Ptolemy[1]

Claudius Ptolemy
(c. 100–170 AD)

Claudius Ptolemaeus (c. 100–170 AD) was a Roman citizen of Greek (or perhaps Egyptian) ancestry. He was born during the reign of Roman Emperor Trajan.

Ptolemy, as he is called in English, was a renowned geographer, a pioneer in music theory, and a highly skilled mathematician — especially in the nascent field of trigonometry. Ptolemy also made major contributions to astrology. It is in astronomy where Ptolemy had his most profound influence. He is recognized as the greatest of his kind in the ancient world.[2]

In c. 150 AD, Ptolemy produced the *Almagest*, a 13-book masterpiece, which did for astronomy what Euclid did for geometry and mathematics. It codified all "extant knowledge in the field of astronomy" known in the West at the time into a single volume.[3]

Ptolemy's sources included the teachings of Aristotle, Hellenistic astronomers such as Apollonius and Hipparchus, as well as Babylonian astronomical data.

Almagest

Ptolemy's "Almagest" shares with Euclid's "Elements" the glory of being the scientific text longest in use. From conception in the second century up to the late Renaissance, this work determined astronomy as a science.

— G. Grasshoff[4]

Ptolemy's masterwork, *Megale Syntaxis* — transformed from *al-majisṭi* or 'The Great' in Arabic to *Almagest* — provided geometric models for calculating celestial motions, extensive tables to compute future or past positions of the five known planets, and a comprehensive star catalogue listing 1028 stars.[5]

How much of Ptolemy's work is based on Hipparchus remains in dispute. Nonetheless, his models and extensive data compilation remained the foundation of astronomical endeavors for the next 1400 years.[6]

In *Almagest*, Ptolemy writes in a scholarly, philosophical fashion. In Book I, he pays tribute to the gods, and later acknowledges the power of mathematics and beauty of astronomy.

"Now the first cause of the first motion of the universe . . . can be thought of as an invisible and motionless deity . . ." Ptolemy writes. He goes on to declare, "Only mathematics can provide sure and unshakable knowledge to its devotees . . . For its kind of proof proceeds by indisputable methods, namely arithmetic and geometry."

He then unites his religious and scientific views: "This science [astronomy], above all things, can make man see clearly from the constancy, order, symmetry, and calm which are associated with the divine. It makes its followers lovers of this divine beauty..."[7]

Ptolemy based his model on the world view of Aristotle. As we learned, Aristotle held that the Earth is *at rest at the center of the universe*. It does not move in any orbit or spin on its axis. Here all celestial objects — the Sun, Moon, five known planets, and stars — orbit the Earth.

In *Almagest*, Ptolemy wrote: "the earth lies in the middle of the heavens... and it has no motion from place to place." He acknowledged that "certain people supposed the heavens to remain motionless, and the earth to revolve from west to east... making approximately one revolution each day. [But] from what would occur here on earth and in the air, one can see that such a notion is ridiculous."[8]

"... the revolving motion of the earth must be the most violent of all motions associated with it; seeing that it makes one revolution in such a short time," Ptolemy argued. "The result would be that all objects not actually standing on the earth would appear to have the same motion; opposite to that of the earth."

"Neither clouds nor other flying or thrown objects would ever be seen moving towards the east, since the earth's motion towards the east would always outrun and overtake them, so that all other objects would seem to move in the direction of the west and the rear."

The idea is that, if the Earth were to rotate, anything not tied down to it would fly off in the other direction. (See Fig. 8.1.) This argument is incorrect. The Earth does, of course, rotate. All earthly objects, whether on the ground or in the atmosphere, share in Earth's motion — thus move at the same general speed as the Earth. More on this in later chapters.

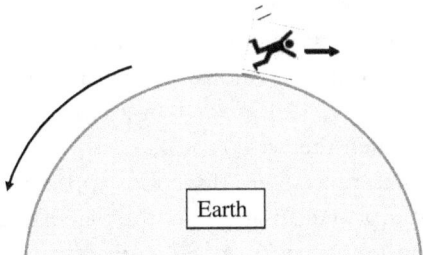

Figure 8.1. Ptolemy's Argument Against a Rotating Earth. Objects would fly off in the other direction. A common mistake in ancient times, since all on the Earth share in its motion. We are all rotating (and orbiting the Sun) along with the Earth, so do not feel its motion.

The twin assumptions that the Earth is stationary and planetary orbits are circular caused no end of difficulties in astronomical models for nearly two millennium. As we shall see in Part IV of this book, this would be finally resolved by Copernicus's Sun-centered system and Kepler's elliptical planetary orbits in the 16th and 17th centuries, respectively.

Ptolemy's Universe

Hereafter, when they come to model Heav'n,
And calculate the Stars; how they will wield
The mighty frame, how built, unbuilt, contrive
To save appearances; how gird the Sphere
With Centric and Eccentric scribbl'd o'er,
Cycle and Epicycle, Orb in Orb. . .

— John Milton, *Paradise Lost*[9]

Ptolemy sought to develop a truly accurate construct on the behavior of heavenly bodies. Here he was faced with his greatest challenge — how to model the odd retrograde motion and apparent distance variation (i.e., changes in brightness) of planets geometrically. And how to restrict this model to *perfect circles* travelling in *uniform* motion ala Plato.

Ptolemy adapted the so-called epicycle model of planetary motion. Apollonius is often credited with its invention. Others say it was probably known all the way back to the Pythagoreans.[10]

The trick is to model the orbits of planets with a combination of circles — a small one rotating around a large one. In this construct, the Earth is placed at or near the center of the large circle. A planet is then placed on the circumference of the small circle. The large circle is called a *deferment*; the small circle an *epicycle*. This arrangement is shown in Fig. 8.2.[11]

The construct is quite ingenious. Notice in the figure how the planet moves towards the Earth, *reverses direction* in a small loop, and then moves away from the Earth. Then the cycle repeats.

The occasional reverse in direction approximates the observed retrograde motion of a planet. In addition, the planet is closest to the Earth when it is in the loop going backwards, i.e., in retrograde motion — thus explaining the observed increase in brightness when in retrograde.[12]

Do you see it? Most of the time the planet moves in one direction. When it is looping, it moves in the other direction. And when it does, it is closer to Earth — so it is brighter!

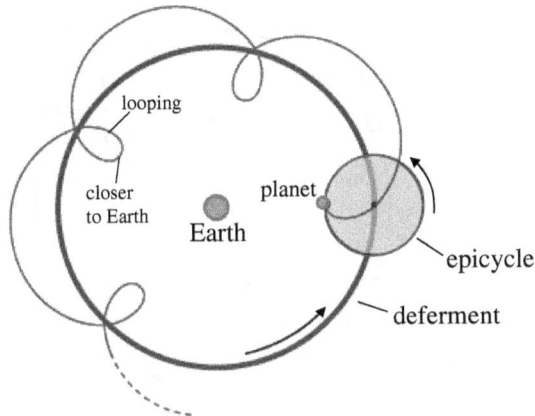

Figure 8.2. The Epicycle/Deferment Model Produces a Looping Planetary Motion. Earth is at the center of the large circle (the deferment). The planet is on the circumference of the small circle (the epicycle). The center of the epicycle rotates around the circumference of the deferment.

How clever. This epicycle construct mimics the retrograde motion and varying brightness of planets.

The Epicycle Ferris Wheel

To better visualize this, imagine yourself seated on the "Epicycle Ferris Wheel." It has a large wheel and a small wheel. You are seated on the edge of the small wheel, as shown in Fig. 8.3.

All is at rest. The attendant presses the button labeled "epicycle." The small wheel begins to rotate counterclockwise. You in your chair rotate in place around its center.

The attendant then pushes the "deferment" button. The giant wheel begins to rotate, also counterclockwise. Now you rotate around the circumference of the small wheel *and* the circumference of the large wheel. You feel a combined motion:

(1) down and up around the small circle;
(2) Simultaneously rising up with the large circle towards the sky, then down to the ground, and up again. You experience the "looping motion" of ancient Greek epicycles.

Fun ride. I should patent it. (Don't tell anybody about it.)

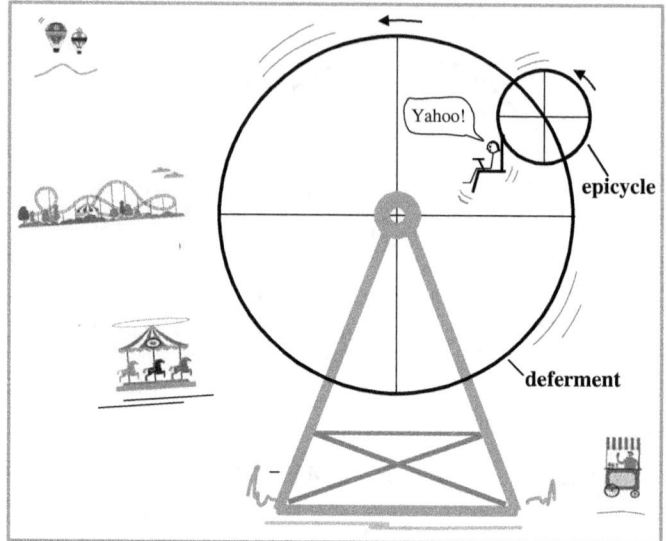

Figure 8.3. Epicycle Ferris Wheel. Large wheel (deferment) rotates around its center. Small wheel (epicycle) rotates around circumference of large wheel. Rider on small wheel represents planet. Center of large wheel represents Earth.

Ptolemy's Breakthrough

In his model, Ptolemy placed the Earth *near but not exactly at* the center of the large circle (the deferment). This is called the "eccentric" method. He then placed the planet in question on the circumference of the small circle, the so-called epicycle, as usual.

Ptolemy then makes a key modification to previous epicycle approaches. With the Earth off to one side of the large circle's center, he adds a point called an *equant* "symmetrically placed on the opposite side."[13] (See Fig. 8.4.)

Ptolemy's breakthrough was to make the motion of the epicycle uniform around the eccentrically-placed equant — rather than the center of the deferment as in prior models. This is shown in Fig. 8.5.[14]

Centered Method — Notice the two dotted lines in Figure 8.5(a). These lines intersect the Earth at the *center* of the deferment. They form two interior angles which are equal. Two large arrows in the figure show the path the epicycle takes over these angles — one arrow on the left and the other on the right.[15]

As usual, the epicycle travels in uniform angular motion; i.e., over equal angles in equal times. The epicycle's motion is with respect to the center

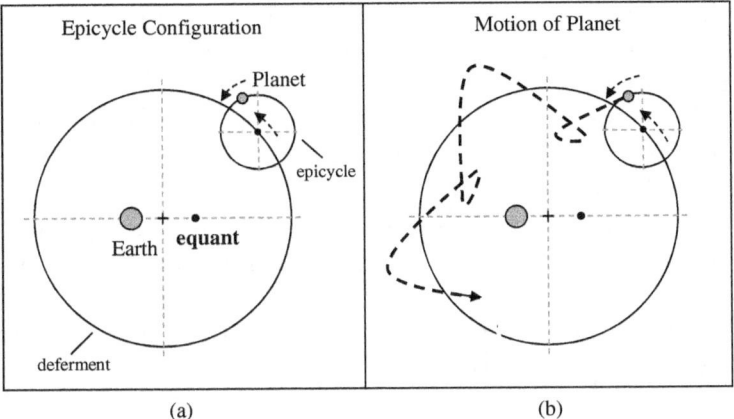

Figure 8.4. Ptolemy's Model of Planetary Motion. Earth is offset from the center of large circle, and the so-called equant is offset the same amount on the opposite side. (Figure not to scale.)

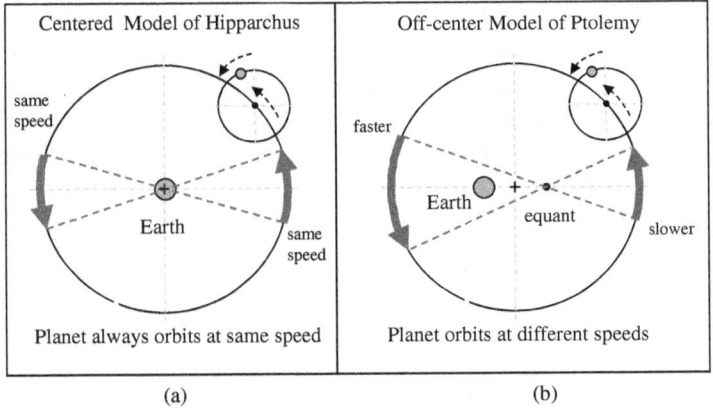

Figure 8.5. Apollonius's Centered vs. Ptolemy's Equant Model of Planetary Motion. (a) Earth is at the center of the deferment. On the left and right side, the epicycle travels over <u>equal angles at the same speed</u>. Thus the planets motion is constant. (b) The epicycle travels in uniform motion with respect to the off-center equant. Thus it moves slower on the left side of the deferment and faster on the right. The planet it is carrying moves at <u>varying speeds</u>. (Figure not to scale.)[15]

of the deferment. So the left and right arrows cover the *same distance* over the same time. Thus, the epicycle's speed is the same for both the left and right paths. And the planet carried by the epicycle travels at the *same speed* as well.

Equant Method — Now look at Ptolemy's construct in Fig. 8.5(b). Here the dotted lines intersect at the *equant*. The two interior angles they make are still equal. And, as before, both arrows travel over the equal angles in equal times. But, due to the offset equant, the left arrow's path is *longer* than right arrow's path.

Do you see it? Longer path on the left. Shorter path on the right. But both travel over the same period of time. So the epicycle must travel faster on the left side and slower on the right.

Thus the epicycle moves *at different speeds* around deferment. And since the epicycle carries the planet with it, the planet also moves at different speeds.

Aha! This is just what planets do — one reason why Ptolemy's model is more accurate than prior constructs.[16]

The Ptolemaic System

Ptolemy's model was flexible. To achieve higher accuracy, all he had to do was add another epicycle. He could place the planet on the circumference of an even smaller circle — with its center on the circumference of the existing small circle. It looks simpler than it sounds, as you can see in Fig. 8.6.

Need even more accuracy? Add another even smaller epicycle. Circle upon circle upon circle; epicycle on epicycle on epicycle. By making the little epicycles rotate slowly, Ptolemy could make tiny corrections to planetary orbits.

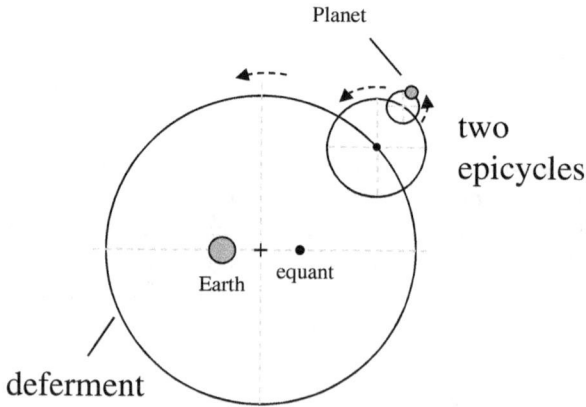

Figure 8.6. Example of a Two Epicycle Model. The addition of more epicycles can provide better agreement between the model and observations.

If the Lord Almighty had consulted me before embarking upon his creation, I should have recommended something simpler.

— Alfonso X of Castile c. 1250 AD, on the complexity of the Ptolemaic model.[17]

Tortuously complicated, yes, and it didn't give a physical picture of what is really going on. But it worked. Ptolemy himself regarded his model as only a mathematical tool and not physical reality.

The Ptolemaic model of the motions for the Sun, Moon, and five known planets required at least 80 epicycles. (Yes, the Sun and Moon had epicycles too.) By adjusting the number of epicycles, their sizes, and their uniform angular speeds, Ptolemy was able to match the observed positions of the known five planets to within 2 degrees — an accuracy never before achieved.[18]

Though no one understood why.[19]

Aristotle versus Ptolemy

Scholars arguing over Aristotle versus Ptolemy faced a dilemma. Aristotle considered his celestial spheres physically real. Ptolemy saw his model of deferments, epicycles, and equants as merely a computational device.

Why was Aristotle's model considered real? First of all, it was developed by the greatest natural philosopher in the ancient world. Secondly, Aristotle had given rational arguments for the reality of his model in *On the Heavens* and other treatises.

Aristotelians argued that Ptolemy's model was an ad hoc attempt to match observations — with no physical, spiritual, or esthetic considerations. Besides, "a planet on an epicycle would crash into" the crystalline sphere that carried it. Plus Ptolemy's model is flawed. Each planet's motion is "not truly centered on the Earth."[20]

Defenders of Ptolemy — generally known as astronomers — argued for mathematical models which best matched observations. Natural philosophers, on the other hand, sought a physical understanding of the heavens ala Aristotle. There was no love lost between the two.

Aristotle's model with "real" celestial spheres was inaccurate. Ptolemy's model was "not real" but highly accurate. Both, as we shall see, played a major role in the clash between science and religion. And both turned out to be wrong.[21]

A New Monotheism

By the time of Ptolemy, a new religion had made major inroads in the Roman empire. With roots in Palestine, its apocalyptic belief system would come to spread across the Roman world and set the stage for the next great conflict between science and religion in the West.

This is the subject of Part III of this book.

PART III
Fall and Rise

Chapter 9

ZEUS VERSUS *YHWH*

Vanity of vanities, all is vanity.

— Ecclesiastes 1:2.

It was a conflict between rich and poor, city and countryside — a struggle between change and tradition, assimilation and exclusion, a clash between West and East, secular and orthodox, the gods and the One God. The very survival of ethical monotheism was at stake.

Hellenization of the Holy Land

When we last talked about Judah, it was the small Persian province of *Yehud*. Centered on Jerusalem, it was roughly a tenth the area of modern Israel. Recall that Persian King Cyrus had issued an edict which permitted Jews to return to their homeland. Several thousand made the initial journey in 539 BC, led by Sheshbazzar, son of Jehoiachin, Judah's last king. They brought with them their new monotheism — the exclusive worship of *YHWH*.[1]

Persian Judah (c. 538–332 BC)

Persian kings let the Jews observe their own local laws — as they did all subject nations. For *Yehud* it was Torah, the laws of Moses given in the first five books of the Hebrew Bible.

Under enlightened Persian rule, *Yehud* was a semi-autonomous theocracy. The *kohen gadol* or high-priest held religious authority as head of a council of elders — the *Sanhedrin*. A Persian appointed Jewish governor attended to internal secular affairs. Over time, a collection of prominent families came to effectively run Jerusalem.[2]

In the cloistered Jewish culture of the time, there were no sculptures or paintings — no doubt due to the biblical prohibition against graven images. Science was not part of orthodox education. The study of foreign tongues was discouraged to avoid the "pernicious influence of heathen philosophy." It seems only the written word was encouraged.[3]

Upon their arrival from Babylon, the first wave of Jews had found the land occupied by Canaanites, as well as a mix of Syrians and Mesopotamians who had been brought to Israel by the Assyrians after they had conquered Israel in 722 BC. A number of returning Jews married some of these local women.[4]

In 445 BC, Persian king Artaxerxes I appointed his Jewish cup-bearer Nehemiah as governor of *Yehud*. Some forty-seven years later, Artaxerxes II sent Torah scholar and priest Ezra to lead the Jews in moral and religious reform.[5]

> *And the seed of Israel* separated themselves from all foreigners,*
> *and stood and confessed their sins, and the iniquities of their fathers.*
>
> — Nehemiah 9:2.

Nehemiah and Ezra vehemently objected to the mixing of Jews with pagan women. As commanded in Deuteronomy (Deut. 7:3–4), they induced the Jews to divorce their Gentile wives and no longer intermarry (Ezra 10:10-11). This further increased the isolation of the Jews and their designation as a separate ethnic group — with ramifications that continue to this day.[6]

The two centuries of Persian rule would be remembered as a time of relative peace. Nonetheless, Judah yearned for independence.

A Turn to the West

On his way to Egypt, Alexander the Great had conquered Tyre and Gaza along the Mediterranean coast in 332 BC. The young Macedonian king did not bother to attack little Judah. With his defeat of the Persians, it came under his rule without a fight. The Holy Land was now Judea or *Ioudas* in Greek — a vassal state of Macedonia.[7]

To the relief of the Jews, Alexander continued the Persian practice of semi-autonomy in Judea. Torah — the laws of Moses — remained the law of the land.[8]

Greek mercenaries, traders, and scholars had visited Judea and its neighbors many years prior to Alexander. Exposure to Greek culture had steadily increased in the region in the first two-thirds of the fourth century BC (c. 399–333 BC). Now Greco-Macedonian culture would seep into Judea

---

*Why does the biblical Book of Nehemiah refer to the people of Judah as "the seed of Israel"? The Jews saw the Samaritans to the north as apostates contaminated by intermarriage to Gentiles brought in by the Assyrians. They saw themselves as the rightful inheritors of *YHWH*'s united kingdom of Israel-Judah. They were the true Israelis, or so they believed. Source: Magness, p. 52.

as never before — along with its pagan ways, high art, and glorification of the human body.[9]

Ptolemaic Judea — 301–198 BC

After the death of Alexander in 323 BC, Jerusalem changed hands six times over twenty-two years. Ptolemy I of Egypt took full control in 301 BC. Judea was now part of the Ptolemaic province of Syria.

There are few extant records of internal events in Judea during this period. We know that Ptolemy sent troops, installed garrisons, and taxed the people in the region. And that Judea remained a semi-autonomous theocracy. Torah continued as the law of the land.[10]

"Hellenic domination of the Near East" and its hybrid Greco-Persian culture would increase trade with the West and spur economic development. The upper classes — particularly the ruling priests of Jerusalem — became "increasingly wealthy and Hellenized."[11]

The majority in Judea — villagers and rural farmers, herdsman and shepherds, weavers and spinners, potters and dyers, tanners and smiths — for the most part did not share in this new wealth. Nor did they adopt the cosmopolitan ways of Jerusalem. This divide would be pivotal in events to follow.[12]

Judea enjoyed a century of relative peace under Ptolemaic rulers. This would come to an end through conquest by the empire to its north.

Seleucid Judea (198–141 BC)

Note to Reader: The primary ancient sources for this section are 1 and 2 Maccabees. They were written in the latter part of the second century (100s) BC by Jewish writers in support of Jewish opposition to the Seleucids. The other major ancient source is the first book of Josephus "Bellum Judaicum" or "Concerning the Jewish War." Josephus was a Jewish military governor of Galilee in the first century AD who defected to Rome. He wrote primarily for a Roman audience. See endnote for more details.[13]

Macedonian general Seleucus (Fig. 9.1) had served as commander of Alexander the Great's "élite infantry corps." He also served under Ptolemy after the death of Alexander. In 312 BC, he founded an empire of his own in Babylon and Northern Syria He took the name Seleucus I Nicator.[14]

Within a decade, the Macedonian general extended the Seleucid Empire from Anatolia in the west and across Asia to the Indus River in

Figure 9.1. Seleucus I Nicator. Roman copy of Greek original – Herculaneum. National Archeological Museum, Naples, Italy. Photographer: Massimo Finiozio.[16]

the east. Mesopotamia, Media, Persia, Parthia, and Bactria were now under his control. His kingdom encompassed the largest territory of Alexander's former empire. (See Fig. 9.2.) The Seleucid capital came to be Antioch in northern Syria — a major center of Hellenistic culture.[15]

In 198 BC, Antiochus III — great-great-grandson of Seleucus I — defeated the forces of boy king Ptolemy V at the battle of Panion in the northern border of present-day Israel. Judea was now part of the Seleucid

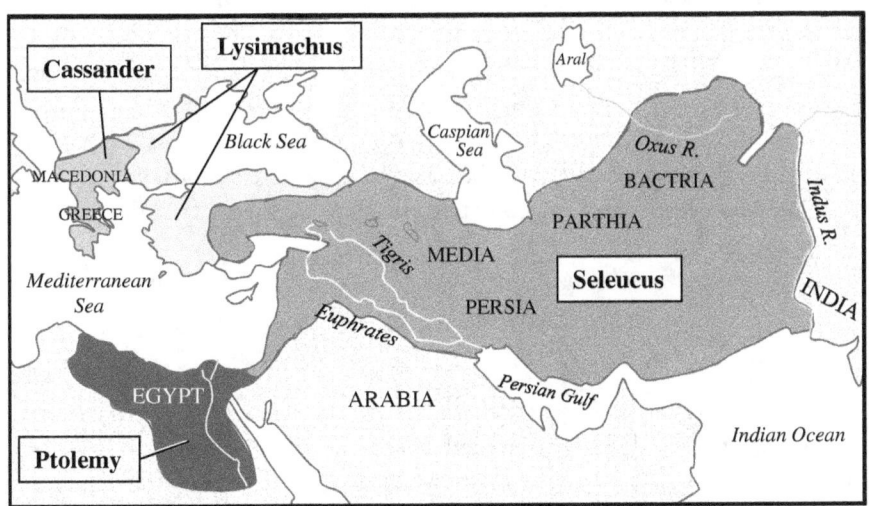

Figure 9.2. The Division of Alexander's Empire — c. 300 BC. The Seleucid kingdom was by far the largest.[17]

Empire. The hundred-year reign of the Ptolemies in the land of the Jews was over.[18]

Antiochus III reaffirmed the doctrine of local religious control. Judea continued to live under Torah law — for a while.[19]

Exposure to Greco-Macedonian customs and religious practices increased dramatically under the Seleucids. Trade also accelerated. Judah enjoyed a prosperity far beyond what it had experienced under Persian and Ptolemaic rule.[20]

This newfound wealth had its impact. Jerusalem's upper classes, its "priests, aristocrats, and government officials," embraced Greek social customs. They assimilated its philosophy, logic, art, and education. Some began to dress in the Greek fashion. Others abandoned circumcision and kosher dietary laws.[21]

The pious masses of Jerusalem and the countryside seethed at these abominations.[22]

Antiochus IV and the Terrors

In 175 BC, a new king ruled in Antioch. In good old Macedonian tradition, he gained the throne by murdering a rival.[23] The new king took the name Antiochus IV Epiphanes — Greek for "God Manifest" — a name, as they say, that would live in infamy.[24]

For Judea, his reign would be one of corruption, rebellion, and civil war.[25]

In c. 170 BC, fighting broke out in Jerusalem between two contenders for the high priesthood, Menelaus and Jason. Both had bribed Antiochus IV for the position.[26] The Seleucid king sent an army to Jerusalem to quell the fighting. His troops stormed the city and massacred thousands of Jews. Survivors were seized and carted off to the slave markets of Antioch.[27]

> *Then there was killing of young and old, destruction of boys, women, and children, and slaughter of virgins and infants.*
>
> — 2 Maccabees 5:13

After the Seleucid troops left, a second rebellion broke out — this time led by pious Jews. Antiochus sent 20,000 troops to the Holy City with orders to slay and destroy. They "plundered Jerusalem and leveled part of the

Holy City," writes Israeli scholar Yisrael Shalem. "The Temple was ritually polluted and abandoned. Jews still alive fled the city."[28]

> *... all the house of Jacob was clothed with shame.*
>
> — 1 Maccabees 1:28

The Edict

In December of 167 BC, Antiochus IV issued a royal decree. It required all inhabitants of his vast empire to adopt Greco-Macedonian practices and customs — including the worship of Greek gods.[29]

Scholars are unclear as to why he did this. Some argue it was an attempt to unite his multi-ethnic empire under a single state religion and culture. Others suggest it was to end the troubles with the Jews and their strange religion.[30]

Whatever the motives, Antiochus had done the one thing that would unite faithful Jews across Judea — he outlawed Judaism.

The Edict made it illegal to own or copy Hebrew scripture, or to sacrifice animals to *YHWH*. Circumcision as well as observance of the Sabbath and other Jewish holidays were also prohibited. Antiochus had pagan altars constructed throughout Judea to conduct pig sacrifices. Jews who refused to participate were tortured and killed.[31]

This religious persecution was unusual for pagan rulers, Chaim Potok points out. With so many gods in their polytheistic pantheon, pagan kings and their subjects tended to be quite tolerant of other gods.[32]

As a final affront to the Jews, Antiochus rededicated the Temple in Jerusalem — the house of *YHWH* — to Zeus. He ordered a statue of the king of the Greek gods be placed on the great rude-stone altar of burnt offerings in the Temple Court.* "Jews were required to make obeisance to it" on pain of death.[33]

> *For the temple was filled with debauchery and reveling by the Gentiles, who dallied with harlots and had intercourse with women within the sacred precincts... The*

---

*1 Maccabees 1:54 says that the Seleucids "erected a desolating sacrilege upon the altar of burnt offering." This was likely a statue of Zeus or some other symbol of the Greek god.

altar was covered with abominable offerings which were forbidden by the laws. A man could neither keep the sabbath, nor observe the feasts of his fathers, nor so much as confess himself to be a Jew.

— 2 Maccabees 6:4-6:7

Antiochus sent troops across Judea to ensure compliance with his royal edict. Enforcement was brutal. 1 Maccabees tells of Seleucid officials who "put to death the women who had their children circumcised, and their families and those who circumcised them; and they hung the infants from their mothers' necks." (1 Maccabees 1:60-63.)

2 Maccabees tells of seven brothers who were "compelled by the king, under torture with whips and cords, to partake of unlawful swine's flesh. One of them . . . said, 'we are ready to die rather than transgress the laws of our fathers.'"

"The king fell into a rage . . . [and] commanded that [his] tongue . . . be cut out and that they scalp him and cut off his hands and feet, while the rest of the brothers and the mother looked on . . . [then] to take him to the fire, still breathing, and to fry him in a pan."

All the brothers were tortured in this way. "Last of all, the mother died, after her sons." (2 Maccabees 7:1 – 42)

We cannot know how accurate these descriptions of Seleucid atrocities are. 1 Maccabees was written in part to stir up hatred against Antiochus and inspire Jewish rebels. The later 2 Maccabees also promoted Jewish support for those martyred by the Seleucids.

Ancient sources tell us a number of Jews "abandoned Judaism and embraced Greek customs," some no doubt to save their lives. Not all succumbed. Resistance spread, especially in the countryside.[34]

The Uprising

And very great wrath came upon Israel.

— 1 Maccabees 1:64

One incident in particular is said to have sparked mass insurrection. A Seleucid officer named Apelles led a patrol to the little village of Modi'in some 20 miles (30 km) west of Jerusalem.[35]

Apelles gathered the people of the village and ordered Mattathias, an elderly priest and head of the Hasmonean clan, to sacrifice to Zeus.

The story is told in 1 Maccabees:

> *"You are a leader, honored and great in this city, and supported by sons and brothers," Appelles said to Mattathias. "Now be the first to come and do what the king commands . . . Then you and your sons will be numbered among the friends of the king, and . . . honored with silver and gold and many gifts."*
>
> *Mattathias answered and said in a loud voice: "Even if all the nations that live under the rule of the king obey him . . . yet I and my sons and my brothers will live by the covenant of our fathers . . . We will not obey the king's words by turning aside from our religion . . ."*
>
> *[Then] a Jew came forward in the sight of all to offer sacrifice upon the altar, according to the king's command. When Mattathias saw it . . . he gave vent to righteous anger; he ran and killed him upon the altar. At the same time, he killed the king's officer who was forcing them to sacrifice, and he tore down the altar . . .*
>
> *Then Mattathias cried out in the city with a loud voice, saying: "Let everyone who is zealous for the law and supports the covenant come out with me!"*
>
> *And he and his sons fled to the hills and left all that they had in the city.*
>
> — 1 Maccabees 2:17-28

Mattathias, his five sons, and their families and followers — a band of some 200 people — ran to the Gophna Hills. In this wilderness some 14 miles (23 km) north of Jerusalem, Mattathias began "training the peasants in guerilla tactics." (See Fig. 9.3.)[36]

A number of rebel groups had formed at this time. Mattathias organized them under his command.[37] He and his forces set up a number of camps in the remote mountain and desert regions of Judea.

From there they launched a series of raids against their own people — Hellenists and other Jews who had acquiesced to Greek practices. They attacked apostate Jewish villages, destroyed homes, and pulled down pagan altars. They forced Jews who had submitted to the Edict to circumcise their children.[38]

The hardships of life in the wilderness took its toll on the elderly Mattathias. In c. 165 BC, less than two years after the start of the revolt, the country priest who some today would call a terrorist passed away. Responsibility for directing the insurrection would pass to one of his sons — a

Figure 9.3. Mattathias appealing to Jewish refugees. Illustration by Gustave Doré from 1866 La Sainte Bible.

military leader whose brilliance on the battle field would be compared to Alexander the Great.[39]

The Hammer

[He] has been a mighty warrior from his youth; he shall command the army for you and fight the battle against the peoples.

— Mattathias (1 Maccabees 2:66)

Mattathias had five sons: Johanan, the eldest called "the saint." Simon, deemed the wisest and most prudent. Judas, the boldest. Eleazer called Avaran perhaps meaning the palest. And Jonathan, the wary and crafty. In his wisdom, the dying Mattathias named Judas as his successor. In their wisdom, his other four sons agreed.[40]

Judas came to be called "the Maccabee," likely meaning the Hammer. Under his leadership, the small band of insurgents quickly grew in numbers.

Judas continued his father's hit-and-run guerilla war against Hellenistic Jews and other apostates. "Coming without warning, he would set fire to towns and villages." (2 Maccabees 8:1, 5-7) This ethnic fratricide was fast becoming a countrywide civil war.[41]

As stories of Seleucid atrocities against religious Jews spread, the revolt grew. The victims of Judas and his rebel attacks in turn "appealed to the king for protection." Antiochus answered their pleas — with a full-out assault on little Judea.[42]

Four Battles

From 167 to 161 BC, Seleucid armies engaged Judas and his rebel forces in a series of four pitched battles. The first was the *Battle of Nahal el-Haramiah* in late November, 167 BC. Here Judas and his brothers set up an ambush at a narrow mountain pass in Wadi Haramiah near modern day Ma'ale Levona in the West Bank. (See Fig. 9.4-1.) With some 600 men, the Maccabees crushed the overconfident Syrian forces and slew its leader, the hated Apollonius.[43]

A year later, Antiochus sent general Seron and 4000 infantrymen from Coele Syria to Judea. The Maccabees attacked at another mountain pass,

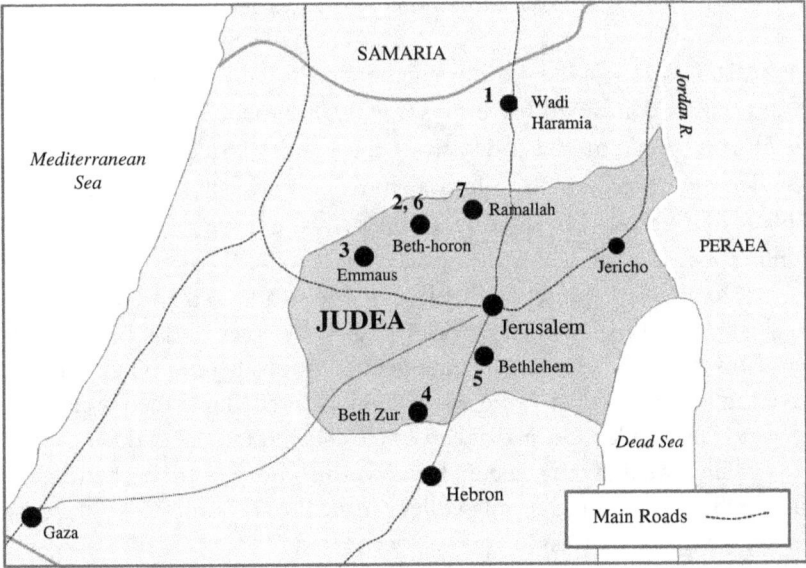

Figure 9.4. The Battles of Judas Maccabee against the Seleucids. (1) Wadi Haramia; (2) Beth-Horon; (3) Emmaus; (4) Beth Zur; (5) Beit Zachariah; (6) Adasa; (7) Elasa.

this time near Beth Horon some 14 miles (22 km) northwest of Jerusalem (Fig. 9.4-2). Here almost on the spot where the biblical Joshua is said to have commanded the Sun to stand still, Judas set his trap.[44]

With some 1000 troops, he sealed the exit to the defile. Another 1000 or so suddenly arose from hiding and attacked from the rear. Jewish archers perched high on both sides of the passageway shot deadly volleys from above. Some 800 Seleucid troops were slain, including Seron. More were killed in their frantic retreat to the Philistine plains.[45]

With the two unexpected victories, volunteers began to "stream in from all parts of Judea." The Maccabean army soon expanded to more than 6000 men. They would need them.[46]

Antiochus IV had had enough. While off raiding the temple of Nanaea in Persia for its treasury, he relayed orders to Lysias, his viceroy in Antioch: Destroy the Jews.[47]

The Battle of Emmaus

> ... send a force ... to wipe out and destroy the strength of Israel and the remnant of Jerusalem ... to banish the memory of them from the place, settle aliens in all their territory, and distribute their land.
>
> – Antiochus IV to Lysias (1 Maccabees 3:35-36.)

Again in 166 BC, Lysias sent his best generals, Ptolemy, Gorgias and Nicanor to Judea with a combined force of some 20,000 men. To avoid the pass of Beth Horon, they entered Judea from the south. The Seleucid army then "ascended by the main road to Jerusalem from the west" and marched along the plain of Sharon. They set up camp near the town of Emmaus in the foothills above the bucolic Ajalon valley (Fig. 9.4-3).[48]

Gorgias and some 6000 Seleucid troops attacked the nearby Jewish camp in the dead of night. Forewarned again by spies, Judas had evacuated most of his army. He left a small contingent of troops behind. They lit bonfires to give the impression the camp was fully populated. They then escaped into the Shaar Hagai Valley. Gorgias followed close behind in the darkness.

The Jews drew the Seleucids into yet another a bottleneck in a narrow defile. Archers again fired deadly volleys from above. Some 1500 Maccabean warriors pounced. Gorgias' army was decimated.[49]

Now it was Judas' turn. With the main body of Jewish troops, he approached the principle enemy encampment at Emmaus from the south. The Maccabean forces arose out of the morning mist into the breaking light of dawn. To their surprise, Seleucid troops were already deployed in classical phalanx formations in front of the camp. They too had their spies.[50]

Curiously, some 12,000 Syrian troops remained in their camp — along with a train of camp followers and slave traders prepared to barter for captured Jews after the expected Seleucid victory. Such was the overconfidence of the Seleucids.[51]

Seeing the Seleucid formation, Judas "divided his command into three 1,000-man battalions and struck the Seleucids in their vulnerable western flank," military historian Kelly Bell writes. "While one of these groups engaged the covering cavalry, the other two assailed the enemy formation from the side."

Then, "to the sound of silver trumpets," a hidden 1500 Maccabean troops attacked from the north. They overran the unarmed and unmounted Seleucid camp and slayed with the wrath of *YHWH*.[52]

Pandemonium broke out. Horses stampeded. Freight-carrying elephants trumpeted and ran amok. Terrified slave traders and their entourages ran for their lives. Bitter hand-to-hand fighting ensued. "Black columns of smoke rose from burning tents."[53]

Nicanor lost some 3000 men in the melee. Surviving Seleucids fled west toward their fortress of Gezer with Maccabean rebels in hot pursuit. Faithful Jews in villages along the way rose up to harass and slay more of the enemy.[54]

Judas Maccabee's rebel forces were victorious over a professional army some three times its size. The Battle of Emmaus has been called "one of ancient history's greatest triumphs."

The Battle of Beth-Zur

In 164 BC, Lysias himself led a Seleucid army to the land of the Jews with some 20,000 infantry troops and a cavalry numbering 4000. They traveled along the coastal route and entered Judea at the southernmost part of the country. Still there was no way to reach Jerusalem without passing through treacherous highlands.

They approached Beth Zur, close to modern Hirbet Beth-heiran in the mountains of Hebron (Fig. 9.4-4). His army trudged uphill into a defile overlooked by high ground. Jewish spies tracked their movements. Three thousand hidden Maccabean warriors attacked from the rear. Two one-thousand-man columns pounced from both sides of the canyon. Some 5000 Seleucid troops were slain. (1 Maccabees 4:28-35.)[55]

On hearing the news, the 8000 or so Syrian troops who had remained at base camp abandoned their posts and fled to the city of Hebron. Why Lysias divided his troops remains a mystery.

Note to Reader: *The authors of 1 and 2 Maccabee, as well as Josephus, likely exaggerated the number of Seleucid troop and casualties. They did the opposite for Maccabean ones, as Jewish historian Bezalel Bar Kochva points out — this to make Judas's feats appear even more remarkable.*[56]

In the meantime, Antiochus IV led the bulk of his army to the east to fend off an invasion by the Parthians. In the lull, Judas occupied Jerusalem.[57]

Festival of Lights

Once in the Holy City, Judas selected orthodox priests untainted by Hellenist heresy to restore and cleanse the Temple. On the 25th day of the month of Kislev in the Hebrew calendar, the Temple was rededicated — an eight-day event celebrated to this day as the Jewish holiday of Chanukah[58]

Judas and his brothers liberated Jewish residents of Hellenized cities in Galilee and in Gilead across the Jordan River. The Maccabee brothers fought by the brutal rules of holy war outlined in Deuteronomy.[59]

> *. . . thou shalt save alive nothing that breathest,*
> *but thou shalt utterly destroy them . . .*

— Deuteronomy 20:16-17

The Maccabee brothers rescued thousands of Jewish men, women, and children. The entourage "marched down the great valley" in triumph "and ascended the hills of Jerusalem with songs and rejoicing," as nineteenth century English antiquarian Claude R. Conder put it."[60]

Deliverance

Antiochus IV died during his campaign against the Parthians in 164 BC. His nine-year-old son Antiochus V Eupator became king. His father's regent Lysias now served as nominal ruler of the empire.

The Seleucid regent gathered a massive army. It was reported to comprise of 30,000 infantry troops, a cavalry of some seven thousand, and thirty elephants "trained for war." Lysias led them from the south towards Jerusalem in the spring of 163 BC.[61]

Judas and his army "set up camp at Beth Zechariah" in an attempt to block the Seleucid advance. (See Fig. 9.4-5.) According to 1 Maccabees, Judas engaged Lysias in a brief skirmish. Seeing the might of Lysias' army, he and his forces retreated to Jerusalem.[62]

It was the sabbath year in Judea. Torah commands that in the seventh year of the agricultural cycle, farmland must not be cultivated.[63] "Whatever grew of itself during that year was for the poor and the stranger and the beasts of the field."[64]

Lysias' troops surround Jerusalem. The Maccabees find themselves trapped in the Holy City with food supplies dangerously low. Then, out of the blue, Lysias receives a message from Antioch. Philip, a rival to the Seleucid throne, is attacking the Seleucid capital.[65]

Just like when Assyrian king Sennacherub rushes back to Babylon to quell an uprising over five centuries earlier, the Holy City is saved. Before departing, Lysias "reaches terms" with Judas. The Edict of Antiochus outlawing Judaism is rescinded! Hallelujah!

Only this time, Jerusalemites do not watch the departing troops from the city walls. In defiance of the peace agreement, Lysias orders the great rampart walls of Jerusalem torn down.[66]

The peace would not last.

The Battle of Adasa

The Seleucid "game of thrones" continues. In 162 BC, Rome frees Seleucid hostage Demetrius, cousin to boy king Antiochus V. He returns to Antioch and kills the boy king and his regent Lysias. Demetrius I Sotor, as he called himself, now rules the empire. (1 Maccabees 7:1-4.)[67]

Civil war again breaks out in Jerusalem. Hellenists fight against the pious Chasidim. Demetrius I sends an army to the Holy City to support the Hellenists, this time under general Nicanor. He arrives in Judea in the spring of 161 BC with orders to hunt down Judas and suppress the party of Chasidim.[68]

This time Judas is ready. Some one-thousand Maccabean troops surprise Nicanor's army near the village of Adasa north of Jerusalem. (See Fig. 9.4-6.) Judas is once again victorious. And this time Nicanor is slain. The flat open valley where the battle is fought would become known as "the Valley of Blood."[69]

Upon returning to Jerusalem, Judas hangs Nicanor's head and right hand from the Temple gate. (1 Maccabees 7:47.)

Note to Reader: This marks the end of 2 Maccabees. Our primary ancient sources going forward are 1 Maccabees and the writings of Josephus.

Judas and the Eight Hundred

Demetrius does not give up. In 160 BC, a massive Seleucid army — some 20,000 infantry and 2000 cavalry — arrives at the northern edge of the "great valley at Berzetho near modern day Ramallah." (See Fig. 9.4-7.)[70]

With the Edict of Antiochus IV rescinded, many who had joined the Maccabean rebellion had returned home. Judas's army had shrunk to some three thousand men. When they see the size of the Seleucid forces, many more flee the battlefield.[71] With a mere eight hundred steadfast warriors, the Maccabees face an enemy over twenty times their size.

"Judas' troops try to dissuade him, saying, 'We are not able. Let us rather save our own lives now, and let us come back with our brethren and fight them; we are too few.'" Judas refuses to concede. "'Our time has come,' he tells his men, 'Let us die bravely for our brethren, and leave no cause to question our honor.'" (1 Maccabees 9:9-10).

Judas orders a "desperate charge on the enemy right flank." The Seleucid left wing wheels and attacks from the rear. The small band of rebels are surrounded. Nearly all are slain, including Judas Maccabee.[72]

Reflections

Despite facing vastly superior numbers, Judas was victorious in all but his last battle. The series of Maccabean victories against "the superpower of the East" was nothing short of extraordinary.

Ironically, it was Lysias's agreement to rescind the religious Edict of Antiochus IV which doomed Judas. Making Judaism legal again took away the primary motive to rebel. In his last battle, when the Seleucids brought their great force against him, Judas found himself effectively without an army.

Still, his ability to train and motivate troops, to develop a nationwide espionage network, to rapidly adapt to battlefield conditions, to again and again outfox and overcome forces many times his size — not to mention his fearlessness in battle — places him amongst history's greatest military leaders.[73]

An apocalyptic message, a brief period of freedom, and conquest by a great power from the West would set the stage for a new version of ethical monotheism — one which would come to spread across the ancient Mediterranean world.

This is the subject of the next chapter.

Chapter 10

FOUND AND LOST

All things have I seen in the days of my vanity; there is a righteous man who perisheth in his righteousness and there is a wicked man that prolongeth his life in his evil-doing.

— Ecclesiastes 7:15

Recall that the prevailing view of the afterlife in the Hebrew Bible is similar to Sumerian belief. All who die, whether righteous or sinner, go to the underworld or *She'ol* in Hebrew.* Out of the Edict of Antiochus IV and its brutal enforcement would come a new vision of Godly justice. And a re-imagining of the afterlife — one which is with us to this day.

The Book of Daniel

Antiochus IV's proclamation outlawing Judaism caused great soul-searching among the Jews. Those who refused to sacrifice to idols, those who continued to obey Torah Law were persecuted, tortured, and even killed. Those who yielded to the demands of Antiochus were not. And Jews who collaborated with the Seleucids were rewarded.

It seemed to faithful Jews that the righteous suffered while the wicked prospered. Where was God's justice?

A new concept of divine retribution arose in first century BC Judea. It is found in the Book of Daniel, the last book of the Hebrew Bible to have been written. Events depicted in Daniel support the argument that it was composed at the time of the Maccabean revolt.[1]

A New Vision of Justice

The Book of Daniel contains the first known detailed biblical account of an Apocalypse. If our timing is correct, it was a reaction to atrocities committed by Seleucid forces during the enforcement of Antiochus's Edict.

The stories in the Book of Daniel are veiled. They contain dreams and visions, fanciful images, and strange symbolism. (See Fig. 10.1.) This ambiguity is likely deliberate. It is an attempt by the author to protect his safety, as much of his writing is directed against Antiochus IV.[2]

The author of Daniel prophesizes a time of great conflict between good and evil. At first, the evil powers of the world will grow stronger. Then *YHWH* will intervene in human affairs. His archangel "Michael shall stand up," and destroy the evil kingdoms of Earth. (Daniel 12:1)[3]

> ... and there shall be a time of trouble, such as never was ... And many of them that sleep in the dust of the earth shall awake, some to everlasting life, and some to reproaches and everlasting abhorrence.

> — Daniel 12:1-2

---

*See for example: Isaiah 14.7-11, Psalms 141:7, Proverbs 7:27, Job 10:21-22, 17:16.

Figure 10.1. Daniel's vision of the four beasts. Woodcut by Hans Holbein the Younger, 16th Century. (My annotations.)

All who have died will be raised from the dead and given corporal bodies, including those who suffered under Antiochus.[4] *YHWH* himself will then come down to Earth. His arrival is described in Daniel:

> *I beheld*
>   *Till thrones were placed,*
>   *And one that was ancient of days did sit:*
>   *His raiment was as white as snow*
>   *And the hair of his head like pure wool;*
>   *His throne was fiery flames . . .*
>
>   — Daniel 7:9

For this is the end time, the Day of Judgment. The resurrected dead and those still living, all "ten thousand times ten thousand," shall stand before the One God. "The judgment [shall be] set, and books opened." (Daniel 7-10.)

Those who had been faithful to *YHWH* will be rewarded with eternal life in a new paradise — here on Earth (*not* in Heaven). Those who were unfaithful will feel the fury of *YHWH*'s wrath and be cursed forever.[5]

An enigmatic passage follows:

> *And behold, there came with the clouds*
> *of heaven, one like unto the son of man.*
>
>   — Daniel 7:13

Who is this "son of man"? An angel of sorts, it is thought. We will meet him again in the chapter on Jesus.[6]

Daniel continues:

> *And there was given him [the son of man] dominion, and glory, and a kingdom, that all the peoples, nations, and languages should serve him.*
>
> — Daniel 7:14

A new kingdom of Israel shall be established here on Earth. Its king shall be the "son of man." All nations shall attend to him. All the world shall worship YHWH, the God of Israel, the God of the Jews, the God of the universe.

These passages in the Book of Daniel were directed towards pious Jews — to give them hope in a time of great suffering. Hold to YHWH and His Laws. He shall soon punish Antiochus and those who obeyed him. And reward you who have remained faithful with eternal life in an Earthly paradise.[7]

Roots of the Apocalypse

There are striking similarities in Daniel's Apocalypse story to Zoroastrianism. This ancient Persian religion also predicted a final battle between good and evil; the coming of *Saoshyant*, a divine king; and the resurrection and judgement of the dead.[8] It seems the Apocalypse story has roots in the Jewish exile in Babylon and its subsequent conquest by Persia.[9]

There are stories of an Apocalypse in "a number of Jewish writings in the centuries after Daniel" — including the Dead Sea Scrolls. Later Pharisee schools would teach the Apocalyptic doctrine of Daniel* with a future earthly paradise for the devout ruled by a Messiah-king in Jerusalem.[10]

Nowhere is its historical impact greater than in the development of Christianity, as we shall see in the following chapters.[11]

---

*The Apocalyptic message of the Book of Daniel is not part of present-day mainstream Judaism — yet it is in the Hebrew bible. A number of devout Jews wrestle with this issue. Source: Freeman, Tzvi. "Is the Book of Daniel authentic?" chabad.org. https://www.chabad.org/library/article_cdo/aid/489751/jewish/Is-the-Book-of-Daniel-authentic.htm. Retrieved Feb 2, 2021.

> **SIdebar: The Three Philosophies**
>
> Around this time, three Jewish sects came to form in Judea — the Sadducees, Pharisees, and Essenes.[12]
>
> *Sadducees** — Primarily drawn from "priestly, aristocratic, and military circles," this group was known for its pursuit of wealth and social standing. Their supporters were generally better educated urbanites.[13]
>
> The Sadducees emphasized worship at the Temple in Jerusalem and felt they alone were suited to rule. They tended to be "haughty and harsh towards common Jews," who in turn disliked them.[14]
>
> *Pharisees* — Pharisees were religious conservatives. They called for strict adherence to Torah law — including regulations on "diet, dress, ceremonies, and the observance of Sabbath." Their primary appeal was in the countryside. Many Sadducees saw them as radicals.[15]
>
> The Pharisees violently opposed Hellenism and emphasized "oral" as well as written Torah Law. They taught that only a Pharisee could correctly interpret Scripture.[16]
>
> *The Essenes* — This devout Jewish sect lived in communal groups, held their worldly goods in common, and lived in shared poverty.[17] They abstained from worship at the Temple in Jerusalem.
>
> According to Josephus, Essene men often abstained from sex as well. They considered "self-control and not succumbing to the passions [a] virtue." Certain things in particular were "matters of personal prerogative: assistance and mercy, and extending food to those who are in want."[18]

Independence!

The death of Antiochus IV in 164 BC begun a century-long decline of the Seleucid kingdom. Incessant competition for the throne, civil wars, and invasions of its territory would lead to the gradual unraveling of its empire.

The Jews found themselves effectively free of Seleucid rule around 100 BC. The exact date is uncertain. Judea was now an independent

---

*The name Sadducees is believed to be derived from Zadok, Solomon's High Priest. He was said to have been a descendent of Eleazar the son of Aaron, who in turn was a priest and brother of Moses. (1 Chron 6:4-8). The name Pharisees comes from the Hebrew *Perushim*, to separate, i.e., to separate from Hellenist influences.

nation — for the first time since before the Babylonian exile over four centuries earlier. Pious Jews called it a miracle from God.[19]

Descendants of country priest and rebel Mattathias would rule the new nation. It would be called the Hasmonean dynasty after the family name Hasmon, the great-grandfather of Mattathias.[20]

By the third generation of Hasmonean rulers, the piety of country priest Mattathias and the family loyalty of his five Maccabean sons would be forgotten. Their Hellenized descendants would act more like pagan kings.

The royal Judean court would be marked by matricide, fratricide, and civil war.[21] Hellenization and impiety increased. The Holy Land would be marred by internal dissention, conflict between Sadducees and Pharisees, and exceptional cruelty.[22]

Mattathias must have been rolling over in his grave.

The Jewish nation would be independent and free to fight amongst themselves for a mere thirty-seven years — only to be conquered by yet another foreign power, the greatest the ancient Western world had known.

Roman Judaea

Gnaeus Pompeius Magnus aka Pompey was a brilliant and ruthless Roman General. He was in Damascus in 64 BC — as part of his Eastern campaign against Armenia and Pontus on the southern coast of the Black Sea. He had seized Syria that same year and placed its king, Philip II as client king under Rome.[23]

At the same time in Judea, Hasmonean high priest Hyrcanus II and his brother Aristobulus II were engaged in a civil war over kingship. Each brother appealed to Pompey for support. The Sanhedrin in Jerusalem also sent a delegation to Pompey. The latter urged the Roman general to abolish the Hasmonean monarchy and restore rule by high priest.[24]

Their collective naivety regarding Rome's intentions was staggering. The Roman general chose the weaker, more malleable Hyrcanus.[25]

In 63 BC, supporters of Hyrcanus opened the gates of Jerusalem for Pompey. The Roman general arrested Aristobulus. His troops then attacked the Fortress of Antonius on the Temple Mount where supporters of Aristobulus had barricaded themselves. Pompey was joined by Hyrcanus and his forces.[26]

After a two-month siege and the massacre of some twelve thousand Jews, Pompey's forces controlled the Holy City. They entered the Temple.

Josephus writes:

> *The [Romans] then fell upon them, and cut the throats of those that were in the temple, yet those that offered the sacrifices [the priests] could not be compelled to run away . . .*
>
> — Josephus, *Antiquities of the Jews* XIV 4.3

Pompey then entered the innermost sanctuary, the Holy of Holies — an act forbidden for all but the High Priest, let alone a pagan. It is said he left its treasures intact.[27]

The following year, the Roman Republic established Syria and lands south as a single Roman province. Judea was now "Judaea," a vassal state of the Roman Empire — its high priest chosen by Rome.[28]

The whale had swallowed the minnow. Rome, the insatiable beast of conquest, now ruled the Holy Land of the Jews. Little did the Romans know that in the following centuries this seemingly insignificant act would come to have a most profound effect on its entire empire.

After Pompey's takeover, the Jews rebelled four times within a period of six years. Rome crushed them all.[29] The Jewish population was far from united, as usual. Some fought against the heathen Romans with fanatical hatred. Romanized Jews on the other hand reportedly swelled the ranks of the Roman army.[30]

Herod

In 37 BC, Herod "the Great," as he would become known, was named Rome's client king in Judaea. He was an Idumean. This neighboring tribe to the south were descendants of biblical Edomites. The Idumeans had been conquered and forced to convert to Judaism by Judean King Hyrcanus I when Judea was independent.[31]

Herod's thirty-three-year reign was one of contrasts. He was a "brilliant strategist and politician" — and a brutal autocrat. His rule was marked by increased prosperity and extensive building projects — with crippling taxes on the populace to pay for them. This included a magnificent rebuilt of the Temple Compound and house of *YHWH* in Jerusalem. Herod then defiled the Temple by placing a Roman eagle at its entrance.[32]

Herod's cruelty was legendary. He executed his second wife Miriamne, granddaughter of Aristobulus II, as well as "her two sons, her brother, and her mother." This effectively ending the last Judean royal line. He later had his first-born son slain. (He had eight wives and fourteen children.)[33]

It is better to be Herod's pig than son.

— Augustus Caesar (allegedly)[34]

Rome's strategy of occupation was as effective as it was insidious. It called for the manipulation of local aristocracy — in this case the Sadducee priesthood — to rule in its name. Prodded by fear, rewarded with riches, already corrupted under prior Jewish rule, the Sadducees kowtowed to the Romans. They became collaborators of the occupation. Under Herod, the Sadducees grew even wealthier. Pharisees despised the Idumean "foreigner." Essenes thought they were all corrupt and kept to themselves.[35]

All this served to exacerbate the historical enmity between the wealthy and the poor, the cosmopolitan city dwellers and the religiously conservative countryside across the Palestine region.

The Next Chapter

Herod died in 4 BC. Riots broke out throughout the land. It took three Roman legions from Syria to restore order. Some Jews yearned once more for a messiah* — a king of the line of David — to overthrow the hated Roman overlords and their collaborators in Jerusalem.

The hope for a messiah king to liberate Judaea coupled with visions of a judgment day to reward the faithful and punish sinners would provide fuel for itinerant preachers across the region. One in particular would come to change history.[36]

This is the subject of the next chapter.

---

*Again, the word "Messiah" or *mashiach* in Hebrew means "the anointed one"—from the coronation ceremony in which a new Hebrew king is anointed with oil to symbolize being blessed by *YHWH*. Messiah can also refer to a high priest or prophet, who were also anointed with oil. Source: Anthony R. Petterson, "The Shape of the Davidic Hope across the Book of the Twelve," *Journal for the Study of the Old Testament* 35 (2010): 225–246. HYPERLINK "https://doi.org/10.1177%2F0309089210386022" https://doi.org/10.1177/0309089210386022

Chapter 11

HE HAS RISEN

. . . ye believe in God, believe also in me.

— John 14:1

<u>Note to Reader</u>: *As noted in the Introduction, I try to present religious figures as human, with desires, doubts, hopes and fears. The "historical" account presented here does not at times agree with Scripture.*

After the death of Herod, Rome divided the Palestine region among his three sons. (Herod didn't kill them all.) Herod Antipas became governor of Galilee and Perea. Philipp II ruled lands east of the Jordan River. Herod Archelaus governed Judaea, Samaria, and Idumea.

The rule of Herod Archelaus was so inept that Jews and Samaritans jointly appealed to Rome for his removal. He was replaced in 6 AD by a Roman prefect who reported directly to emperor Augustus. Judaea was now an autonomous Roman province, its capital in the "splendid sea resort Caesarea."[1]

In these troubled times, Jewish "preachers and prophets roamed the countryside." Some called for the end of the world. A new teacher in particular would preach a fiery Apocalyptic message straight out of the biblical Book of Daniel.[2]

No Greater Prophet

In the time when Herod was king of Judea . . . an angel of the Lord appeared . . . "your wife will bear you a son," [he said] . . . "and many will rejoice because of his birth, for he will be great in the sight of the Lord . . . and he shall be filled with the Holy Ghost even from his mother's womb . . . I am Gabriel . . . and I have been sent . . . to tell you this good news."

— Luke 1:5-19

It was, they say, a miraculous birth.

He was born at the time of Herod "the Great," Rome's cruel and oppressive puppet king of Judaea. This charismatic Jewish preacher would come to wander the countryside and preach especially to the poor, the outcast, the reviled — and criticize the religious aristocracy for its hypocrisy.[3]

He warned all who would listen that the end of the world is imminent. *YHWH* is coming to overthrow the forces of evil — the Romans and their Jewish collaborators. The One God will resurrect the dead and judge us all. Repent and pray for forgiveness. Sin no more and walk the path of righteousness.[4]

The Birth

Like Sarah, the wife of the Hebrew Patriarch Abraham, his mother had been barren and was thought too old to bear children. (Luke 1:7; Genesis 21:1) Her name was Elizabeth, "of the daughters of Aaron."

Her husband Zechariah was a Jewish country priest of the order of Abijah. "Fear not" the angel Gabriel told him, "for thy prayer is answered . . . Elizabeth shall bear thee a son, and thou shalt name him John." (Luke 1:5-13).

His Hebrew name was *Yochanan*, meaning "*YHWH* is gracious."[5]

What little we know of the life of John is given in the New Testament and the writings of Josephus.[6]

His Youth

The New Testament tells us John was raised as a Nazirite.* Under this strict code, John was forbidden to "drink neither wine nor strong drink," nor cut his hair or touch a corpse.[7]

John spent his formative years in the solitude of the Judean desert. Some scholars say while there he may have been taught by the Essenes. Perhaps, they suggest, it was from them that John acquired his Apocalyptic message on the end of the world.[8]

It is believed that John began his ministry in 27 or 28 AD. Tiberius was now emperor in imperial Rome. The infamous Pontius Pilate served as his governor in Judaea. Herod Antipas still ruled for Rome in Galilee and Perea.[9]

Itinerant Preacher

> *In those days came John the Baptist, preaching in the wilderness of Judea. And saying, "Repent ye: for the kingdom of heaven is at hand."*
>
> — Matthew 3:1

They come in droves to the wild country of the lower Jordan River valley. This "barren region of rugged hills and valleys" east of Jerusalem was

---

*The biblical Samson of "Samson and Delilah" and Hebrew judge-prophet Samuel were also said to be Nazirites. (Luke 1:15, 12:15, Numbers 6.)

alleviated only by the muddy Jordan River as it wound its way south to the Dead Sea.[10]

Here stands John, an odd "wiry ascetic." He is dressed in the traditional garb of a Jewish prophet: a garment of "camel's hair, and a leather girdle about his loins." Men and women of the countryside listen with rapt attention. "It is said," a member of the crowd whispers, "he subsists on a diet of 'locusts and wild honey.'" (Matthew 3:4.)[11]

He tells his audience it is not enough to "obey Torah Law and honor YHWH with one's lips. The heart must change." He urges them to personal action, to acts of charity and justice.[12]

> *He who has two coats, let him share with him who has none,*
> *and he who has food, let him do likewise.*
>
> — John the Baptist (Luke 3:11)

Sadducees and Pharisees come to see what the fuss is all about. John rails "at their smugness." You breed of vipers!" he tells them. "Who hath warned you to flee from the wrath to come? Bear fruit that befits repentance . . ." (Matthew 3:7-9.)

The Baptist

> *John, that was called the Baptist was a good man, and commanded the*
> *Jews to exercise virtue, both as to righteousness towards one another, and*
> *piety towards God . . .*
>
> — Josephus (Jos. *Antiquities of the Jews* 18.5.2)

John became known for the practice of baptism — an ancient Jewish ritual of purification through immersion in running water. A number of Jewish teachers practiced ritual washing, including the Qumran community of the Dead Sea Scrolls. John would submerge repentant Jews in the waters of the Jordan River.[13]

According to early Christian writings, John's circle of devotees is said to have included thirty apostles and some three thousand followers. We cannot know if this is an exaggeration.[14]

The Death of John

According to Josephus, Herod Antipas had John executed because he feared his popularity would spark a revolt (Jos. *Antiquities of the Jews* 18.5.2). The Gospel of Matthew tells a different, more scandalous story (Matthew 14:3-11):

Herod Antipas had been married to the daughter of Nabatean king Aretas IV. While in Rome, he was smitten with his half-brother Philip's wife, Herodias. Herod divorced his wife and began "living in sin" with Philip's wife.[15]

The Baptist told Herod "it is against God's law for you to marry her. So Herod had [him] arrested and imprisoned as a favor to his wife Herodias. [Then] at a birthday party for Herod, Herodias's daughter [the infamous Salome] performed a dance that greatly pleased him. So he promised . . . to give her anything she wanted."[16]

"At her mother's urging, the girl said, 'I want the head of John the Baptist on a tray!' John was beheaded in the prison, and his head brought on a tray and given to the girl, who took it to her mother."[17]

Whatever the story, John is believed to have been executed somewhere between 29-30 AD. One of his disciples would take on John's mantle and change history.

Y'shua

At his circumcision at the local synagogue, his father Joseph named his newborn son *Yehoshua* in Hebrew, meaning "*YHWH* helps." The name comes from Moses' successor Joshua in English. It was likely shortened to *Y'shua*. It was a common Jewish name.[18]

We will refer to him as Jesus, the name by which he has come to be known in English.

Who was this Jesus? We know little for sure from a strict historical perspective. He was a "Jew who led a popular Jewish movement in Palestine in the beginning of the first century AD," writes scholar of religions Resa Aslan. "The Romans crucified him for doing so."[19]

Most historical scholars tell us that Jesus did not intend to found a new religion, nor did he claim to be the divine Son of God. It is "pretty certain" he was associated with and baptized by John the Baptist.[20]

Reconstructing Jesus

> *[High priest] Albinus . . . assembled the Sanhedrin . . . and brought before them the brother of Jesus, who was called Christ, whose name was James . . .*
>
> — Josephus, *Antiquities of the Jews* (20, 9.1)

Outside the New Testament, there is little mention of Jesus. Josephus briefly mentions him twice in *Antiquities of the Jews*, written around 93–94 AD. Roman historian Tacitus remarks on his execution by Pontius Pilate in *Annals*, written c. 116 AD.[21]

In religious writings, the letters of Paul — the first Christian writings, generated between 40 to 50 AD — say very little about the life of Jesus.

The four Gospels of the New Testament — perhaps written between 70 and 100 AD, some 40 to 70 years after the death of Jesus — do tell the story of Jesus' birth, ministry, and death, albeit with significant differences among them. They are "neither biographies nor objective historical accounts," writes religious scholar Paula Fredriksen. She refers to the them as "a kind of religious advertisement." In addition, the Gospels we read today were recopied and edited over the centuries. Nonetheless, scholars generally feel that the "meaning and intent of the New Testament has been preserved."[22]

Somewhat of a consensus has emerged among secular scholars on the life of the "historical" Jesus. Let us attempt to briefly reconstruct his life as a human being — based on these mainstream views, as well as Gospel writings,* and, for what it's worth (very little), my own interpretations. This is a highly controversial subject and I understand there will be those who disagree.

---

* Here we access the so-called Synoptic Gospels of Mark, Matthew, and Luke. Secular scholars generally consider the later Gospel of John less historically reliable. Source: Ehrman, Bart. *Jesus: Apocalyptic Prophet of the New Millennium.* (Oxford Univ. Press, 2001) p. 88. Bible scholar Dr. John Stevenson, adjunct professor at Trinity International University, points out that the lack of historicity of John "has come under fire as a result of Dead Sea Scroll studies. They show that much of the theology of the Gospel of John is couched in terms that was specifically found in the first century prior to the destruction of Jerusalem." Source: Stevenson, John. "Review of Cosmic Roots draft", email to author, Aug 23, 2020.

Jesus the Man

Jesus was perhaps born somewhere between 4 BC and 6 AD. He was from the village of Nazareth in lower Galilee. The small conservative village stood in stark contrast to the sophistication and wealth of the Hellenistic city of Sepphoris a mere four miles away.[23]

Known as a "hotbed of radicalism," the Galilean region was separated from Judaea by Samaria to its south. Jesus likely came from a population of Jews who were ethnically mixed with those who had come to the area after the Assyrian conquest in the eighth century BC.[24]

He was apparently from a large family. It included his father Joseph; mother Mary; brothers James, Joseph, Simon, and Judas; and unknown sisters. We know almost nothing about his early years.[25]

At some point in his late twenties, Jesus left Nazareth and went off to hear the teachings of John the Baptist.* He was spell-bound. It seems the call to repent one's sins, to follow the path of righteousness, to act with kindness and charity towards the downtrodden moved him as nothing had before.

> *Among those that are born of women, there is not a greater prophet than John the Baptist.*
>
> — Jesus (Luke 7:28)

Jesus became a follower, then a disciple, and finally a protégée of John.** "That Jesus was baptized by John the Baptist is as certain as anything historians know about him," writes New Testament scholar John Dominic Crossan.[26]

The Wilderness

Jesus was likely devastated by John's death. It surely filled him with great sorrow — and fear. We imagine him asking, "Why did *YHWH* allow this to

---

*According to Scripture, Jesus and John were cousins through their mothers Mary and Elizabeth, who were sisters. (Luke 1:36)

**The gospels say that John the Baptist predicted the arrival of Jesus, who was superior to John. (Matthew 11:9–11.)

happen? If I now go out on my own preaching the word of God, will I also be killed?

The Gospel of Matthew tells us Jesus went off to "a desert place apart" after John's death. When the people heard of this, "they followed him on foot out of the cities. And Jesus went forth, and saw a great multitude, and was moved with compassion toward them, and he healed their sick." (Matthew 14:13-14.)

What drew Jesus out of his despondency and grief? Compassion — a lesson for us all.

The Ministry of Jesus

I was sent only to the lost sheep of the house of Israel.

— Matthew 15:24

Jesus returned to Galilee and took up John's mantle. Avoiding gentiles and the Hellenized cities of the region, it seems that Jesus preached to his fellow Jews exclusively — mostly farmers and fishermen of the countryside. (Mathew 10:5-6; 15:24)[27]

He preached what John had preached. He told his fellow Jews the end of the world is near. *YHWH* is coming. Any day now. He will resurrect the dead and judge us all. Repent and pray for forgiveness.

In this Apocalyptic vision, the God of Abraham would soon break the yoke of Roman occupation and establish a new Kingdom of Israel here on Earth. A cosmic judge — the biblical Son of Man of the Book of Daniel — would come down, raise all who had died, reward the righteous and punish the wicked.

This End of Days was imminent. "Truly I tell you," Jesus said, "this generation will not pass away before all these things take place." (Mark 13:30.)

Like the fiery biblical prophet Amos, Jesus railed against the rich, called for social justice, and urged compassion for the poor.[28] A frequent target of his wrath were the Temple priests in Jerusalem, the Sadducees — these "lovers of luxury" as Josephus called them, who had grown wealthy from collaboration with the Roman occupiers.[29]

<u>Jesus and the Law</u>

> *For I desire mercy, not sacrifice [burnt offerings].*
>
> — Jesus, citing Hebrew prophet Hosea.
> (Hosea 6:6; Matthew 9:13)[30]

Jesus attacked the religious establishment for their veneer of righteousness. He argued that Pharisee ritual observance of the minutia of Torah Law was not the key to serving YHWH. It was the spirit and intent behind the Law which was paramount. Like his mentor John, he urged a return to what he saw as the core values of Judaism — empathy for the socially powerless, the poor, the oppressed, the outcast.[31]

"When you give a banquet," Jesus said, "invite the poor, the crippled, the lame, the blind, and you will be blessed. Since they cannot repay you, you will be repaid at the resurrection of the righteous." (Luke 14:13-14.)

In my favorite New Testament passage, the Son of Man says to the those judged virtuous:[32]

> *Come, ye blessed of my Father, inherit the kingdom prepared for you from the foundations of the world:*
>
> *Because I was hungry and you gave me food, I was thirsty and you gave me drink, I was a stranger and you welcomed me, I was naked and you clothed me, I was sick and you visited me, I was in prison and you came to me . . .*
>
> *As you did it to one of the least of these, my brethren, you did it to me.*
>
> — Matthew 25:34-40

Jesus expands on these words in the Sermon on the Mount or Beatitudes. Like a new Moses, he addresses the multitudes from a mountainside. Jesus says in part:

> *Blessed are the poor in spirit:*
> *for theirs is the kingdom of heaven.*
> *Blessed are they that mourn:*
> *for they shall be comforted.*

Blessed are the meek:
*   for they shall inherit the earth . . .*
Blessed are the merciful:
*   for they shall obtain mercy.*
. . . whoever shall smite thee on thy right cheek, turn to him the other one.
Love your enemies, bless them that curse you . . . and pray for them which . . . persecute you.

— Matthew 5:1-47

It is as though Jesus had reached back to the founding of the Hebrew nation and its national god, *Ēl* the Kindly, *Ēl* the Merciful. The preacher from Nazareth had raised the principles of compassion and grace to unprecedented heights.

The Apostles

The Gospels tell us that Jesus had seventy disciples, including women. (Luke 8:1-3; 10:1-12.) It seems a number of them had followed John.[33]

Chief among them were twelve Apostles — all men from Galilee.[34] Most were poor fishermen, one a tax collector and another a revolutionary. They were recruited to represent the "long since destroyed and scattered" twelve tribes of Israel.[35]

Jesus told them that in the future kingdom of *YHWH*, "you who have followed Me will also sit on twelve thrones, judging the twelve tribes." (Matthew 19:28.) This belief that all Hebrew tribes would be reconstituted in post-apocalyptic Israel was also from the Book of Daniel.[36]

Apparently the first apostles to join Jesus were Andrew and John, fisherman from the village of Bethsaida on the shores of the Sea of Galilee. (See Fig. 11.1.) They had been devoted followers of John the Baptist. Through them Jesus found Simon (*Shim'on* in Hebrew), later called Peter. He was considered leader of the apostles.

Jesus told them to "go, sell your possessions and give to the poor, and you will have treasure in heaven. Then come, follow me." (Matthew 19:21.) He trained the apostles to spread his apocalyptic message and preach independently. This would be key to the future growth of the movement.[37]

Troubles in Galilee

The Gospels tell us enthusiasm for Jesus began to wane in the north country (Mark 6:3-6). He found himself rejected by religious authorities and, most

Figure 11.1   Jesus Preaching on the Shores of the Sea of Galilee while Sitting in a Boat. Matthew 13:1-3 ASV. Right Rev. Richard Gilmour, D. D. Bible History, 1904.

disturbing, the common people as well. To make matters worse, Pharisees warned him to "get away, for Herod [Antipas] wants to kill you." (Luke 13:31-33.)[38]

It seems the people of the lake towns tired of Jesus' constant badgering on righteous behavior. This reminds one of Socrates. The outcome also had similarities.

The Messiah Secret

At some point Jesus took the teachings of John the Baptist a step further, a huge step. It seems he came to believe that he, Jesus was the Messiah — the future king to be appointed by *YHWH* to rule the restored kingdom of Israel after the Apocalypse.

As noted, Jesus did not contend that he was divine. This was an assertion made after his death, as we shall see.[39]

On the way to villages around Caesarea Philippi, Jesus asked his apostles:

> *"Who do people say I am?"*
> *"Some say John the Baptist; others say Elijah; and still others, one of the prophets,"* they replied.

"What about you?" he asked Peter, *"Who do you say I am?"*
"You are the Messiah," Peter answered. (Mark 8:27-29.)

Jesus warned his apostles not to tell anyone else he was the Messiah. (Mark 8:30.)

Why this "Messiah secret," as scholars call it? Bible experts debate as to the theological reason for this secrecy. Some argue it is a mistranslation. Others say it is a later add by the author of Mark.[40]

Perhaps it was simply voiced by Jesus out of fear for his life. To claim kingship was treason in the eyes of the Roman overlords. If it became known to them, he would be crucified.

[YHWH's] message becomes a fire burning in my heart, shut up in my bones . . .

— Jeremiah 20.9[41]

Like the Hebrew prophet Jeremiah, the fire burning in Jesus' heart would not subside. He had to go to the Holy City, to the One Temple of the One God, to confront the hypocrisy there.

Jesus led his followers to Jerusalem during the time of the Passover festival.

Jews from Palestine and across the Mediterranean came to the Holy City to commemorate Moses' liberation of their ancestors from bondage in Egypt. During Passover, "the city's population could swell to over a million people."[42]

To the House of YHWH

<u>Note to Reader:</u> *The description here is based on the Gospels and in part on the writings of Reza Aslan in his excellent book, Zealot: The Life and Times of Jesus of Nazareth.*[43]

A motley gang of men and women in ragged cloaks and worn sandals march in joyful anticipation towards the Holy City. As they approach, Jesus instructs two of them to go the village ahead and borrow a donkey colt. They bring the colt to him and throw their cloaks over it. He sits upon it.

Other followers spread their clocks on the road before him. Still others lay branches they had cut in the fields. "Hosanna!" they shout, "Blessed is he who comes in the name of the Lord. Blessed is the coming kingdom of our father David. Hosanna in the highest heaven!" (Mark 11:1-18.)[44]

Why did Jesus enter Jerusalem riding a donkey colt? Apparently because it was a symbol for the arrival of a king of the Jews — as prophesized by the biblical prophet Zechariah:

Behold, thy king cometh unto thee,
He is triumphant, and victorious,
Lowly and riding upon a donkey,
Even upon a colt the foal of a donkey . . .

— Zechariah 9:9

Did it really happen? I mean given Jesus' reluctance to publicize his claim of being the future Messiah? We cannot know. If it did, why was he not arrested? Perhaps in the city teeming with pilgrims, a Galilean riding a donkey colt was not noticed. Or the authorities noticed but did not take it seriously.[45]

The Jews of Jerusalem would likely have seen Jesus and his ragged band of followers from Galilee as country bumpkins. When the rural Galileans spoke, their crude northern dialect would have confirmed the impression. And the notion of the Galilean Jesus as the Messiah-king would have seemed laughable. Everyone knew he would come from Judea. (cf. John 7:40-42.)[46]

According to the Gospels, Jesus went to the Temple the next day with his followers. They joined "the crush of pilgrims passing through the Huldah Gates at the Temple's southern wall" and entered the public plaza — the Court of the Gentiles. (See Fig. 11.2.) Roman soldiers were likely "standing among the colonnade of the Temple looking down at Jews celebrating," writes Fredriksen.[47]

Oh, how Jesus and his followers must have felt when they viewed the Temple Sanctuary off in the distance. The House of Almighty *YHWH* "seemed like a mountain covered with snow; for any part not covered with gold was dazzling white," Josephus would later write. "In front of [its] 82½-foot-high and 24-foot-wide golden doors was a curtain of the same length — a Babylonian tapestry embroidered with blue, white, and linen thread, scarlet and purple . . . the scarlet symbolized fire, linen the earth, blue the air, and purple the sea . . . Worked on the tapestry was the whole vista of heavens except for the signs of the Zodiac."[49]

Now in the Court of the Gentiles, Jesus in a fit of righteous anger "cast out them that sold and bought in the temple, and overthrew the tables of the money changers, and the seats of them that sold doves." (Matthew 21:12.) He chased away the oxen and sheep to be sold for sacrifice. "With the

152   *Cosmic Roots*

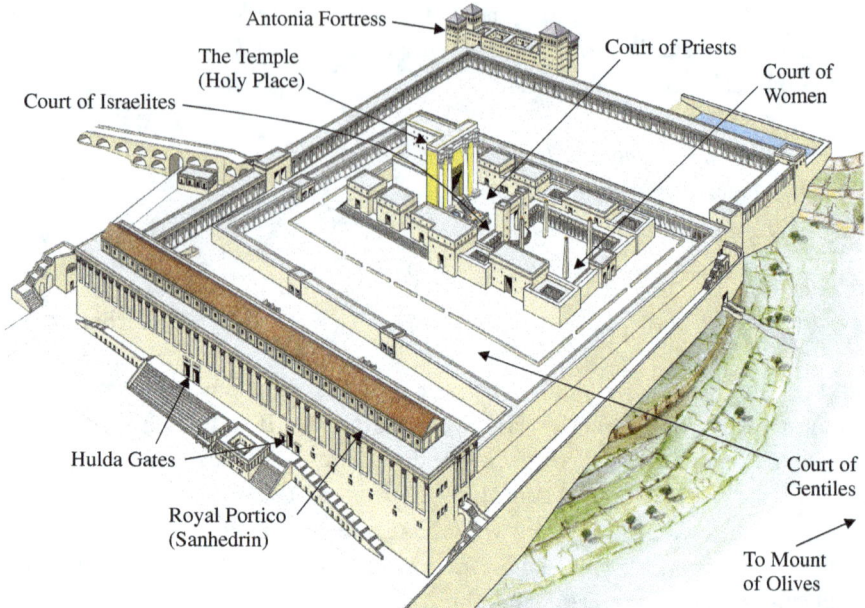

Figure 11.2.   The Herodian Temple Complex. The Court of Gentiles was open to all. The sacred enclosure within was for Jews only. It was forbidden to Gentiles under penalty of death. Within its walls was the Court of Women, Court of Israel (men only), Court of Priests, and the "Holy Place." In front of this was the Altar where sacrifices took place. Within was the Holy of Holies where the High Priest entered once a year on Yom Kipper, the Day of Atonement. It is said to have once contained the Ark of the Covenant and the Commandments.[48]

help of his disciples, he blocked the courtyard entrance forbidding anyone carrying goods from entering the Temple."[50]

The rage expressed by Jesus was likely rooted in the age-old animus of the poor of the Jewish countryside towards the wealthy of Jerusalem. As an impoverished Jew from rural Galilee, Jesus had a natural repugnance for the priestly Jewish aristocracy who ran the Temple* and grew rich in their collaboration with Rome.

---

*Typical of oriental temples (ala Sumer), the Jerusalem Temple was "the hub of economic and commercial activity, and housed the national treasury." Source: Tabor, James "The Jewish Roman World of Jesus" uncc.edu. https://pages.uncc.edu/james-tabor/the-jewish-world-of-jesus-an-overview/. Retrieved April 7, 2019.

It is easier for a camel to go through the eye of a needle, than for a rich man to enter into the kingdom of God.

— Matthew 19:24

It seems Jesus's protest was not so much against commercial practices being carried out so near to the Holy of Holies, as is commonly thought. It was in defense of the poor, who could not afford to pay for the animals being sold for sacrifice to Almighty *YHWY*. (See endnote for other explanations.)[51]

Why was Jesus not arrested for this provocation at the Temple? Surely the Romans had implemented heightened security in the city during Passover to put down disturbances — especially at the entrance to the Temple.

Perhaps the stories of Jesus's triumphant entry on the first day and near riot on the second are exaggerations.[52]

The Betrayal

The Gospels tell us Jesus continued to go to the Temple to preach for a week. Then, it seems, he got wind of a plot to arrest him. He took his disciples and hid in the dark of night in the Garden of Gethsemane at the foot of the Mount of Olives just east of the Holy City.* (Mark 14:32-42; Matthew 26:36-46.)[53]

Then "Judas Iscariot, one of the Twelve, appeared." With him were a "great multitude armed with swords and clubs." They were sent from the "chief priests, the teachers of the law, and the elders." (Mark 14:43.)

Judas walked over and kissed Jesus. That was the signal. The soldiers and officers arrested Jesus and took him away. The remaining disciples "deserted him and fled." (Mark 14:50.)

Why did Temple authorities arrest Jesus? What did Judas tell them? Bible scholar Bart Ehrman suggests that Judas told the authorities Jesus' secret: he claimed to be the Messiah, the future King of the Jews.[54]

Jesus was handed over to Pontius Pilate. It seems the Roman governor had a standing order that the High Priest (appointed by and subservient to Rome) "arrest and turn over any Jews seen as agitators or subversives." Pilate

---

* According to Luke 22:39, Jesus goes to a place that was known to Judas Iscariot. He comes here so he can be apprehended by those authorities that he has predicted would arrest him (Matthew 16:21; 17:22-23; 20:17-19).

ordered Jesus be crucified — the standard Roman punishment for treason.[*] It was the most painful and humiliating form of execution.[55]

Jesus was flogged and stripped naked, a crown of thorns mocking his royal aspirations placed on his head. He was beaten and spit upon, his arms tied to a cross, and made to carry it to the crucifixion site.[56]

My God, my God, why hast thou forsaken me . . .

— Psalm 22:2, Hebrew Bible

The Romans placed a plaque above his head to identify his crime: "Jesus of Nazareth, King of the Jews." (Matthew 27:37.) Jesus suffered death by suffocation on the cross — a slow, excruciating torture that could last for days.[57]

Alongside Jesus, two other Jewish "rebels" were crucified by Rome that day. (Matthew 27:38.)[58] People passing by could hear their screams of agony. They were among the likely hundreds of Jews nailed to the cross by Pontius Pilate — not to mention those crucified by other Roman governors.[59]

Although we cannot know for sure, the majority of scholars estimate the date of Jesus' crucifixion was around the year 30 AD.[60]

He Has Risen

Other first century Jewish preachers and zealots had sought to cast out the pagan conquerors. Some also claimed the title of Messiah. They included the prophet Theudas with some 400 disciples; "the Egyptian" who raised an army in the desert; Athronges the Sheppard boy who crowned himself "King of the Jews." There were also "the Samaritan"; "Hezekiah the bandit chief; Simon of Peraea; Judas the Galilean," who led an armed rebellion against Roman taxes; his grandson Menahem, leader of the *Sicarii* or Daggermen; and Simon son of Giora. (More on this in the next chapter.)[61]

---

[*]The fact that Jesus was turned over to and crucified by the Romans tells us he did not claim to be the divine Messiah or Son of God of later Christianity. If so, Jewish authorities would have charged him with blasphemy. If found guilty, he would have been stoned.

Whether strictly religious messiahs like Jesus or armed warriors, they were all executed by Rome. But, unlike the others, Jesus' story did not end there.

The Resurrection

Not much changed after the death of Jesus. The Holy Land still suffered under the Roman boot. The imminent Day of Judgment had not come.[62]

Then three days after his death, Jesus appeared alive and *in the flesh* to his followers:

> *He [Jesus] appeared first to Mary Magdalene, from whom he had cast out seven demons. She went and told those who had been with him, as they mourned and wept. But when they heard that he was alive and had been seen by her, they would not believe it . . .*
>
> *Afterward he appeared to the Eleven themselves as they were reclining at table, and he rebuked them for their unbelief and hardness of heart . . .*
>
> — Mark 16:9-20

Was the miraculous return of Jesus from the dead a vision, a hallucination, or perhaps group hysteria of his followers borne out of guilt for having left their master to die? Did it really happen? The resurrection — and other miracles — cannot, of course, be considered historical fact. They are matters of faith.[63]

> *. . . this is the night when Christ broke the prison-bars*
> *of death and rose victorious from the underworld.*
>
> — *Exultet*, Ancient Easter Chant[64]

Whatever its reality, many followers of Jesus came to believe it happened. Out of this event came the belief that Jesus was a divine being — one who would return again soon in a second coming.[65]

Reflections

Of all the Jewish rebels, prophets, and messiah claimants of the first century AD, why is Jesus alone remembered and worshipped? The Resurrection. This event is without precedent in the Hebrew Bible or anywhere else in Jewish history.[66]

At least this is what scholars tell us. I think there is more: the character, the righteousness, the charisma of Jesus himself. Surely his passion, compassion, devotion to *YHWH*, inspiration, holiness, integrity, righteousness, force of personality, and ultimate courage were major factors in why his followers remembered him, spoke of him, wrote about him, and worshipped him long after his demise. He touched their hearts.

Jesus had no political power, no army, no huge following in his lifetime. It is amazing that he wasn't lost to history. He preached for perhaps two years and is remembered for two thousand. That to me is the real miracle.

The Way

The "growing understanding that Jesus had been raised from the dead," writes Bible scholar L. Michael White, "seems to have spread very quickly among his followers." A movement began to form, a sect of Judaism which called itself "the Way." (Acts 9:2, 19:9, 19:23, 24:22)[67]

A number of followers of the Way dared to openly claim Jesus' divinity. This led to "unofficial, sporadic, and local opposition, sometimes violent — at the hands of" some mainstream Jews, writes Ehrman. In retaliation, Jesus followers declared *YHWH* "had rejected his own people, but only because they had rejected Him first."[68]

According to the Book of Acts, the first martyr of the Way is said to have been Stephen — a Greek-speaking Hellenistic Jew.[69] He reportedly gave a speech to Diaspora Jews living in Jerusalem at the Synagogue of Freedmen. Stephen tried to convince them that, in denying Jesus, Jews were rejecting the *divine* Messiah. For this blasphemy, they "cast him out of the city, and stoned him" to death. (Acts 6:8-10; 7:59)[70]

Why couldn't pious Jews accept the beliefs of Jesus followers, who were also Jewish? To claim a human being who walked the earth as divine, let alone the Son of Almighty God, was the ultimate blasphemy. Thus, the justification for the persecution of Jesus followers.

Perhaps no behavior towards our fellow humans is so cruel, so delusional, so self-righteous as to persecute, even spill blood in the name of God. If Satan does exist, such acts surely give him the greatest pleasure.

> *Whoever keeps the whole law but stumbles*
> *at just one point is guilty of breaking all of it.*

— Book of James 2-10

With the death of Jesus, his older brother James (*Ya'akov* in Hebrew) now led the Way. He was both a believer in his brother's divinity and a strict follower of Torah Law. "James the Just" as he was known was especially devoted to the poor. He and his community would for the most part remain in Jerusalem awaiting the second coming of Jesus.[71]

The Next Chapter

Despite a mere one-to-two-year ministry, Jesus' power to inspire and his message of Godly hope and righteousness remained with followers long after his death. Memories of his words and deeds would sustain and nurture the movement.[72]

Traveling preachers — "wandering charismatics" as they are called — would spread Jesus' humanitarian message, tell stories of his life, death, and resurrection, and speak of the coming kingdom of *YHWH*.[73]

In the meantime, tensions in Palestine would continue to escalate. Ever more oppressive Roman rule would soon lead to open revolt. It would alter both mainstream Judaism and the nascent Jesus movement forever.

This is the subject of the next chapter.

Chapter 12

JERUSALEM IN FLAMES

And I will stir up thy sons, O Zion ...
And the Lord shall be seen over them,
And His arrow shall go forth as the lightning,

— Zechariah 9:13-14, Hebrew Bible

It seems inevitable in hindsight. Oppressive taxation by greedy Roman governors. Repeated Roman insults to the God of Abraham. Collusion and avarice of Temple leadership. Land loss and unemployment exacerbated by severe drought and famine.[1]

It was just a matter of time.

On the Brink

<u>Note to Reader</u>: The primary source for this period is again Josephus aka Yosef ben Matityahu. As noted, he was a Jewish general in Galilee who had gone over to the Roman side. While he attempts to explain the motivations and actions of his Jewish brethren, he exhibits a far greater compulsion to place his Roman benefactors in a positive light. Though modern scholars consider his history generally accurate, his portrayal is deeply colored by his fawning to Rome.

After the death of Jesus, the Holy Land continued to boil with messianic fever. In 36 AD, a holy man called "The Samaritan" declared himself messiah king to a crowd of devotees on Mount Gerizim. Under orders from prefect Pontius Pilate, Roman soldiers "cut his followers to pieces." Samaritan leaders appealed to Rome. For this final cruelty, emperor Tiberius exiled Pilate to Gaul.[2]

In 44 AD, Theudas stood on the banks of the Jordan River and crowned himself messiah. With arms stretched, he proclaimed he would part its waters like Moses and liberate the Holy Land from the Roman infidels. They chopped off his head.[3]

Two years later, Jacob and Simon, sons of failed messiah Judas the Galilean, launched their own rebellion against Rome. Both were crucified.[4]

As revolutionary bandit gangs began terrorizing the countryside, a new threat formed which would strike at the heart of Judaea.[5]

The Sicarii

No god but YHWH

— *Sicarii* Motto[6]

In mid-century (50s AD), a "shadowy" band of Jewish fanatics began to terrorize the Holy City. The Romans called them *Sicarii* or daggermen, from the Latin *sica* for curved dagger.[7] With daggers concealed in their cloaks, these religious fanatics assassinated any and all fellow Jews they saw as enemies of *YHWH*.[8]

Propelled by an apocalyptic worldview, their main target was the wealthy priestly aristocracy — the Sadducees and other Roman collaborators. Their reign of terror began with the assassination of High Priest Jonathan during Passover of 56 AD.[9]

In the meantime, Rome continued its ruthless crackdown. A Jewish false prophet mysteriously called "the Egyptian" drew thousands of followers. They were massacred by Roman troops.[10]

The final affront came in 66 AD. Gessius Florus, an Ionian Greek, was Rome's procurator of Judaea. Perhaps the most corrupt of all Roman governors, Florus seized "vast quantities of silver from the Temple treasury." He claimed it was on orders from Emperor Nero.[11]

> *The most heartless of men, [he] indulged in every kind of robbery . . . everyone might be a bandit so long as he himself received a rake-off.*
>
> — Josephus on Florus[12]

Some Jerusalemites mockingly passed the hat around for "that poor procurator Florus." Outraged, he had his soldiers "sack the city's Upper Market-place and kill all they met there." Some 3600 Jews are said to have perished in the onslaught, including women and children.[13]

Pent-up for over half a century, Jewish rage exploded into a full-out riot. In August of 66 AD, a Jerusalem mob overwhelmed the small Roman garrison stationed in the Holy City. Under guarantee of safe passage, the Roman troops surrendered and laid down their arms. The Jewish rebels slew them all.[14]

Eleazar ben Ananias, captain of the Temple and son of the high priest, persuaded Temple priests to end daily sacrifices to the Roman Emperor and state. This was tantamount to a declaration of independence from Rome.

Florus failed to quell the uprising. The responsibility went up the Roman chain of command.

Cestius

Gaius Cestius Gallus was Rome's Legate of Syria—under which Judaea fell. He was a civil administrator reportedly "with no battle field experience."

In the fall of 66 AD, Cestius marched from Antioch into Palestine at the head of the Twelfth Legion, *Fulminata*. It was "reinforced with units of III *Gallica*, IIII *Scythica*, and VI *Ferrata*, plus auxiliaries and allies." All in all, it was a force of over 30,000 Roman troops.[15]

The legions wiped out the inhabitants of Chabulon in Galilee southeast of Ptolemais. They burnt the city and countryside villages to the ground. Cestius' troops "liberated" the Hellenistic city of Sepphoris, Roman capital of Galilee. They killed 8400 Jews in Joppa on the Mediterranean coast and burnt down surrounding villages.[16]

Cestius and his troops then advanced on Jerusalem. Leaders of the pro-Roman Jewish faction in the Holy City tried to negotiate peace. Jewish rebels murdered them.[17]

Roman forces "stormed the city's inner wall for five days, and undermined the northern wall protecting the Temple." Then "without a reason in the world," as Josephus put it, Cestius ordered a retreat.[18]

His army marched in haste away from the Holy City. Jewish rebels followed in hot pursuit. The insurgents trapped the Romans in the mountain passes of Beth-horon. Some 5300 Roman infantry and 480 cavalry were slain — in the same area where Judas Maccabee had ambushed the Seleucids over two centuries earlier.

That night, Cestius and his remaining forces slipped away to the north. Elated rebels sung "hymns of victory" on their way back to Jerusalem.[19]

Cestius was an honorable Roman. He returned to Antioch and committed suicide.[20]

Fearing Roman reprisals, a number of "distinguished Jews abandoned the city," Josephus tells us, "like swimmers from a sinking ship."[21]

Zealots* took advantage of the power vacuum and seized the Temple compound in Jerusalem. The Zealots possessed an uncompromising devotion to YHWH and the Law and raging hatred towards the heathen Romans. Their minions included Pharisees, some priests, and common people.[22]

With this unlikely victory over the Romans, the "Great Revolt," as it would come to be called, soon spread to the rest of Judaea. Its tendrils of rage and frustration enflamed Galilee, Idumaea, Peraea, and even Samaria.[23]

---

*The Zealots were a separate Jewish nationalist sect far greater in number than the *Sicarii*. They date back to at least the time of Judas Maccabee. Source: Padfield, David. "Jewish Sects of the Second Temple Period" 2017. https://www.padfield.com/acrobat/sermons/jewish-sects.pdf. Retrieved July 27, 2021.

Rebel forces swelled across the land. Soon they will face the full wrath of Rome — the greatest military power the ancient world had ever known.[24]

It was hopeless.

Vespasian

The route of Cestius and the twelfth legion by a ragtag group of Jewish rebels threatened the perceived invincibility of Rome. Emperor Nero recognized that his very throne was in jeopardy. He commissioned his best general, Titus Flavius Vespasianus, to crush the uprising without mercy.

In the spring of 67 AD, on came the old war-horse with an army of some 60,000 men. Under his command were the famed Fifth and Tenth Legions from Syria, the Fifteenth Legion from Alexandria, and some twenty-three cohorts. To this were added "large allied contingent contributed by nearby kings under the Roman yoke: Antiochus, Agrippa, and Sohaemus*."[25]

March Across Galilee

Vespasian's forces rampage through Galilee with his son Titus at his side. At the same time, civil war breaks out throughout Judaea between those who want to fight the Romans and those who seek peace.[26]

The Romans decimate town after town and wipe out adjacent villages. Japha, Gerizim, Jotapata, and Tarichaeae fall like dominoes. Roman forces then lay siege to the large fortified city of Gamala, "capital of the Golan district of the Galilee."[27]

The city is "perched on a ridge that falls off steeply on the north, south and west," Josephus tells us. It is very difficult to attack. For three months, the Romans try to break through its fortifications. They finally succeed on November 10, 67 AD.[28] Roman forces flood into the city.

"Despairing of their lives and hemmed in on every side," Josephus writes, "multitudes plunge headlong with their wives and children into the ravine which had been excavated to a vast depth beneath the [city's] citadel." According to Josephus, nine thousand Jews perished that day. Over half are suicides.

---

*Antiochus ruled the Armenian kingdom of Commagene. Sohaemus ruled the Emesene kingdom in Syria.

It is perhaps difficult for us to today fully comprehend the depth of hatred and fear the Jews felt for the occupying Romans. Unlike others under the Roman yoke, the Jews had *religious* objections to their pagan overlords. Many believers in the One God would rather give up their lives than succumb. In any event, the captured would likely be put to death, sold as slaves, or for those in arms, crucified.

By late 67 AD, Vespasian's legions had subdued the whole of Galilee.[29]

Roman forces then enter Judaea proper. City after city are destroyed.[30] By June of 68 AD, they subjugate most of Judaea, as well as Idumaea to the south and Peraea across the Jordan River to the east.[31]

Only Jerusalem, the Holy City of *YHWH*, remains.

In the meantime, civil war breaks out in Jerusalem. Jewish refugees from Galilee join with the Zealots and fight the ruling moderates. Aided by some 20,000 Idumaeans, they slaughter the moderates and kill their leader, former high priest Ananus ben Ananus.[32] According to Josephus, "the entire outer court of the Temple was deluged with blood, and 8500 corpses greeted the rising sun." The revolutionaries now control the Temple.[33]

The Pause

In late summer, Vespasian receives word that Emperor Nero had committed suicide. The Roman general ceases all military actions and awaits orders from a new emperor. His troops would remain idle for a year.[34]

What do the rebels in Jerusalem do in the interim? Do they organize and train a fighting force under a single command? Do they build up their defenses? Do they attack the Romans in Maccabean-style guerilla war? No, they do not. They continue to fight among themselves.

With the death of Nero, Rome finds itself engulfed in its own civil war.

The Year of Four Emperors — 69 AD

Nero's early rule had been marked with sensibility and good judgment. It seems his behavior turned to megalomania with his executions of his mother and wife in 59 and 62 AD, respectively.[35]

The Emperor of Imperial Rome then took to reciting poetry and playing the lyre — publicly. He acted in plays and took the roles of "pregnant women and slaves about to be executed." This behavior was seen as a scandalous breach of public decorum.[36]

Nero raised taxes to exorbitant levels to fund extravagant expenditures. To make matter worse, Britain revolted in 60-61 AD. This was followed by an uprising in Gaul, a revolt in Spain, and the insurgency in Judaea.[37]

By June of 68 AD, Nero had lost the support of the military, the upper class, and the common people. The Roman Senate pronounced him a "public enemy." Rather than face the proscribed Roman punishment of being beaten to death by rods, the Emperor killed himself.[38]

After the death of Nero, three emperors ruled in quick succession. Galba followed Nero as emperor in June of 68. He was assassinated by the Praetorian Guard the following January. The Senate then recognized Otho as emperor. He was defeated by the forces of Vitellius in April of 69. Otho committed suicide and Vitellius now sat on the throne in the Imperial City.[39]

Then Vespasian's supporters made their move.[40] Roman legions in Egypt, Syria, and Palestine swore allegiance to Vespasian in July of 69. Danube legions followed a month later and invaded Italy. On December 20, they killed the unpopular Vitellius in the Imperial Palace. The following day, the Roman Senate recognized Vespasian as emperor.[41]

In the interim, a new Jewish insurgent would arise and take the mantle of leadership in the Holy City.

Simon

<u>Note to Reader</u>: *This section is primarily based on the writings of Josephus and Oxford historian Cecil Roth.*[42]

His name was Simon bar Giora. He was a Jewish radical who despised the Jerusalem ruling class and their collusion with Rome. Like the *Sicarii*, he "attacked the wealthy, sacked their houses, and molested their persons." Like the Hebrew prophet Isaiah, he called for economic justice for the poor and the freeing of slaves.[43]

> *Because the Lord hath anointed me*
> *To bring good tidings unto the humble . . .*
> *To proclaim liberty to the captives . . .*
> *And the day of vengeance of our God . . .*

— Messiah prophecy: Isaiah 61:1-2

Simon was among the Jewish rebel leaders who had routed Cestius at Beth-Horon in the fall of 66.[44] He and his followers had likely fought alongside the Zealots against the moderates in Jerusalem.[45]

When Vespasian had become inactive in 68 AD, Simon seized the opportunity. With some 40,000 men, he conquered Idumaea, and "the entire south of the country, including Hebron," Cecil Roth tells us."[46] Simon's conquests were short lived. The Roman general returned to action in the Spring of 69 and retook the region for Rome.[47]

Simon retreated to Jerusalem and camped outside the city walls. Tired of rebel atrocities, former High Priest Matthias opened the gates and pleaded with Simon to overthrow the Zealots and Galileans.

Simon in Jerusalem

In April of 69 AD, Simon enters the Holy City with the remnants of his southern army — some 10,000 Judeans and 5000 Idumaean allies. It is the largest, "most disciplined and best organized force of all his rivals."[48] They find a Jerusalem swelled with refugees from Galilee and Jewish pilgrims celebrating Passover. His troops rapidly take control of almost the entire city.[49]

The Galileans under John of Gischala hold the Temple Mount and its outer court with some 6000 men. Now estranged from the Galileans, Zealots under Eleazer ben Simon are ensconced in the Temple Sanctuary or Inner Court.

The rest of the city — the Upper and most of the Lower City "with its bazaars and shops and palaces and fortresses, and all uncommitted citizens, their numbers swollen by tens of thousands of refugees — are now in the hands of Simon," writes Roth.[50]

The madness inside Jerusalem continues. In attempts to starve each other out, Simon and John "set fire to the houses that were supplied with grain and supplies of every kind," Josephus writes. "Almost all the grain — enough to support them through many years of siege — went up in flames."[51]

Why haven't we heard more of Simon? Largely because of Josephus. A member of the Jewish aristocracy, Josephus significantly downplayed the contributions of low-born revolutionaries like Simon bar Giora. We yearn to know more of Simon, his origins, his motivations — but beyond the writings of Josephus, the trail grows cold.

Upon his return to Rome in December of 69 to serve as emperor, Vespasian gives command of the Judaean campaign to his son. This is Titus's opportunity to make his own mark on history — to crush the troublesome people with their strange beliefs who dared rebel against Imperial Rome.[52]

The Siege of Jerusalem

What corner of the earth had escaped the Romans, unless heat or cold made it of no value to them? . . . God was on the Roman side.

— Josephus, *The Jewish War*.[53]

Titus reaches the outskirts of the Holy City in April of 70 AD. Within its walls, the Zealots form under John. Simon and John then make an "unholy alliance," as Josephus put it. Simon is recognized as the overall leader of the rebellion.[54]

Under Titus' command are four legions of "superbly trained and magnificently equipped" professional soldiers. They serve under a highly disciplined command structure. Along with auxiliaries and local troops, they number a combined force of some 60,000 men.[55]

The vaunted Roman war machine now faces a large, well-fortified city. Jerusalem is protected by three successive layers of stout walls. Within are "a maze of narrow streets." Numerous underground water and sewage passages provide hide-outs "from which to launch surprise attacks."

Urban warfare would force the Romans to fight within these cramped quarters. This would mitigate to some extent their 3 to 1 advantage in numbers over the Jewish rebels.[56] Rome's siege capabilities along with its famed military engineering would more than made up for this.

Roman engines were "masterpieces of construction," Josephus writes.[57] They included catapults, battering rams with protective sheds, rolling iron-clad siege towers 75 feet (23 meters) high — as well as iron-pointed wall-borers, stone-throwers which could fling 55-pound (25-kilgrams) stones some 700 yards (640 meters), and "iron-plated mobile screens to protect those bringing up the engines"[58]

Simon on the other hand had no cavalry and some artillery pieces and other gear taken from the Romans at Beth-Horon. The latter included 300 bull-throwers and 40 stone throwers

In his account, *The Jewish War*, Josephus refers to Simon as a "tyrant," a "terrorist," a "monster."[59] Still, he could not help admit:

> "[Simon bar Giora] was regarded with reverence and awe, and such was the esteem in which he was held by all under his command, that each man was prepared even to take his own life had he given the order."[60]

Though outmanned and ill-equipped, Jewish rebels under Simon fought back with "ferocious hand-to hand-combat," Cecil Roth writes, "They conducted a series of bold sorties, including a pitched battle which drove the Romans all the way back to their camps, a brilliantly organized raid which nearly led to the destruction of Roman camp on Mount of Olives, and a surprise assault that almost crushed Titus himself."[61]

Let us take a look at the great historic conflict.

*The Siege Begins**

On May 10, 70 AD, the Romans attack the third and outermost wall of Jerusalem. (See Fig. 12.1 – Arrow A).

The thud of Roman battering rams echoes throughout the terrified Holy City. The pounding is relentless.

Jewish rebels attack from atop the walls with fire and projectiles. Jewish ground sorties under Simon set fire to Roman works. Titus and his cavalry drive them off. A rebel is captured. He is crucified in front of the wall for all to see.

The Romans break through on the fifteenth day.[62]

Four days later, Roman battering rams breach the second wall at the central tower gate. (See Fig. 12.1 – Arrow B.)[63] Legionaries pour into the breach. They are met by a barrage of rebel missiles. Savage fighting continues "all day and into the night." The Jews drive the Romans back and mend the break.[64]

The Romans resume their attack and rams break through again. This time they advance. Simon orders the rebels to pull back to the Antonia Fortress.[65]

With grain supplies destroyed, famine begins to threaten the populace inside the City. Hungry Jews sneak out in search of food along the valleys. Roman soldiers seize up to 500 starving people a day. Titus orders them scourged and crucified in sight of the wall for all inside the City to see.[66]

The fight for the great Antonia Fortress begins. (See Fig. 12.1 – Arrow C.)

---

*For an excellent four-part video set depicting the siege of Jerusalem, please see: "The Siege of Jerusalem–70AD" youtube.com. https://www.youtube.com/watch?v=Hen0wmoj5RU&list=PLkOo_Hy3liEL-DFpCESTbk_fC2RVUFU5o

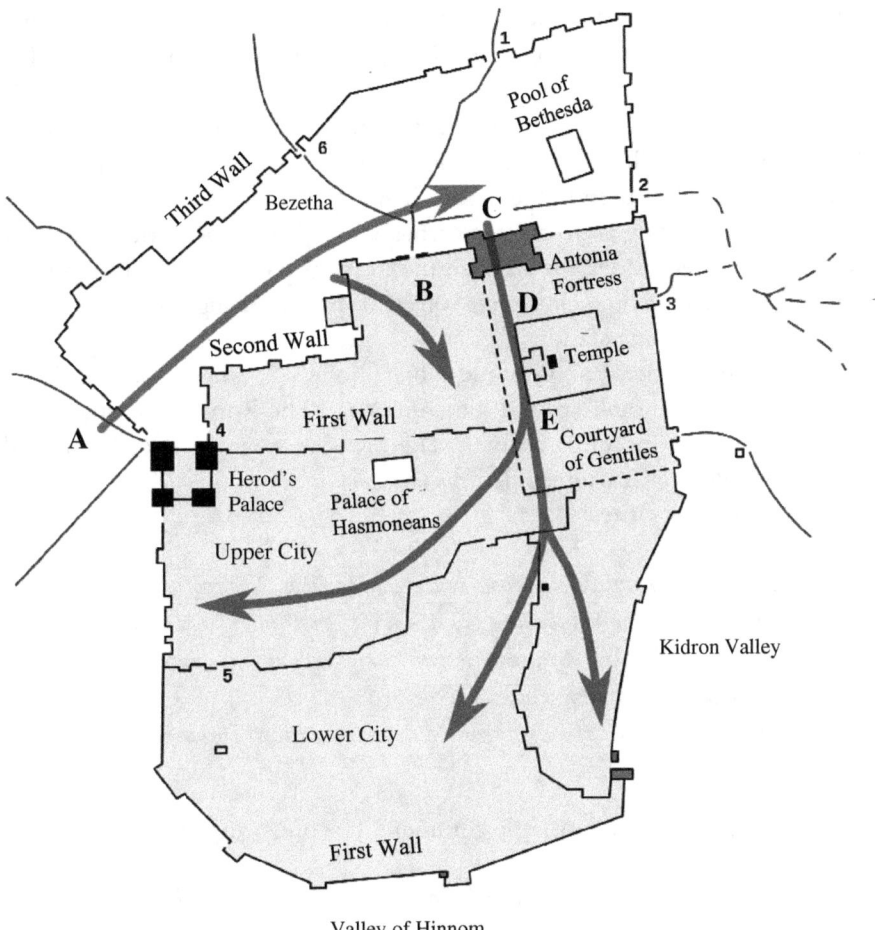

Figure 12.1. The Siege of Jerusalem. Arrow A: Romans break through third wall. Arrow B: Second wall is breached. Arrow C: Romans stall at Antonia Fortress. Arrow D: Battles of the Temple. Arrow E: Romans attack Upper and Lower City.

The Battle for Antonia

[The Romans discover] the Jews have an inner courage that rises superior to faction, famine, war, and disasters beyond number.

— Josephus, The Jewish War.[67]

Constructed by Herod the Great, the Antonia Fortress rises up against the northwest corner of the Temple Enclosure. It is "built on rock 75 feet high and precipitous on every side," Josephus tells us. "Before the actual tower is a 4½ foot wall, and inside this the whole elevation of Antonia rises 60 feet in the air."[68]

The Romans "lock their shields together and charge in their units." Outmanned and ill-equipped, Jewish rebels counterattack with blood-curdling yells. They fight in "ferocious hand-to hand-combat."[69]

The fighting is brutal, the carnage terrible on both sides. Time and time again, rebels under Simon push the Romans back. Still the legionnaires come, gradually pushing closer like waves of an ever-rising tide.[70]

Still, Antonia stands.

Titus orders a four-and-a-half mile (7 km) stone wall of circumvallation be built around the entire City. The Romans complete it in three days, or so Josephus claims.[71] The Jews are now completely trapped. Famine within the City intensified.[72]

Josephus writes:

"[Famine begins to] devour whole houses and families. The roofs are covered with women and babies too weak to stand, the streets full of old men already dead. Young men and boys, swollen with hunger, haunt the squares like ghosts and fall wherever faintness overcame them . . .

[When] they can no longer bear the stench, [survivors] throw bodies from the wall.

In perhaps his most vile comment, Josephus writes:

[The rebels] welcomed the destruction of the people: it left more for them.[73]

Under heavy rebel fire from the wall, Romans bring up siege engines. A series of attempts to take Antonia again fail.

Then on July 24 at 2 AM, Roman soldiers scale the wall on their own initiative. In the dead of night, "twenty of the men guarding the Roman platforms and one trumpeter silently move forward." They slit the throats of rebel sentries in their sleep. The Roman trumpeter blows his horn. Thinking the mass of the Roman army is upon them, the remaining Jewish guards panic and run away.

Hearing the trumpet sound, Titus "arms his forces with all speed, and with his generals and picked troops leads the way to the top." Antonia is theirs at last.[74]

By late July 70 AD, Roman forces reach the grounds of the Holy Temple. (See Fig. 12.1 — Arrow D above.)[75]

The House of YHWH

In three pitched battles on blood-stained Temple grounds, the "exhausted and half-starved" rebels manage to hold off the Romans.[76]

On August 5, priests halt daily sacrifices to *YHWH* at the Temple altar. They have run out of sacrificial lambs.[77]

On August 10, the west portico of the Temple wall appears undefended. Legionnaires rush in. It's a trap. Rebels have filled the rafters with dry wood and bitumen. They set it alight. Roman troops are burned alive in the inferno.

Two days later, Titus orders another storming of the Temple wall, this time in the north. It again fails. The Roman general has the entire northern colonnade torn down. Simon and his forces pull back to the center of the Temple mount.

Titus orders the Temple gates set on fire. The flames spread to the Temple porticos. Roofs and columns attached to buildings are now ablaze. The next day, the Roman general orders the fire extinguished. Still, it continues to burn.[78]

On August 27, the fourth and final battle of the Temple begins. Jews and Romans "grapple in a life and death struggle round the entrances" to the Temple grounds, Josephus writes, "For six days . . . the most powerful battering-ram of all pounds the wall" of the Temple Enclosure — without result. Roman siege engines and mines also have little effect.

At dawn, rebels pour out of the Eastern gate of the Temple and attack. The air is thick with hissing arrows from archers from both sides on opposite walls. The Roman counterattack pushes Jewish forces up against the Temple walls.

The following day, Simon orders an attack. Legionnaires counterattack and drive the rebels up to the walls of the Temple enclosure.[79]

Titus retires to Antonio to consider his next move. In his absence, rebels again attack. They clash with legionnaires who are putting out the fire in the inner court. The Romans turn and chase the Jews to the Sanctuary.[80]

On impulse, a Roman soldier "snatches up a blazing piece of wood," Josephus writes. "[He] hurls the brand through the golden aperture [which has] access to . . . the chambers built round the Holy Sanctuary.*"

"As the flames shoot into the air the Jews send up a cry that matches the calamity . . . for that which hitherto they had guarded so devotedly is disappearing before their eyes."[81]

As the Temple Sanctuary, the Holy of Holies, the *Kodesh HaKodashim* burns, Roman soldiers put all who are caught to the sword. "Little children and old men, laymen and priests alike are butchered."[82]

"Cries from the hill are answered from the crowded streets [below]," Josephus writes, "and now many who are wasted with hunger and beyond speech find strength to moan and wail when they see the Sanctuary in flames."

The Lower and Upper City Ablaze

Under orders from Titus, Roman forces set fire to the Lower City. (See Fig. 12.1 – Arrow E above.) The rebels flee to Herod's Palace stronghold in the Upper City.[83]

On September 7, the Romans lay siege to Herod's Palace from the west outside the walls. Sections of the wall crumble under the massive firing of Roman projectiles. The wall is breeched on September 25. The rebels panic and flee and leave its massive towers undefended.[84]

Roman forces "pour into the streets sword in hand, cut down without mercy all who come within reach, and burn the houses of any who took refuge indoors" Josephus writes ". . . in search of plunder, they find whole families dead and rooms full of victims of starvation."[85]

During the night, "the fire gains mastery." On the eighth of Gorpiaios (September 26) — some four and a half months after the siege had begun — "the sun rises over Jerusalem in flames."[86]

Simon Appears

As Roman troops sack the City, Simon "takes his most trustworthy friends and some tunnellers" and hides in "one of the concealed sewers," Josephus

---

*Modern scholars question Josephus' claim that Titus sought to prevent the burning of the Temple. This was more likely another effort to present his Roman sponsor in a favorable light. Source: Josephus. *The Jewish War*. Trans. by G. A. Williamson. (New York: Penguin pbk. 1981) Ch. 21, endnote 10, p. 452.

tells us. Where the tunnel ends, they try to hack through solid rock, "hoping to emerge at a safe spot and get clear away."[87]

"Progress is slow . . . food is at the point of giving out." Simon "dresses himself up in several short white tunics with a crimson cape fastened over them" — like a Biblical king of Israel, like a Messiah. "At the very spot where the Temple had once stood, he appears out of the ground."[88]

Some scholars suggest he did this to confuse Roman soldiers guarding the area in order to make his getaway. Others say it was to indicate he was indeed the Messiah. Either way, he was recognized, captured and sent to Caesarea in chains.[89]

Remaining men of Jerusalem over the age of sixteen are "sentenced to death in the mines of Egypt or gladiator arenas* across the Roman empire." Women and children are shipped off to slave markets.[90]

The Procession

The following year, Emperor Vespasian and son Titus hold a Triumph in Rome to celebrate their hard-fought victory over the Jews. The procession winds its way through the streets of ancient Rome with some 700 captured prisoners, along with booty from the Temple. (See Fig. 12.2.)

A series of stages of immense size follow — many "three or even four storeys high," Josephus writes. Each platform represents a city or town defeated in the Jewish war. On each is its captured rebel commander. The final platform represents the fallen Jerusalem. On it stands a chained Simon bar Giora still arrayed in white and crimson.[91]

As the principal leader of the revolt, Simon is given the honor of being scourged all the way to the great Temple of Jupiter Capitolinus. Then "a halter is thrown around his neck," writes Roth, "and Simon is dragged to the Mamertine Prison, where he is strangled."[92]

When the signal is given that he has been executed, cheers rise from the crowd. Vespasian and Titus end the Triumph with offers of sacrifices to the chief god of Rome.[93]

Thus died the last messiah aspirant of the Second Temple period.

---

*It was primarily Jewish slaves captured in the 67-70 AD revolt who built the Colosseum in Rome. "Inscriptions also attest to the existence of Jewish gladiators. . . They were common victims of Roman brutality" in this and other arenas across the empire. Source: Salah, historum.com.

Figure 12.2. The Arch of Titus in Rome — South inner panel. Depicted are treasures taken from the Temple during the siege of Jerusalem, including the Menorah and trumpets of Jericho. The Arch of Titus was erected by Domitian in c. 81 AD at the foot of the Palatine hill on the Via Sacra in the Roman Forum. It stands to this day.

Aftermath

The siege of Jerusalem lasted nearly five months. It has been called "the most massive Roman military endeavor since the fall of Carthage two centuries earlier."[94]

There were no miracles, no justice for the Jews. The faithful died by the starvation, fire or sword, cut down by the heathen along with their wives and children. Those of lesser faith who had fled the Holy City earlier survived.

Almighty God was held blameless in the slaughter. Surely this catastrophe was due to some human failings. Pious Jews would say it was punishment for not being observant enough to Torah Law. Christians would come to say it was because Jews had rejected Jesus, the Son of God. Non-believers would say it was the inherent cruelty of our species once more on display.

An estimated 1.1 million Jews died in the Great Revolt.* Jerusalem and its Temple were destroyed. Ethical monotheism was not.

---

*This from Josephus. Roman historian Tacitus puts the number at 600,000. Source: Josephus. *The Jewish War*, Ch. 21, endnote 40, p. 454.

The Sadducees, Essenes, and Sicarii were wiped out by the Romans. The Pharisees alone survived. It would be they who would take sole possession of the Jewish religion.

Judaism would become a religion without a Temple, without priests, without animal sacrifices — a people in exile led by Pharisees.* These teachers of Torah would become known as rabbis. They would lead services at synagogues throughout the Roman Empire and beyond to this day.[95]

The Next Chapter

The slaughter in Jerusalem wiped out both Jews and Jewish Christians. The small sect of Jesus followers under the leadership of brother James** were no more.

The center of the Christian movement would shift from Jerusalem to Antioch, Alexandria, and Rome.[96] Its chief proponent would be a Jew, a Pharisee from the Diaspora. He would come to lead this nascent movement in the Hellenistic world outside the Holy Land. It would change everything.

This is the subject of the next chapter.

---

*Archeological evidence indicates that Jews continued to live in Jerusalem as well as other parts of the Holy Land after the war with Rome. Source: Ehrman, response in "Aslan's Zealot: Historical Mistakes" ehrmanblog.org.

**In an arbitrary and unpopular act, James the Just had been executed by High Priest Ananus in 62 AD. (Jos. *Jewish Ant.* 10:9)

Chapter 13

THE THIRTEENTH APOSTLE

> ... *circumcised on the eighth day, of the people of Israel, of the tribe of Benjamin, a Hebrew of Hebrews; as to the law, a Pharisee;*
>
> — Philippians 3:5

In appearance he was reportedly "a man of small stature, with a bald head and crooked legs, a pale countenance with ruddiness of skin, with eyebrows meeting and nose somewhat hooked."[1]

In character, he was head-strong, persistent, some might say obsessive, fierce in his loyalties, and equally fierce in the denunciation of those he saw as his enemies. He was a skilled organizer, persuasive debater, compelling writer, of extraordinary stamina, and a natural leader.

A contemporary of Jesus, he was likely born somewhere between 5 BC and 5 AD. He was a devoutly religious, educated Greek-speaking Jew. He lived in the Hellenized city of Tarsus, capital of the Roman province of Cilicia on the southern coast of present-day Turkey. His Hebrew name was *Sha'ul* or Saul in English, after the first king of Israel.[2]

As a youth, Saul was educated in Jerusalem at the school of esteemed rabbi Gamaliel, grandson of the great Jewish teacher Hillel the Elder (Acts 22:3). Here, no doubt, Saul learned of Hillel's famous words: "What is hateful to yourself, do not do to your fellow man. That is the whole Torah; the rest is commentary."[3]

As an adult, he worked for the Sadducees and became a zealous persecutor of Jesus followers. "I persecuted the church of God and tried to destroy it," he later wrote. (Galatians 1:13.)[4]

In a notable example, Saul approved of the stoning of Stephen in Jerusalem (Acts 8:1). Some suggest he "may have been the ranking Jewish leader (an elder) who participated."[5]

The Conversion

The year is c. 34 AD. According to the Book of Acts, Saul is on the road to Damascus to pick up captured members of the Way to bring them before the High Priest in Jerusalem for punishment.

"As he neared Damascus, suddenly a light from heaven flashed around him. He fell to the ground and heard a voice say, 'Saul, Saul, why do you persecute me?'" (Acts 9.3-6.) See Fig. 13.1.

"'Who are you, Lord?' Saul asked.

"'I am Jesus, whom you are persecuting,' he replied. 'Now get up and go into the city, and you will be told what you must do.'"

Saul proceeded to Damascus, where a Jesus disciple named Ananias restored his sight and baptized him into the Way. The newly converted Saul then went to Arabia and returned to Damascus. After three years, he returned to Jerusalem. Why Arabia remains a mystery. (Gal 1:17.)[6]

Figure 13.1. **The Conversion of Saul.** Fresco by Michelangelo, c.1542–45.

This story is told in the Book of the Acts of the Apostles. It was written by an unknown author perhaps some fifteen to thirty-five years after the death of Saul. Curiously, Saul himself never wrote about his conversion.[7]

What made persecutor Saul join the persecuted? A divine epiphany? The humanitarian teachings of Jesus? The miraculous story of the resurrection? The enduring faith of Jesus' followers? Whatever it was, it was arguably the most significant religious conversion in human history.

Even though he had never known the living Jesus, Saul declared he too was an Apostle. Although, he said, "I am the least of the apostles and am unworthy to be called an apostle because I persecuted the church of God." (1 Corinthians, 15:9.)

The man who had hunted the followers of Jesus now applied the same fanatic zeal to the conversion of Jews and Gentiles to the Way.

Saul began preaching the "good news" of Jesus's resurrection in cities across the eastern Mediterranean. They often held large populations of Hellenized Jews. Nonetheless, he found it easier to convert Gentiles. (See Fig. 13-2.)

Figure 13.2. Saul delivering the Areopagus sermon in Athens. By Raphael, 1515. Royal Collection of the United Kingdom.

Why? For one thing, deification of a living person was a common pagan concept. As we have seen, a number of rulers had proclaimed themselves gods or sons of gods and were worshipped accordingly.[8]

In the Greco-Roman world, gods at times "are sent from heaven to earth on some redemptive mission on behalf of humankind," writes James Tabor, esteemed scholar on Christian origins. "There are also human figures from history or legend believed to be endowed with divinity to perform superhuman feats . . . such as miracles. . . [Titles are] often bestowed on an emperor such as 'Lord,' 'God,' 'Son of God,' and 'Savior.'"[9]

The story of Jesus, on the other hand, was a problem for most Jews. The Hebrew Bible never said "the Messiah would be born of a virgin," as Bart Ehrman points out. Or that he would be "raised from the dead." Plus it was believed that the Messiah would liberate the Holy Land, which did not happen. These were likely the foremost reasons why the vast majority of Jews rejected Christianity.[10]

Saul had been brought up in a Hellenistic city among Gentiles. He spoke their language (Greek), understood their culture, and was comfortable interacting with them. Somewhere along the line, he found he was more effective in converting non-Jews, as noted.

It would make all the difference.

Is God the god of Jews only? Is He not the God of Gentiles too?

— Saul (Romans 3:29.)

God-Fearers

Of particular interest to Saul were Gentiles who were attracted to the high morality of the Jewish religion. "We have evidence that in every synagogue, especially in the Diaspora," James Tabor writes, "there were groups of Gentiles" who had converted to the faith — as well as those "sympathetic to the God of Israel" but were not formal converts.[11]

Saul saw these "God-fearers," as they were called, as the seeds of Gentile conversion to the Way. For Saul, this was an act of salvation. According to the book of Acts, he believed they "needed to be separated from the wicked, so they will be spared during the End Times culling," writes Ehrman.[12]

To be more palatable to Greek-speaking Gentiles, Saul began to use his Latin name Paulus or Paul in English. The Hebrew *Y'shua* became the Greek *Iesous* or Jesus in English. The Hebrew *Mashach* or Messiah in English became the Greek *Christós* (the anointed one) or Christ in English. Thus, to the Greek-speaking population of Hellenistic eastern Europe and the Near East, Jesus the Messiah became *Iesous Christós* or Jesus Christ in English.[13]

Over time, followers of the Way would come to be called *Christianos* in Greek, meaning "Followers of the Anointed" or Christians in English.[14]

And *YHWH*, the God of Abraham, the God of Israel, the God of the Universe would become God the Father — a God for all nations, as the Hebrew Bible had predicted.

All nations whom You have made shall come and worship before You,
O Lord. And they shall glorify Your name.

— Psalm 86:9

It is difficult to know how much of Paul's message came from prior oral tradition or early written works now lost. Certainly, some of his views came from earlier followers of the Way, especially his beliefs on the death and resurrection of Jesus, The rest? This question "has long puzzled scholars."[15]

Let's take a look at his doctrine.

The Canon of Jesus Christ According to Paul

There is no longer Jew or Greek, there is no longer slave or free, there is no longer male and female; for all of you are one [person] in Christ Jesus. And if you belong to Christ, then you are Abraham's seed and heirs according to the promise....

— Paul (Galatians 3:28-29.)

Paul declared that to become one in Christ, Gentiles need not follow the Torah, i.e., Jewish law. He taught that salvation was not based on the Torah but on faith in Jesus. He said Gentiles did not have to first convert to Judaism. If you believe in Christ. you become heir to God's promise of the Covenant.

According to Paul, a Gentile who comes to Christ does not have to become circumcised — a mutilation seen as repulsive in Hellenistic culture (not to mention painful).[16] A Gentile male does not have to wear a *Yumalka* (head covering). He or she does not have to observe Jewish festivals (holidays) or practice kosher dietary laws. (Galatians 5:1-5; 1 Corinthians 9:20-23; Romans 14:2-3; Colossians 2:16-17.)

Paul was ambiguous as to exactly how much of the rest of Jewish law and the Hebrew Bible is to be followed. This issue is disputed to this day by Christian scholars.[17]

Paul continued to teach Torah traditions of "compassion, ethics, and righteous behavior." He wrote in Romans: "The commandments 'Do not commit adultery.' 'Do not murder.' 'Do not steal.' 'Do not covet.' and any other commandments are summed up in this decree: 'Love thy neighbor as yourself.'" The latter phrase is taken from the Hebrew Bible. (Romans 13:9; Leviticus 19:18.)[18]

Paul spoke out against prostitution, sex before marriage, divorce, and homosexuality. Jesus reportedly said nothing about any of these except divorce. To be more palatable to Rome, Paul also called for obedience to secular rulers. This would be later used to justify the divine rights of Kings. (1 Corinthians 6:9-20, 7:11-27; Romans 13:1-7; Matthew 19:1-12.)

The Cross

The crucifixion of Jesus must have been a difficult issue for the followers of Jesus. To be stripped naked for all to see and be crucified by agonizingly slow torture, one's bowels loosened at the moment of death, was an act of supreme humiliation. That was Rome's intent. How could Jesus Christ, the embodiment of "glory, divinity, and triumph" be a victim of such base degradation?[19]

> *... Christ died for our sins ...*
>
> — 1 Corinthians 15:3

The rationale goes something like this: Yes, Jesus died a humiliating death. But he was resurrected. This attests to his divine nature. His suffering and death had to have meaning. It could not have been for *his* sins — he was the purest of souls. It was an intentional sacrifice for the sake of his followers. It was to atone for their sins, to reconcile their transgressions with God. This had always been God's plan — to have Jesus die on the cross to save humanity, to bring "salvation to the world." (Galatians 1:3; 1 Corinthians 15:3; Romans 5:6).[20]

Paul emphasized this concept in his writings — though Bible scholars tell us the idea predated Paul. Perhaps it originated with the eleven Apostles who fled when Jesus was arrested. Maybe it was to assuage their guilt. "Jesus died for *our* sins," they thought; i.e., the sins of the Apostles.

In time, the cross would come to be *the symbol* of the Jesus movement. Human suffering would be at the core of its canon — an act shared with Jesus in preparation for entrance into the kingdom of God. "Momentary affliction is preparing for us an eternal weight of glory beyond all comparison," Paul wrote. (2 Corinthians 4:17.)[21]

> *Christ is the end of the [Torah] Law, in order to bring righteousness to everyone who believes.*
>
> — Paul (Romans 10:4)

What Paul taught was a new religion: Belief in Jesus supersedes Torah Law. For this and reasons mentioned earlier, "most Jews did not join movement. It became largely a Gentile religion."[22]

Other Gentiles Converts?

Was Paul the first to convert Gentiles to the Way? No, at least according to the book of Acts. The first convert was a "God-fearing" Roman centurion stationed in Caesarea named Cornelius. He was converted in c. 36 AD by Peter. The apostle then preached and converted other Gentiles gathered there. (Acts 10:1-13.)[23]

Of course, we cannot know the historicity of this story. Either way, it appears the number of Gentiles conversions were at best minimal before the coming of Paul. In fact, secular scholars tell us "there is no [historical] evidence of Gentile converts" before Paul.[24]

It seems at one point, Jewish followers of the Way tried to tell Gentile followers of Paul's doctrine that he was wrong. One must observe Torah law, including circumcision, to be true followers of Jesus.

Incensed, Paul declared that his version of Christianity was the only legitimate one. "As we have said before, so now I say it again," he wrote in Galatians, "If anyone is preaching to you a gospel contrary to the one you embraced, let him be under a divine curse!" (Galatians 1:9.)

Sidebar: The Good Samaritan

> *. . . thou shalt love thy neighbor as thyself.*
>
> — Leviticus 19:18

On its way to modern Jericho in the West Bank, a two-lane country highway leaves the outskirts of Jerusalem and weaves its way through hills and dales down into the Jordan Valley. The ochre landscape is arid and empty, except for occasional Bedouin camps with tents and flocks of sheep and goats.[25]

The 18-mile road descends from some 2500 feet above sea level to about 800 feet below sea level.* It passes Bethany, where Jesus is said to have stayed as a guest of Lazarus. It goes by a tomb called *Nebi Musa* in Arabic where Moslems believe the Prophet Moses is buried. Continuing downhill past a sign marking sea level is a small, dreary, isolated building set alone along the roadside — allegedly "The Inn of the Good Samaritan."[26]

---

*This is a drop of about 1010 meters. Located some 7 miles (11 km) north of the Dead Sea, Jericho is the lowest city in the world. It is also arguably the oldest continuously inhabited settlement, dating back to 9000 BC. Source: Compton, Nick. "What is the oldest city in the world?" The Guardian Feb 16, 2015. https://www.theguardian.com/cities/2015/feb/16/whats-the-oldest-city-in-the-world. Retrieved Jul 27, 2021.

(Continued)

Two thousand years ago, this road was a dangerous trail notorious for bandits. It was known as "The Ascent of Blood" for its red rocks and outlaw assaults.[27]

The road is the setting for the Good Samaritan story, as told in the Gospel of Luke in the New Testament — likely written between 80 to 100 AD. Christian tradition says that "Luke" was a companion of Paul, an assertion doubted by most historical scholars. It is generally recognized that the author of Luke also wrote the Book of Acts.[28]

The tale begins with a Jewish lawyer who asks Jesus,

"Master, what shall I do to inherit eternal life?"

"'What is written in the law [of Torah]? How readest thou?' Jesus says.

"'Thou shalt love the Lord thy God with all thy heart and with all thy soul, and with all thy strength and with all thy mind,' the lawyer answers, 'and thy neighbor as thyself.'

"'Thou hast answered right . . .' Jesus says.

"Then the lawyer asks, 'And who is my neighbor?'"

He is asking whether his neighbor refers to his fellow Jews or to Gentiles as well.

Jesus answers with a parable:

"'A certain man went down from Jerusalem to Jericho, and fell among thieves, which stripped him of his raiment, and wounded him, and departed, leaving him half-dead.'

"'And by chance there came down a certain [Jewish] priest . . . and when he saw [the wounded man], he passed by on the other side.'

"'And likewise, a Levite [one who assists priests in ritual services], when he was at the place, came and looked on him, and passed by on the other side.'[29]

"'But a certain Samaritan . . . came where he was, and when he saw him, he had compassion on him. He bound his wounds, pouring oil and wine, and set him on his beast, and brought him to an inn, and took care of him.'

"When [the Samaritan left the next day], he gave money to the host [of the inn] to take care of [the injured man.]"

Jesus then asks the lawyer. "'Which of these three, thinkest thou, was neighbor unto him that fell among thieves?'"

"'He that showed mercy on him,' the lawyer answers.

"'Then,' says Jesus, 'go and do likewise.'" (Luke 10:25-37.)

Why did the author of Luke make a Samaritan (Fig. 13.3) the hero of the story? At the time of Jesus, Jews and Samaritans generally despised each other. As we learned, this animosity went back some ten centuries before Christ to the biblical rivalry between the kingdom of Judah and the kingdom of Israel to the north.

Figure 13.3. The Good Samaritan. Stops to help a Jewish man accosted by thieves on the road to Jericho. Art by Richard Andre for The Coloured Picture Bible for Children, London, 1884.

Again, despite mass deportation under Assyrian king Sargon II in 721 BC, the "Lost Tribes of Israel" were not all lost. Many, especially the poor, remained in their native land. They merged with foreign refugees brought in by the Assyrians and became known as the Samaritans, after their capital city of Samaria. (2 Kings 17:1-40.)

Jews of the time considered the Samaritans a "mixed-race" contaminated by intermarriage to pagans. Yet it was a Samaritan who stopped to help the

> *(Continued)*
>
> injured man. The parable tells us that it is one's character rather than one's race, nationality, or ethnicity which is important in the eyes of God.[30]
>
> The "Good Samaritan" story is only found in the Gospel of Luke, as are a number of its stories. The narrative serves several purposes. It demonstrates the compassion of Jesus towards those who society shuns. It promotes Apostle Paul's vision that to be saved in the Second Coming, one must show kindness and mercy to all humanity.
>
> And, by elevating their sworn enemy over a priest and Levite, it serves as a not-so-subtle rebuke of the Jewish religious establishment.
>
> The "Good Samaritan" story in the Gospel of Luke has transformed the view of a Samaritan from an idol-worshiper cursed by *YHWH* to a person who comes to the aid of another. Whatever his intensions, the author of Luke has made the name "Samaritan" synonymous with kindness and good deeds — a legacy which holds to this day.

The Ministry of Paul

Apostle Paul preached his version of the Jesus message to Gentiles over a period of some 20 years. He did this alone or with companions Barnabas, John David, Silas or Timothy. Often "hungry, ill, and cold," he is said to have traveled over 10,000 miles through the Eastern Mediterranean — most times on foot and occasionally by ship.[31]

He suffered criticism, abuse, punishment, and threats on his life. Gentiles attacked him for steering people away from their pagan gods. Jews persecuted him for preaching against Torah Law.[32]

The Word

Paul is believed to be the first known Christian author.[*] He began composing his famous letters to Jesus followers — often while in

---

[*] Scholars generally agree that Paul wrote the following New Testament books, between c. 52 and 67 AD: 1 Thessalonians, Galatians, 1 and 2 Corinthians, Romans, Philippians, and Philemon. There is remarkably "little overlap between Paul's epistles and the stories told in the Gospels." Scholars wonder why. Sources: Tabor, James. "The quest for the Historical Paul" biblicalarchaeology.org. Nov, 2012. https://www.biblicalarchaeology.org/daily/people-cultures-in-the-bible/people-in-the-bible/the-quest-for-the-historical-paul/; Comment and response, seward414. Ehrman, "James the Brother of the Lord" ehrmanblog.org.

prison — over a decade before the first Gospel, Mark, is believed to have been written.³³

In Second Corinthians, he describes the trials and tribulations of his mission:

"I have worked much harder [than other missionaries], been in prison more frequently, been flogged more severely, and been exposed to death again and again. Five times I received from the Jews the forty lashes minus one. Three times I was beaten with rods, once I was pelted with stones, three times I was shipwrecked, I spent a night and a day in the open sea, I have been constantly on the move.

"I have been in danger from rivers, in danger from bandits, in danger from my fellow Jews, in danger from Gentiles, in danger in the city, in danger in the country, in danger at sea, and in danger from false believers. I have labored and toiled and have often gone without sleep; I have known hunger and thirst and have often gone without food; I have been cold and naked." (2 Corinthians 11:22-27.)³⁴

In his travels through Asia Minor, Greece, Macedonia, and Cyprus, Paul established as many as twenty Christian churches in his lifetime. "What is far more impressive," writes religious author Neil Cole, "is how many daughter, grand-daughter and great-grand-daughter churches were birthed from those."³⁵

To the Holy City

Now, however, I am on my way to Jerusalem to serve the saints there.

— Romans 15:25

Paul went to Jerusalem three times. He went there to meet James and the Council of elders in c. 37 AD, some three years after his conversion. He stayed with Simon Peter for a fortnight. (Galatians 1:13-24.)

He returned to the Holy City in c. 48 AD — a period of increasing unrest in Palestine. It was the time of Roman procurator Cumanus. Under his rule, Roman troops slaughtered 20,000 Jerusalemites, a Roman soldier near Beth-Horon tore up the Torah, and a Jewish "bandit gang" invaded Samaria.

Paul came to Jerusalem in the midst of these troubles to announce his mission to Gentiles. According to Paul's account, James and the Council accepted that circumcision would not be required for Gentile converts. They also agreed that Paul would be principal apostle to Gentiles while Peter would minister to Jews. (Galatians 2:1-10.)[36]

In light of the reported piety of James toward Torah law, not to mention the apparent reluctance of Jesus to preach outside Jewish circles, the Council's acquiescence to Paul's requests was remarkable — if it truly happened.

Paul's Final Return

According to Acts, Paul came again to the Holy City in c. 57 AD.

This was around the time when the *Sicarii* had begun their reign of terror with the assassination of High Priest Jonathan. Perhaps what happened next was influenced by the uneasy tension in the City.

The brother of Jesus warned Paul that he had a reputation for being against Torah Law:[37]

> *They have been told that you teach all the Jews who are among the Gentiles to forsake Moses, telling them not to circumcise their children or live according to our customs. (Acts 21:21.)*[38]

James had Paul take part in a week-long cleansing ritual in the Temple along with four other men.

"The seven days were almost ended when some Jews from the province of Asia [Minor] saw Paul in the Temple and roused a mob against him."

> "They grabbed him, yelling, *'Men of Israel, help us! This is the man who preaches against our people everywhere and tells everybody to disobey the Jewish laws. He speaks against the Temple — and even defiles this holy place by bringing in Gentiles.'"* (Acts 21:27-28)[39]

Apparently, a riot broke out. Roman soldiers broke it up and arrested Paul. (See Fig. 13-4.) According to Acts, he was held for two years as a prisoner in the Roman port of Caesarea Maritima on the Mediterranean coast (south of Haifa in modern day Israel).

The new Roman governor reopened his case in 59 AD. At Paul's request as a Roman citizen, he was sent to Rome to stand trial. This is also from the Book of Acts.[40]

Figure 13.4. Paul Arrested. Bible illustration, early 1900s.

Paul is said to have arrived in Rome in c. 60 AD and placed under house arrest — at least "officially." He preached in the city for two years and was released. Second century Christian tradition tell us that the apostle Peter was in Rome at the same time.[41]

A terrible fire swept the imperial city in c. 64 AD. The conflagration raged for nine days and destroyed some two-thirds of the city. Rumors circulated that Emperor Nero ordered it set because he wanted to rebuild the city. "To deflect suspicion on himself," he blamed the Christians.[42]

According to late first century Roman historian Tacitus, "an immense multitude [of Christians] was convicted. Covered with the skins of beasts, they were torn by dogs and perished, or were nailed to crosses." Others were "rolled in pitch and set aflame" to serve as a nightly illumination in [Nero's] gardens."[43]

Allegedly, among those executed were Paul and Peter.[44]

Paul's Legacy

More than any other individual, Paul (Fig. 13.5) took a local Jewish phenomenon — belief in the life, death and resurrection of Jesus — and transformed it into a primarily Gentile movement.

In terms of what Jesus had taught and what Paul had preached, the differences are stark. Jesus preached exclusively to Jews. He taught that observance of Torah law was important but not enough. One must put into

Figure 13.5. The Ecumenical Saint Paul. Depictions by six Christian artists.

action its core values of love of neighbor and charity towards the poor, the outcast. Jesus warned of the imminent Apocalypse and coming of the Son of Man to judge us all.[45]

Paul spoke primarily to Gentiles. He de-emphasized Torah law. For him, belief in Jesus was the path to God, the only path. Paul spoke of the imminent second coming of Jesus to judge us all.

Largely through Paul's efforts, the central principles of ethical monotheism would spread beyond the confines of a small country in the Levant to Gentile believers across the Roman world and beyond. Others followers of Jesus would espouse different interpretations and schools of thought. It would be Paul's conceptions on the life, death, resurrection, and proper worship of Jesus which would come to dominate Western civilization.

Outside of Jesus, Paul is the man most responsible for Christianity as we know it today. For his seminal contributions to the nascent religion, Saint Paul is celebrated throughout Christendom.

The Next Chapter

The Romans crushed the Jewish revolt in Palestine soon after the death of Paul. As we learned, the fall of Jerusalem in 70 AD meant the end of the Jewish state and Temple Judaism. The destruction of the Holy City also freed the now international Jesus movement from Jerusalem oversite.[46]

The second coming of Jesus did not happen. What did occur was an epochal religious transformation no one had predicted. This is the subject of the next chapter.[47]

Chapter 14

THE BLOSSOMING

*I planted the seed,
Apollos* watered it,
but God has been making it grow.*

— Paul (Corinthians 3:6)[1]

---

*Apollos was a fellow Jew from Alexandria who, along with Paul, preached on Jesus at the synagogue in Ephesus. Source: Kerr, C. M. "Apollos" biblestudytools.com. https://www.biblestudytools.com/dictionary/apollos/. Retrieved Aug 22, 2020.

After the passing of Paul, belief in the divinity of Jesus infused with apocalyptic Judaism would blossom into a new religion of ethical monotheism. Its tenets of righteousness and charity, good versus evil, angels and devils, and life after death would give hope to the poor and downtrodden — as well as others in the Roman world who sympathized with their plight.

This new religion of Jesus Christ would spread across the Roman world and, within a mere four centuries, conquer it.

A believer might say the idea of a single compassionate God was too vital, too important, too sacred to be held by just a small group of people in the Levant — that the spread of Christianity was guided by a Divine hand so His word would come to all nations.

Whatever one's beliefs, this seminal historical and religious phenomenon is a story for the ages.

The Spread of Christianity

Historians estimate that at the time of Paul's death in c. 66 AD there were thousands of Jesus followers in the Roman empire. By the end of the fourth century, they would represent some 30 million people — about half the empire's population.[2]

How did this happen? How did the Way grow from a small community of Jesus followers in the eastern Mediterranean to become the dominant religion of the Roman empire?

Surely the appeal of Jesus' moral teachings on charity and kindness were its primary attraction. Its emphasis on compassion and mercy was unprecedented in pagan religions and philosophies.

There were, it seems, a number of other factors in the growth of Christianity. They include its appeal to the lower classes, Christian fellowship, the zeal of its followers, the attraction of its miracle stories, Jesus as a personal God, and the promise of immortality for believers. Let's take a look.

Christianity and the Lower Classes

There was no single pagan religion within the empire. As in Sumer, polytheism invoked a multitude of gods. Worship and practices varied by region and within regions.

Still, they all generally included rituals and cultic acts of "sacrificial offerings, prayer, and divination." This to show respect for and gain the favor of the gods. Ethics tended to be a separate issue, primarily dealt with by philosophers.

Some examples: The philosophy of Stoicism espoused "practical morality and an ordered society." This appealed to the stability-loving and pragmatic Romans. Neoplatonism appealed to reason, along with the "desire for a mystical experience of an Absolute beyond the grasp of human thought," as British philosopher Philip Sherrard put it.[3]

These philosophies appealed particularly to intellectuals. They offered little for slaves and the poor. They and others low on the economic ladder were often drawn to "mystery" religions. This included cults of North Africa and the East with "exotic and bizarre rites."

Mithraism, which featured the worship of the Indo-Persian Sun-god Mithra, was an example. It spread across the empire in the second and third centuries AD. Rome originally regarded Christianity — with its resurrection of Jesus, strange rites, and promise of immortality — as another mystery religion."[4]

Christianity also drew converts especially from the poor, uneducated lower classes — the vast majority of the Roman empire's population. Indications are that the religion tended to attract more women than men, perhaps due to its emphasis on morality and compassion.[5]

> *Brothers and sisters, think of what you were when you were called. Not many of you were wise by human standards; not many were influential; not many were of noble birth. But God chose the foolish things of the world to shame the wise; God chose the weak things of the world to shame the strong.*
>
> — Apostle Paul (1 Corinthians 1:26-27)[6]

Christian Fellowship

Though paganism was fundamentally tolerant, there was "no sense of a community of worshippers throughout the empire to which one could 'belong,'" writes Bart Ehrman.[7] Christianity, on the other hand, offered a community of followers linked across the empire. The righteous behavior of its members drew converts. Pliny the Younger, Roman governor of Bithynia in Asia Minor, wrote around 122 AD:[8]

> ... *meeting on a certain fixed day before it was light, [Christians] bound themselves by a solemn oath, not to do any wicked deeds, never to*

> *commit any fraud, theft or adultery, never to falsify their word, nor deny a trust when they should be called upon to deliver it up . . .*

Christianity was commonly called atheism as it rejected the gods. Pagan emperor Julian later lamented that it was the kindness of these "atheists" to both fellow Church members and pagans which fostered the growth of Christianity.[9]

> *. . . it is their benevolence to the stranger, their care for the graves of the dead, and the pretend holiness of their lives that have done most to increase atheism [Christianity]. . . the impious Galileans [Christians] support not only their own poor but ours as well . . .*
>
> — Emperor Julian (ruled 361–363)

Miracle Stories

According to esteemed 18th century English historian Edward Gibbon, another factor in the growth of Christianity was the unabashed zeal of its followers — a strength of conviction "unheard of in pagan antiquity."[10] Their stories of miracles are seen as catalysts for the growth of the new religion. After all, the very foundations of Christianity are based on a miracle: the resurrection.

This and other miraculous acts by Jesus and others demonstrated the validity and power of the Christian God to potential converts. It was not, as Ehrman points out, that they had actually seen these miracles for themselves. Some heard them repeated again and again — as told by Christians with great conviction. They came to believe them to be true.[11]

A Personal God

Jesus was the ultimate personal God. He was a God who walked the Earth and suffered in human form, who died a most painful death — for you, to wash away your sins. Belief in Jesus provided solace in the often-brutal Roman world.[12]

Other Factors

The spread of Christianity was aided by the vast network of Roman roads, as well as shipping routes which had been designed to facilitate rapid movement of the military and foster trade.[13]

Ehrman contends that pagan "henotheism" — belief in a single ultimate divinity who ruled over the other gods — paved the way for Christian monotheism. This belief had become popular, especially among the educated.[14]

What was the perhaps the biggest factor in the spread of Christianity? Ehrman argues that it was its "exclusive, not inclusive" nature. Unlike polytheism, conversion to the new religion required the abandonment of all other gods. Thus, its continued growth tended to reduce the pagan population.[15]

What about competition from the millions of Jews* in the Roman world? They too declared the existence of a single, all-powerful God to the exclusion of all other gods. They too emphasized morality and kindness. Didn't this attract sympathetic pagan converts? To some degree it did, such as the "God-fearers" we discussed in Chapter 13. Still, Jews saw themselves as a nationality, an ethnic group — one which opposed intermarriage. They generally held themselves aloof from the Gentile world. Christians on the other hand sought it out.[16]

Christians felt a compulsion to evangelize. The world would soon face the Apocalypse. It was a moral imperative. It was urgent. We must convert non-believers to save them from God's wrath to come. The "evangelizing mission of the Christian church was unparalleled and unprecedented in the ancient world," Ehrman points out."[17]

How did they do this? Not so much through organized missionary efforts. They "principally used everyday social networks — family, friends, co-workers, acquaintances, and other people they met — to convert people simply by word of mouth." It was this simple approach which over time gained the new religion its "massive following."[18]

The growth of Christianity was further enhanced by the stories and teachings of Jesus, as documented in the latter half of the first century AD.

---

*Jews had a "special status in the Greco-Roman world," James Tabor tells us, "They were [generally] exempt from emperor worship and were permitted a number of special privileges based on their observance of the Sabbath and the festivals." This included "exemption from military service," appearing in court on the Sabbath, and "certain business arrangements." They were also allowed to "settle inter-Jewish legal disputes according to their Law and traditions." In addition, Jews were permitted to "administer their own funds and send money to Jerusalem, especially the Temple tax. (before the fall of Jerusalem)" Source: Tabor, "The Jewish Roman World of Jesu

The Gospels

After the death of Jesus, tales of his birth, life, crucifixion and resurrection would spread by word of mouth across the Roman world. After Paul's death and perhaps because of it, Jesus followers began to record these stories in what has become known as the Gospels or "Good news."

Based on pre-Gospel oral histories and perhaps earlier lost writings, the Gospels would provide early Christian communities with written stories on Jesus' life and teachings. Read aloud in church services, they served to unify and spread the "good news" deeper into the Roman Empire.[19]

These are four "canonical" gospels: Mark, Matthew, Luke, and John. We drew much of the story of Jesus in Chapter 11 on these writings. Here we find the words attributed to Jesus — his humanitarian message of love and kindness, his compassion for the outcast and the poor, his forgiveness for repentant sinners.

No one knows when these four Gospels were written or by whom. Scholars generally suggest they were perhaps written between c. 65 and 110 AD — some thirty to eighty years after the death of Jesus. They were likely composed by people who never met Jesus.[20]

Though they tell of the life of Jesus, the Gospels are not biographies or for that matter historical documents. They are religious works. Each gospel promotes its author's "interpretation of the Christian message," scholar of early Christianity Paula Fredriksen writes.[21]

Like most New Testament books, the Gospels appear to have been written "largely if not exclusively" for Gentile audiences. Gospel writers in particular sought to depict Rome in a favorable light.

Jesus had been crucified by a Roman prefect for sedition. Jews had conducted a failed armed rebellion against imperial Rome. Gospel writers wanted to convince imperial Rome that it was all "the Jews" fault. We, the followers of Jesus are benign and peace-loving — in concert with the great emperor Augustus and Pax Romana.[22]

An example:

Render to Caesar the things that are Caesar's

— Jesus (Mark 12:17)

I cannot imagine a Jewish holy man who preached that *YHWH* will soon liberate the Holy Land from Roman rule would say such a thing.

This seems to be yet another attempt by Gospel writers to appease Roman authorities. The nascent Christian movement wanted Rome to recognize it as a legitimate religion within the empire.[23]

A Shadow on Christianity

In his book, *A Moral Reckoning*, Daniel Goldhagen, former Associate Professor of Political Science at Harvard University, "counts some 450 problematic passages regarding Jews in the Synoptic Gospels (Mark, Matthew, Luke) and Acts of the Apostles alone."[24]

This animosity is most evident in the Book of John. In John 8:44, for example, the author calls Jews the offspring of Satan:[25]

> *Ye [Jews] are of your father the devil.*
>
> — John 8:44

According to the New Testament, it is not just Sadducee priests corrupted by Rome but all Jews who are responsible for the death of Jesus. The Romans — who crucified any Jew who dared challenge their rule — are held blameless.[26]

Why this attitude? One factor was the continued refusal of the vast majority of Jews to believe in the divinity of Jesus. In addition, Jewish followers of "the Way" had been "expelled from local synagogues" and shunned, even persecuted. Some likely added "anti-Jewish elements to the stories." Later Christian scribes copied, recopied, and edited Gospel scrolls. Over the years they added to the denigration of Jews.[27]

Pontius Pilate — Perhaps the most blatant vitriol against Jews is in the Gospels' depiction of Pontius Pilate. The Roman Prefect had "routinely crucified Jews without so much as a trial."[28]

Philo of Alexander, Hellenistic Jewish philosopher, wrote of Pilate's "venality, violence, thefts, assaults, abusive behavior, frequent executions of untried prisoners, and endless savage ferocity."[29] Agrippa I, later Roman ruler of Judaea, confirmed that he was "unbending and severe," a man accused of "cruelty, bribery, and countless murders."[30]

So egregious was Pilate's rule that "the people of Jerusalem lodged a formal complaint with" the Roman emperor Tiberius. As noted, he was later recalled to Rome for overly harsh suppression of a Samaritan rebellion. Pilate's brutality and greed were too venal even for the Romans.[31]

Yet the Gospels portray Pilate as arguing for the innocence of Jesus, while "the Jews" called for his crucifixion. "'We have a law,' answered the Jews, 'and according to that law He must die.'" (John 19.5-7.)

The notion that any Roman governor — let alone the vicious Pilate — would plead mercy for a zealot who called himself King of the Jews is ludicrous.[32]

Deicide — The New Testament blames "the Jews" for the execution of Jesus, fostering the slur "God-killers." In Thessalonians, it says:

> *[The Jews] who killed the Lord Jesus and the prophets and also drove us out . . . displease God and are hostile . . . in their effort to keep us from speaking to the Gentiles so that they may be saved . . .*
>
> — 1 Thessalonians 2:14-16

Scholars generally agree that this passage was not written by the Apostle Paul, but by a later editor.

Today's Christian authorities have typically spoken out against anti-Semitism. Nonetheless, this hate speech is still in the New Testament and read aloud in churches all over the world.

The writings of Paul, the Gospels, and later New Testament works would be key to the spread of the new religion. Stories of Jesus and "instructions on what to believe and how to act" served to unify, inspire, and grow the nascent Christian movement. The pagan world had nothing like it.[33]

The Divinity of Christ

The term "Christology" refers to the divine nature of Jesus. The first gospel, the Gospel of Mark is said to have low Christology. Though it points to his divinity, secular scholars generally contend that Mark portrays Jesus very much as a human being. This is "far from a consensus."[34]

The Gospel of Matthew depicts Jesus as the Son of God from birth.[35] Nonetheless, nowhere in the Gospels of Mark and Matthew — or Luke for that matter — does Jesus himself say he is God."

The Gospel of John has high Christology. Its central theme is Jesus as the Son of God. Here Jesus openly proclaims his divinity. (John 20:30-31.) The author(s) of John also proclaim Jesus as the preexistent Son of God, one who existed *before* creation itself:[36]

*And now, O Father, glorify thou me . . . with the glory
which I had with thee before the world was.*

— John 17:5

The notion of Jesus as the divine preexistent Messiah-God of Christianity did not necessarily represent the beliefs of all followers of Jesus, as we shall see.

One precept of Christianity in particular would promote its growth: the promise of life after death.

The Afterlife

*For the Lord himself shall descend from heaven . . .
and the dead in Christ shall rise first.*

— 1 Thessalonians 4:16

For most pagan religions and mainstream Judaism, the afterlife tended to be of little or no concern. Christianity, on the other hand, promised a spectacular reward for faith in Jesus — immortality.

According to the New Testament, after death, you will be resurrected when Jesus returns in the second coming. You will then "ever be with the Lord." (1 Thessalonians 4:17). Unrepentant sinners and those who deny Jesus will also be resurrected but "to shame and everlasting contempt." (Daniel 12:2.)[37]

Heaven and Hell

Perhaps because the second coming never came, this apocalyptic belief would morph into an idea beyond New Testament writings: the notion of Heaven and Hell. No waiting for the ever-postponed second coming. You will receive your reward or punishment *immediately* after death.[38]

This Christian notion of heaven turned the prospect of death from one of dread to one of joy for a true believer. You will be free from all pain and suffering, illness and hunger, tragedy and loss. You will be reunited with your loved ones and live in bliss with Jesus for all eternity.

The concept of Hell is another story. The Apocalypse of Peter, likely composed between 125 and 150 AD, tells of "a variety of punishments awaiting sinners in Hell." Its descriptions were clearly influenced by "Homer, Virgil, Plato, and Orphic and Pythagorean traditions." Plato's *Gorgias* (c. 400 BC), for example, tells of Tartarus, the underworld where all souls go after death to be judged and where "the wicked receive divine punishment."[39]

The Apocalypse of Peter features graphic descriptions of the eternal tortures awaiting sinners in Hell:

Persecutors of Christians are forever "beaten by evil spirits, their inwards eaten by restless worms." Those who bore false witness are punished with "flaming fire in their mouths." Murderers are "cast into a . . . place full of evil snakes." Women who had undergone abortion are placed "up to their necks [in the] gore and the filth of those who were being punished."[40]

A particularly harsh punishment is reserved for blasphemers. They are "hanged by the tongue over a burning fire" or tortured with "a red-hot iron in their eyes" for eternity.

Famed Carthaginian theologian Quintus Tertullianus (c. 160–220 AD) aka Tertullian also wrote of the horrors of Hell. He took special delight at the plight awaiting those who had persecuted Christians:

> *What sight shall wake my wonder, what my laughter, my joy and exultation? As I see all those kings, those great kings . . . and the magistrates who persecuted the name of Jesus, liquefying in fiercer flames than they kindled in their rage against Christianity . . .*[41]

Called the "the only sadistic literature in the Roman world," these early Christian descriptions of Hell were as disturbing as they were effective. Evidence suggests that more than the promise of heaven, it was the eternal terrors of Hell "which convinced potential converts" to join the new religion, according to Ehrman.[42]

All soul is immortal.

— Plato[43]

Christianity also came to embrace the Greek concept of an immortal soul — an essence within you that lives on after you die. It is your immortal soul which goes to Heaven or, if you have been naughty, to the eternal torments of Hell.

Yet another feature instrumental in the survival and growth of Christianity was its evolving organization structure.

Church Hierarchy

After the death of Paul, the boundary between Christianity and Judaism remained fuzzy. Not all agreed on how Jews fit into the new religion. Over the second half of the first century AD, Christianity gradually broke away from its Jewish roots. Somewhere near the end of the century, "Jesus followers began to refer to themselves as Christians."[44]

> ... and the disciples were called Christians first in Antioch.
>
> — Acts 11:26

The new religion established a hierarchical structure across the Roman world. It included deacons, priests, and bishops — all under the "Vicar of Christ" or Pope in Rome. This centralized organization sought "oversite over religious practices throughout the empire." It would constitute a major strength in the Church going forward.[45]

In the following century, Christianity would begin to establish a single doctrine of faith.

A Single Christian Canon

Christianity was far from a homogenous religion of like beliefs. Throughout its first three centuries or so of existence, it was a "diverse phenomenon," writes Ehrman, "with different Christians advocating an enormous range of beliefs and engaging in strikingly different practices."[46]

A number of small movements, sects, schools, and splinter groups had formed in the early days of the Way. Some held Jesus was a man who had become divine at death, others as divine at birth, still others as divine from the beginning of time. Yet others said, "Jesus was only a spirit manifestation" or "only a man."[47]

In the latter half of the second century, Irenaeus (Fig. 14.1) — bishop of Lugdunum in Gaul, now Lyon in France — sought to unify the

Figure 14.1. St. Irenaeus (c. 130–202). Engraving by unknown artist.

many disparate Christian views. By this time a number of gospels had been written, possibly as many as thirty.

Shortly after the death of astronomer Ptolemy in Alexandria in c. 175–185, Irenaeus declared Mark, Matthew, Luke, and John as the "only Gospels that Christians should read." The four Canonical Gospels were the only ones which told of the life of Jesus. They were also the most popular.[48]

Irenaeus further asserted that all who did not follow his views were heretics who should be expelled from the Church.* He proclaimed only Church bishops could provide the correct interpretation of Scripture — an edict which would come to affect the great Galileo some fourteen centuries later.[49]

Irenaeus emphasized the letters of Paul in his canon. He also stressed the "unity of the Old Testament and the Gospels." His views eventually came to dominate church theology. In great part due to Irenaeus, the Hebrew Bible — in its Greek Septuagint form — would become part of Christian canon and an integral part of Western culture.[50]

---

*Heretics included the then widespread religious movement called Gnosticism. "Gnosticism was not a single religion," James Tabor tells us, "but a diversified and complex religious phenomenon both independent of, and interacting with Judaism and early Christianity." Gnostics wrote the "Gospel of Thomas, the Epistle of Luke, and the gospel behind the Toldoth" likely in the second century AD. Source: Tabor, James. "The Roman World of Jesus: An Overview," uncc.edu.

> **Sidebar: The Septuagint**
>
> As we learned, a number of Jews had returned to Jerusalem under Persian king Cyrus in the sixth century BC. Still, most Judean refugees remained outside the Holy Land. This included Jewish communities in Babylonia, Sardis (in Asia Minor), and Lower (northern) Egypt.[51]
>
> Jewish emigration to Egypt increased dramatically under the Ptolemies — especially to Alexandria. Some Gentiles in the city attended Synagogues and a number of them converted to Judaism, the so called "God fearers."[52]
>
> Here the Hebrew Bible was translated into Greek — most likely in the Ptolemaic period (third century or 200s BC). According to first century Jewish historian Josephus, it was produced in Alexandria to take its place among the books of its great library.
>
> It is called the Septuagint from the Greek *septuaginta* or "seventy" for the number of Jewish scholars allegedly employed in its translation.[53] The Greek translation of the Hebrew Bible was the major literary contribution of Hellenic Judaism and a historic step for ethical monotheism.
>
> For the first time, the Hebrew Bible was accessible to Greek-speaking Jews and the rest of the vast Greek-speaking world. "The Hebrew view of God, of history, of law, and of the human condition, in all its magnificence would spread around the world," writes Washington State University professor Richard Hooker.[54]
>
> With the imprimatur of Irenaeus, the Septuagint would later become the basis for the "Old Testament" of Christianity. It would come to include a few books not found in the Hebrew Bible, including First and Second Maccabees.[55] It has since been translated into "Latin, Coptic, Ethiopian, Armenian, Georgian and Slavonic" languages, and more. King James of England would commission its translation into English in 1604 AD.[56]

Early Christian Cosmology

It seems Greek advances in cosmology did not reach very early Christian writers. The New Testament reflects the flat earth cosmology of the Sumerians/Hebrew Bible. Paul speaks of multiple heavens and an underworld. (2 Corinthians 12:1-4; Philippians 2:10)[57]

The author of the Book of Revelations writes: "*And after these things I saw four angels standing on the four corners of the earth.*" (Revelation 7:1.) The Gospel of John says: "*. . . thou lovest me before the foundation of the world.*" (John 17:24.) (My underlines)

In time, the early Christian Church would become Hellenized and adopt the spherical Earth model of pagan Greece. There were several holdouts. Tertullian, 5th century Alexandrian monk Cosmas Indicopleustes, and others argued, based on Scriptures, that Earth is indeed flat. Though it was "never a majority or official position of the early church," writes Donald Simanek, emeritus professor of physics at Lock Haven University of Pennsylvania."[58]

Like the Jews before them, Christians would pay a heavy price for their strange monotheistic religious beliefs — as we shall see in the next chapter.

Chapter 15

THE AGE OF MARTYRS

*You will be handed over to be persecuted and put to death . . .
the one who stands firm to the end will be saved.*

— Jesus (Matthew 24:9-13).

208   *Cosmic Roots*

<u>Note to Reader</u>: *The primary source for the following section is Notre Dame scholar of early Christianity Candida Moss's book: The Myth of Persecution: How Early Christians Invented A Story of Martyrdom.*

The Colosseum, Region III: Isis and Serapis, Imperial Rome.[1]

It is mid-day break. Slaves remove the bloody carcasses of animals slain during the staged morning hunts. The smell of death lingers in the "sweltering mid-afternoon heat." The crowd of some 50,000 — men, women, and children — grow restive.[2]

Let the lunch-time show begin.

Damnatio ad bestias
Condemnation to beasts

A dozen male criminals, naked, their hands tied behind their backs, are brought into the arena and chained to poles. Their crime? Being Christian. More precisely, adherence to an unauthorized religion, refusal to honor the image of the emperor, failure to pay homage the Roman gods, and unlawful assembly.

Trap-doors in the wooden floor open. Lions and leopards that have been fed on human flesh and then starved rush out from below. The beasts lunge and tear their victims from limb to limb. (See Figure 15.1.) Cheers rise

Figure 15.1.   Leopards Attacking a Victim in Arena. Roman floor mosaic, 3rd century AD, Archaeological Museum of Tunisia. A lovely addition to any discerning Roman home.

up from the crowd. Some collect bets on who would be eaten first. Victims who have managed to survive are dispatched by gladiators.[3]

A group of Christian women, also naked and bound behind their backs, are brought in. Slaughter by beast repeats. Modest entertainment on an otherwise dull Sunday.[4]

Hopefully, the afternoon gladiator fights will be livelier.

Is this horrific scene true? We know that those charged with breaking the law or captured in armed revolts were thrown to lions and other beasts on a regular basis. Were Christians? No historical records or physical evidence has been found to verify this notion. The scholarly consensus is that the number of Christians condemned to death in *any* Roman arena was exaggerated by later Christian writers.[5]

There is no doubt that Roman persecution of Christians did occur. It is the extent and frequency that are in question. It appears that over the first three centuries AD, persecutions were sporadic, isolated incidents — mostly local affairs often "initiated by provincial mobs."[6]

"There are literally hundreds of stories describing the deaths of thousands of early Christian martyrs," writes Candida Moss, "but almost every one of them is legendary."[7]

Still there were persecutions. Six events in particular are considered the most historically accurate. They are: *The Martyrdom of Polycarp* in c. 155 or c.166; *Ptolemy and Lucius* in 165 AD; *Justin Martyr* shortly after; *Lyon and Vienne* some sixteen years later; the *Scillitan Martyrs* around 180, and *Perpetua and Felicia* in 203. For more details, please see endnote.[8]

Though rare, these events were deeply traumatic for the small percentage of the empire's population who followed Jesus. Christians* were socially shunned, discriminated against, and at times persecuted. One only has to experience being a persecuted minority to understand.[9]

Let's take a look at possible motives.

Dislike of Christians

> *A class hated for their abominations . . .*
> *a most mischievous superstition*
>
> — Roman historian Tacitus on Christians[10]

---

*Ehrman estimates there were some 40,000 Christians in the Roman empire by 150 AD. Source: Ehrman, *The Triumph of Christianity*, p. 294.

In their own way, Romans were deeply religious. Rooted in Etruscan and Greek culture, they honored a great number of gods. The ever-practical Romans based their religious practices on the principle of *do ut des* or "I give that you might give." They performed their rituals, prayers, and sacrifices to the gods with this in mind.[11]

Christians refused to publicly honor the Roman gods. The pagan population saw this as impious and unpatriotic — not to mention dangerous. One had to sacrifice and show respect for the gods. To do otherwise was to invite calamity on all.[12]

Certain "aspects of Christianity sounded like treason or revolt." Jesus as King, one who is superior to the emperor, is one example.[13] Christians "abstained from volunteering for active duty in the Legions." This caused even more resentment among pagans.[14]

Christians met privately. This spawned rumors and suspicion. A second century rumor, perhaps initiated by a personal grudge, alleged that they "practiced incest and other sexual perversions."

They were also suspected of cannibalism. This may have been a misinterpretation of the Eucharist ceremony where Christians eat bread and drink wine as symbolically eating the flesh and drinking the blood of Christ.[15]

> *Whoever eats my flesh and drinks my blood has*
> *eternal life, and I will raise them up at the last day.*
>
> — John 6: 54.

As Christian numbers grew, persecutions became more prevalent. Imperial persecutions in particular would become a recurring theme. They too were sporadic and rare, and depended very much on who was emperor.

State-sponsored Persecutions

The mere name ['Christian'] is assailed.[16]

— Tertullian, famed third century Christian writer.

We discussed the first known instance of imperial persecution in Chapter 13: the torture and murder of Christians by Emperor Nero in 64 AD for allegedly starting a fire in Rome. The second such incident involved Emperor Trajan.

Trajan (ruled 98–117 AD)

He was known for his "high-mindedness, generosity," and love of conquest. Under Trajan, the Roman Empire achieved its maximum territorial extent — stretching from Armenia in the East to Britain in the West.[17]

In c. 112, Trajan's governor of Bithynia on the Black Sea was Pliny the Younger. He began executing anyone who had been publicly denounced as a Christian. He wrote to Trajan that "the contagion of this superstition [Christianity] has spread not only to the cities, but also to the villages and farms."[18]

> *I interrogated these as to whether they were Christians; those who confessed I interrogated a second and a third time, threatening them with punishment; those who persisted I ordered executed.*
>
> — Pliny the Younger in letter to Emperor Trajan.[19]

Trajan wrote back: If a person "denies that he is a Christian and really proves it — that is, by worshipping our gods, then he will be pardoned." Otherwise, he (or she) is to be executed. Apparently, some Christians were killed, though the exact number is lost to history.[20]

Hadrian (ruled 117–138)

A decade later, the emperor was now Hadrian. Known for "high-mindedness and prudence," he would give up the eastern territories conquered by Trajan. Though no one realized it at the time, the Roman Empire had begun a slow, inexorable decline.[21]

Anti-Christian riots broke out in western Asian Minor in c. 122. Hadrian wrote to his provincial governor that being a Christian was not enough to justify execution. They "had to be proven guilty of illegal acts before they could be condemned." We do not know how many were executed or saved under this order.[22]

Sidebar: The Later Jewish Wars — *Bellum Judaicum*
After the first Jewish War and the destruction of Jerusalem and its Temple in 70 AD, tension remained high between the Jewish and Gentile populations of the Roman Empire — especially in North Africa. A second uprising would begin in 115 AD under Emperor Trajan.

The Kitos War (115–117 AD)

Unorganized and apparently spontaneous, the Jewish rebellion started in Cyrene, capital of the Roman province of Cyrenaica on the Mediterranean coast of North Africa (now Algeria). A very large community of Jews had resided there since pre-Roman times.[23]

As children, they "had heard tearful recountings of the Roman barbarities in Judea a generation before," writes historian Saleh. Most of Rome's army were in the East with Trajan on his war of conquest against Armenia and the Parthians. This was their chance for revenge.[24]

Jewish mobs, apparently mostly young and lower class, went on an "orgy of bloodshed and arson." They attacked lightly manned Roman garrisons, destroyed Roman government buildings, and bath houses; razed pagan temples and massacred Gentiles.

The uprising spread eastward to Egypt and the island of Cyprus, where there were also large Jewish minorities. Alexandria alone had a Jewish community of some 150,000.

"Embittered by prejudice endured at the hands of Alexandria's Greek majority," Saleh writes, "local Jewish communities joined in the rebellion." A significant part of Alexandria was set aflame. Jewish mobs also destroyed Pompey's tomb. In Cyprus, Jewish rebels "burned the Roman provincial capital of Salamis to the ground."[25]

The uprising spread to Judaea itself and recently conquered areas of Mesopotamia. Trajan sent three separate forces to quash the rebellion: Two legions to Egypt and Cyrene; the Seventh Legion *Claudia* to Cyprus; and sadistic Moorish prince Lucius Quietus to Mesopotamia and Judaea.

It took "two years of bitter fighting to suppress the revolts." The campaigns are said to have "wiped out the ancient Jewish community on Cyprus" and "destroyed and drove out nearly all the Jews of Mesopotamia."[26] According to second century Roman historian Cassius Dio, "220,000 Greeks and Romans were murdered in Egypt and Cyrene, and another 240,000 perished in Cyprus. How many Jews were killed, Dio did not bother to record."[27]

Lucius Quietus was subsequently named Roman governor of Judaea. By Jewish tradition, it is Quietus (pronounced kweetus) from whom the uprising derives its name — hence the "Kitos" War in English.[28]

Another Jewish revolt broke out in Judaea fifteen years later. Its cause also remains unclear. Hadrian was now emperor in Rome.[29]

The Bar-Kokhba Revolt (132–135 AD)

Unlike the Kitos War, the third and last Jewish rebellion was well-planned and well-organized. It was led by a single individual — the charismatic Simon Ben Kosiva aka Bar-Kokhba. Apparently, many Jews hoped, nay believed that he was the Messiah promised by the Hebrew Bible who would restore Jewish independence in the Holy Land.

The appellation "Bar-Kokhba" means "Son of Star" in Aramaic. It is apparently a messianic reference.

The Bar-Kokhba uprising began with guerilla tactics and later built up into an armed fighting force reminiscent of Judas Maccabee. Surprisingly successful, the rebels came to hold large parts of Judaea, perhaps including Jerusalem.[30]

Bar-Kokhba established an independent state of Israel that endured for over two years. Archeologists have discovered coins of the period which corroborate "the existence of an independent Jewish state for a brief period."[31]

Emperor Hadrian sent nearly a third of his forces in the Roman Empire to quell the uprising — an estimated 60,000 to 120,000 troops. He charged his best general, Julius Severus, Roman governor of Britain, to lead the Roman juggernaut.[32]

Severus implemented a slow, brutal "scorched-earth" strategy on the Jewish population of Judaea. It was as deadly as it was effective.[33]

Jewish sources tell of "Roman soldiers smashing babies against rocks" and the "mass slaughter of civilians," writes American-Israeli historian Benjamin Kerstein. It seems "the majority of the Jewish population of the province were killed, enslaved, or exiled."[34]

After the rebellion, Hadrian wiped the name Judaea off the map. The region would now be called Palaestina. Hadrian also banned Jewish law and rituals within the newly named region.[35]

The city of Jerusalem was plowed under. In its place, Hadrian built the pagan city of Aelia Capitolina, with a temple to Jupiter placed on the site of the Second Temple. Jews were "forbidden to live within sight of" the new city.[36]

After the reign of Hadrian, over a hundred years would pass with no known new Christian persecution across the empire. Then great troubles would arise for both Christians and the empire itself.

The Crisis of the Third Century

In c. 230 AD, Germanic tribes crossed the northern borders of the empire and vanquished Roman legions there. Around the same time, forces of the Sassanid (neo-Persian) Empire invaded eastern Roman provinces and overran Mesopotamia.[37]

Roman emperor Severus Alexander (ruled 222–235) chose to confront the Sassanids in a campaign which he led personally. To pacify Germanic chieftains, he paid them tribute rather than fight. His troops were outraged. In 235, they murdered the nineteen-year-old emperor and his mother and threw their remains in the Tiber River.[38]

Thus began the Roman era of "imperial assassinations and usurpations" known as the "Crisis of the Third Century." It spanned from 235 to 285. Within this period of fifty years, some twenty men would claim the title of emperor. In comparison, from the beginning of the empire in 27 BC to 235 AD, there had been only twenty-six emperors. And this was a span of over 250 years.[39]

Frequent border raids and invasion would mark the Crisis era. Carpians, Goths, and Alamanni raided from the north. Sassanid armies invaded in the east. All this while Roman generals fought amongst themselves for supremacy.

Rome's vast trading network broke down. Masses of people moved from cities to the countryside for food and protection. Severe economic hardship gripped the land (in some but not all provinces). To make matters worse, the empire suffered outbreaks of the Plague of Cyprian, possibly smallpox.[40]

The empire fractured for a time into a Roman core and "two breakaway states, one in the far west and one in the east." In a series of eastern and western campaigns — culminating in the Battle of Châlons in 274 — Emperor Aurelian (ruled 270-275) would restore the empire.[41]

At times like these, we humans often lash out at "the other," the minority within our midst. So began a new round of imperial persecutions of the Christian population.

Decius (ruled 249–251)*

> *The oftener we are mown down by you, the more in number we grow; the blood of Christians is seed.*
>
> — Tertullian (Apology 50).[42]

---

*The six *local* (non-imperial) persecutions reported earlier occurred after the reign of Hadrian and before the reign of Decius — that is from 155 to 203 AD.

It was still a time of social, economic, and civic discord. Decius tried to rally public support with a call for a return to the traditional values of the Roman Republic. In January 250, he issued a decree that everyone must "sacrifice to the gods, taste the sacrifice, and swear they had always done so." This act must be performed "in the presence of a Roman magistrate." Only Jews were exempt.

Thus began the first empire-wide Christian persecutions.[43]

Christians summoned before a magistrate were caught in a trap. If they refused to sacrifice, they would be executed. If they acquiesced, according to their beliefs they "faced eternal damnation."[44]

A number of Christians gave in and sacrificed to save their lives. Others refused and were martyred in horrific deaths. Pope Fabian (served 236–250) himself was among those executed. Still others avoided the magistrates through bribery or exile.[45]

The intensity of Christian oppression varied greatly with location. "It is impossible to know the extent of the persecutions," Moss tells us. Whatever the amount of suffering, Decius is a name cursed in Christian circles to this day.[46]

Valerian (ruled 253–260)

In 257, Emperor Valerian "demanded Christian leaders participate in pagan rituals and stop meeting *en mass* in cemeteries"[47] Valerian's follow-up letter in 258 directed that "bishops, priests, deacons be put to death at once," writes Moss. In addition, "Christian senators and high-ranking officials were to lose status and property, and if they didn't apostatize, be executed."[48] Nonetheless, "only a handful of Christians seem to have died as a result of Valerian's second letter."[49]

Historians consider the end of the Crisis of the Third Century to be when Diocles, a military commander from the Baltic province of Dalmatia, ascended to the imperial throne. As Emperor Diocletian (ruled 284 – 305), he proved to be a visionary leader, remarkable civic organizer, and a nightmare for Christians.[50]

The Great Diocletian

In his famous reform program, Diocletian (Fig. 15.2) established a new tax system, reorganized provincial administration, and improved military readiness. Most important, he assigned and promoted personnel based on merit and trust.

Figure 15.2 Head of a statue of the Roman emperor Diocletian. Archaeological Museum of Istanbul.

Diocletian believed the empire was too vast for a single emperor to control. In late 285 he split the empire in two. He took charge of the East and named his son-in-law Maximian, an Illyrian officer, as co-emperor in the West. Diocletian remained the senior emperor.[51]

Diocletian placed the senior capital of the Eastern Roman Empire at Nicomedia, a Greek city on the Gulf of İzmit in the Sea of Marmara in Asia Minor. Consequently, the balance of power shifted to the East and its Greek-dominated culture.

Things were looking up for the Roman Empire — and, it seemed, for the Christians. Believers in Jesus Christ had survived persecution and even thrived. Christianity had grown to perhaps as many as 2.5 million followers by 300 AD. It had become "the most powerful force within the empire."[52]

Despite this or perhaps because of it, things were about to get worse for the faithful, much worse.

The Great Persecutions

Emperor Diocletian was deeply devoted to the Roman gods. He saw "the rise of Christianity as a threat to the empire," as Ehrman puts it. In the beginning of the fourth century, he issued a series of edicts intended to wipe out this "atheist" religion once and for all.[53]

Christians had recently erected a new church across from the imperial palace in Nicomedia — an indication of their growing power. In early 303, Diocletian ordered the church razed, its scriptures burned, and its treasures seized.[54]

The next day he issued an edict which ordered the destruction of all Christian churches throughout the empire — as well as confiscation of all Christian scriptures. Christian meetings were declared illegal. Christians of "high social status were to lose their rank. Christian freedmen in the imperial service" were to be enslaved, Ehrman tells us.[55]

There was no imperial force to ensure compliance across the vast empire's regional and municipal levels. Implementation of the edict was spotty in the East and of little effect in the West.

Diocletian issued a second edict in the summer of 303. It called for the empire-wide arrest of all Christian bishops, priests, lectors, and deacons. Apparently, this edict had greater impact — and problems. Jails became overcrowded with imprisoned clergy.[56]

With the "twentieth anniversary of his reign on November 20, 303" imminent, Diocletian announced his third edict: a general amnesty for imprisoned clergy — if they sacrificed to the gods.[57]

In the spring of 304, Diocletian issued his fourth, last, and most severe edict: "Everyone — men, women, children — must gather in a public space to offer sacrifice" or be executed.

Prosecution was again sporadic and varied greatly by region. "Some Christians were tortured and burned to death," Moss writes. "Others were mutilated and sentenced to the copper mines of Egypt."[58]

It is impossible to know how many people were imprisoned or died during the Great Persecutions. Bart Ehrman estimates "possibly hundreds, although almost certainly not many thousands." Despite the efforts of Diocletian, Christianity not only survived, but continued to grow. The Great Persecutions had failed.

It seems there were two key factors for the relatively small number of casualties — and for Christianity's survival. The first was the entrenched growth of the religion across the empire. It was simply too popular to wipe out. Besides, it is notoriously difficult to eliminate deeply held religious convictions.

The second factor was the generally non-violent response of the Christians to persecution. They had been taught to "turn the other cheek," that it was noble to die for their beliefs. There was an eternal reward in heaven for doing so. Unlike the Jews, Christians never took up arms against Rome — as far as we know. This passive Christian response avoided annihilation by the Roman war machine.

Dominance

The Great Persecutions under Diocletian was the climax of Christian suffering under Roman rule. In less than ten years, Christianity would become the preferred religion of the empire.[59] It would be given the imperial imprimatur by an unexpected source, a brilliant Roman general from Serbia — and again change history. This is the subject of the next chapter.

Chapter 16

BY THIS CONQUER

For to whatever quarter I direct my view, whether to the east, or to the west, or over the whole world, or toward heaven itself, everywhere and always I see the blessed one yet administering the self-same empire.

— Eusebius of Caesarea[1]

In 293, Emperor Diocletian attempted to resolve the ancient Roman problem of imperial succession once and for all. He established a "Tetrarchy" or rule of four. To the two ruling co-emperors, now called *Augusti*, he added two vice-emperors or *Caesars*.

Each co-emperor would have his own vice-emperor. Should the co-emperor abdicate or die, his vice-emperor would become the new co-emperor. He in turn would appoint a new vice-emperor. And so on.

Plus, vice-emperors were to be chosen based on merit rather than inheritance — a radical departure from prior imperial practice.

Under Diocletian's idealistic plan, co-emperors would happily share power — and each vice-emperor would be content to wait for their respective co-emperor to resign or die before taking over. No infighting. No more generals constantly battling each other for the emperorship. No more civil wars.

Diocletian named Galerius, a distinguished soldier, as his vice-emperor in the East. His co-emperor Maximian chose Praetorian commander Constantius Chlorus as vice-emperor in the West. The arrangement seemed to work. Diocletian led successful campaigns in Egypt. Galerius did the same against Persia and Germanic tribes along the Danube. Maximian secured victories in North Africa, as did Constantius Chlorus in Britain. Their efforts brought a new *Pax Romana* to the empire.[2]

Like most utopian ideas, it did not last.

After twenty years of rule, the exhausted and ill Diocletian abdicated in 305 — the first Roman emperor to do so. He convinced co-emperor Maximian to step down at the same time.

In the second Tetrarchy, Galerius moved up to emperor in the East with his nephew Maximinus Daia as his vice-emperor. Constantius Chlorus moved up to emperor of the West with Severus as vice-emperor. So far so good — except that Daia was a relative of Galerius, violating Diocletian's tenet of selection by merit.

Constantius Chlorus died in July 306. Severus moved up to emperor of the West and Galerius reluctantly named Chlorus' son Constantine as his vice-emperor — another election by heredity violation.

Diocletian's dream of a smooth transition of power soon fell apart. Starting in 307, the empire began a series of civil wars. Six rulers battled for imperial power: Galerius, Severus, Maximinus Daia, Constantine, and two new contenders: Maximian's son Maxentius, and Licinius, supported by Galerius.[3]

Of the six, it would be Constantine who would rise to ultimate victory. He would initiate a religious and cultural revolution which would change the very character of the empire and the Western world forever.

Who was this Constantine? How did he come to change history? What were his religious beliefs? Let us take a brief look — and in doing so, try as best we can to separate fact from fable.

Constantine the Great

Born in 272 or 273 in the northern Balkans in today's Serbia, Constantine was the son of Constantius Chlorus. He was educated at the court of Diocletian in Nicomedia, where his father served as tribune. His mother was Constantius' consort Helena — an innkeeper's daughter who it seems was sympathetic to Christianity.[4]

Constantine served as "junior officer in Diocletian's court" when the Great Persecutions began. There is no evidence that he "expressed any disapproval." He also fought alongside his father in Britain, where he proved to be a highly skilled military commander. He also "campaigned against barbarians on the Danube, fought the Persians in Syria, and went to battle under Galerius in Mesopotamia."[5]

Constantine was an adherent of Apollo, yet once he became co-emperor he continued his father Constantius Chlorus' policy of tolerance towards Christianity. His father had "shut down some churches, but did not arrest, torture, or martyr any Christians." His motives for this rare open-mindedness are unknown.[6] Perhaps it was in deference to his consort, Helena.

Maximinus Daia on the other hand continued Christian persecutions in the East — especially in Egypt.

> *"We ourselves beheld ... many [Copts] all at once in a single day, some of whom suffered beheading, others punishment by fire ... as soon as sentence was given against the first, some from one quarter and others from another would leap up to the tribunal before the judge and confess themselves Christians...."*[7]
>
> — Eusebius, 4th century Christian historian and bishop of Caesarea

The Conversion

In October of 312, Constantine marched his legions to the outskirts of Rome to face his rival in the West, Maxentius. The night before the great battle for the eternal city, he is said to have had an epiphany.

The story is briefly told a few years later by Christian theologian Lactantius, advisor to Constantine and tutor to his eldest son Crispus:[8]

> *Constantine was directed in a dream to cause the heavenly sign [of God] to be delineated on the shields of his soldiers, and so to proceed to battle. He did as he had been commanded, and he marked on their shields the letter X, with a perpendicular line drawn through it and turned round thus at the top, being the cipher of CHRIST. Having this sign (XP), his troops stood to arms.*[9]

"XP" refers to the Greek letters chi and rho, the first two letters of the word *Christos* or Christ in English. (See Fig. 16.1)

With the symbol of Christ as its battle standard, Constantine engaged Maxentius at the Milvian Bridge. This stone bridge (today's *Ponte Milvio*) crosses the Tiber River at a northern entrance to Rome. The army of Constantine crushed the forces of Maxentius — who drowned in the Tiber while attempting a retreat.

Nearly three decades later, famed Christian historian and bishop Eusebius told a more detailed and somewhat different story. In his biography, *Life of Constantine,* he wrote about the appearance of an image in the sky:[10]

Figure 16.1. The Chi-Rho Sign.

About the time of the midday Sun, a cross-shaped trophy formed from light, and a text attached to it that said, "BY THIS CONQUER." He [Constantine] and the whole company of soldiers that was then accompanying him witnessed this miracle and were gripped with amazement.[11]

Eusebius was vague as to where and when the event occurred. He said only that it was "on a military campaign he was conducting somewhere." Eusebius claimed his version was told to him directly by Constantine.[12]

Constantine would later say that he had experienced his epiphany on the march to Rome, and that he "considered himself a Christian after the Battle of the Milvian Bridge." The location and timing of Constantine's vision(s) and conversion to Christianity remain subjects of scholarly debate.[13]

Constantine used the cross or chi-rho Christogram — symbols for Jesus, preacher of non-violence, compassion, and mercy — to fight and kill. This would be cited throughout Christian history, along with stories in the Hebrew Bible, to justify war and the slaughter of "infidels" and fellow Christians as well. This religious militancy would be glorified in the 19th-century English hymn "Onward Christian Soldiers," an oxymoron if ever there was one.

Free at Last

In April 311 — a year prior to the Battle of Milvian Bridge — Galerius had issued the Edict of Toleration in the East. It allowed Christians to practice their religion without persecution. It also pardoned them from past disobedience to imperial laws and customs.[14]

> *... in view of our most mild clemency and the constant habit by which we are accustomed to grant indulgence to all, we thought that we ought to grant our most prompt indulgence also to these, so that they may again be Christians and may hold their conventicles, provided they do nothing contrary to good order.*
>
> — Excerpt, Edict of Toleration by Galerius (311 AD)[15]

Why did Galerius do such a thing? Edward Gibbon argued that it was prompted by the failure of imperial persecutions. They had not persuaded Christians to turn to worship of the pagan gods. Other scholars suggest Galerius wished to placate Constantine, his co-emperor in the West at the time.

Whatever his motives, Galerius died of illness shortly after.

In February of 313, some five months after the Battle of the Milvian Bridge, Constantine persuaded now co-emperor Licinius to join him in issuing a similar edict. With its pro-Christian tone, the so-called "Edict of Milan" confirmed freedom of religion across the empire. It officially recognized Christianity and declared the end of persecution of its followers. It also ordered the return of Christian property confiscated by Rome.[16]

> *I, Constantine Augustus, as well as I Licinius Augustus . . . saw [that it] would be for the good of many . . . that we might grant to the Christians and others full authority to observe that religion which each preferred . . . no one whatsoever should be denied the opportunity to give his heart to the observance of the Christian religion, of that religion which he should think best for himself, so that the Supreme Deity, to whose worship we freely yield our hearts may show in all things His usual favor and benevolence . . .*
>
> — Excerpt, Edict of Milan by Constantine and Lucinius (313 AD)[17]

After nearly three centuries of on-and-off persecution and discrimination, Christians were now free to practice their religion without harassment throughout the Roman empire. Constantine scholar Harold Drake called this "the first official government document in the Western world to recognize the principle of freedom of belief."[18]

I imagine church bells ringing, people of the faith gathered in the streets embracing each other, tears forming in their eyes, their arms raised wide in supplication, looking upward to the heavens, shouting "thank you, Jesus." I have found no such reference. Nonetheless, perhaps celebrations did break out across the empire.

The Final Battles

In April 313, Licinius crushed the forces of Maximinus Daia at the Battle of Tzirallum in eastern Thrace.[19] The latter died four months later. All that remained of the six emperor wanna-be's were Licinius in the East and Constantine in the West. They co-ruled the empire in an uneasy truce for ten years.

Tensions rose again in 323 over border issues beyond the Danube. With the excuse he had been looking for, Constantine attacked. He outmaneuvered, outsmarted, and defeated the forces of Licinius on land and at sea. His final victory came in the Battle of Chrysopolis on the Asiatic shore of the Bosporus in 324 AD.[20]

Constantine was now sole emperor of the Roman Empire. (See Fig. 16.2)

Figure 16.2. Constantine the Great. Inscription at base reads "Constantine. By This Sign Conquer." The statue rests on the site in 306 AD where he was first declared an emperor: Eboracum, capital of Roman province Britannia Secunda (now York, England)

God's Emperor

Constantine saw his victories as evidence for and validation of the Christian God. With his new powers, Constantine accelerated the granting of imperial favors to the Church. They included financial support for the Church, promotion of "Christians to high office," and exemption of Christian priests from paying taxes just like pagan priests. He ordered a number of new churches built, including the Lateran Basilica in Rome, which was to become the official cathedral of the Pope. He "urged his fellow Romans (e.g., soldiers) to adopt Christianity," writes Bart Ehrman, and had his "children instructed in the Christian faith."[21]

Upper class Romans saw the writing on the wall. Imperial favors towards Christians induced a number of the aristocracy to convert. They began shifting local patronage from pagan temples to Christian churches. This accelerated the rise of Christianity and decline of polytheism. Follow the money.[22]

Under Constantine's new church-state paradigm, Apostle Paul's call to obey the commands of secular rulers would take on a deeper meaning. The Eastern notion of the king as a god and the ancient Jewish tenet of the Judean king as God's corporal instrument would merge — to give

Constantine and later Christian sovereigns the imprimatur of the One God in all they did. Constantine was God's vice-regent on Earth. The divine right of Christian kings had been established.[23]

Though some secular scholars argue to the contrary, it seems that Constantine's devotion to Christianity was sincere. People of the time commonly saw the gods bound up with all other aspects of life. Constantine saw his battlefield victories as coming from the Christian God

The Council of Nicaea

> *I and my Father are one*
>
> — Jesus Christ (John 10:30)

In the summer of 325 AD, Constantine called a meeting of church bishops from across the empire — most from eastern provinces. Its purpose was to establish a single doctrine on the divinity of Jesus. Though the emperor likely did not grasp the nuances of the obscure theological issues, he participated in the debate and enforced the final decision.[24]

> *Though extremely trivial and quite unworthy of so much controversy . . . my first concern was that the attitude towards the Divinity of all the provinces should be united in one consistent view.*
>
> — Constantine[25]

The Council was held in the Bithynian city of Nicaea, now İznik in Turkey. Christian teacher and priest Arius of Alexandria held that Christ was the divine Son of God — but *subordinate* to the Father. His bishop, Alexander, a follower of Paul, argued that Christ was "fully equal with God the Father."[26]

Arias argued that God the father was Almighty. There cannot be "two beings who are both almighty," as Ehrman puts it. Thus the logical conundrum that God the father and Christ the Son are co-equal.[27]

Despite the arguments of Arias, Alexander and other followers of Paul won over the council. Under pressure from Constantine, 315 of the 318 bishops voted for the "fully equal" precept of Alexander. Arius and two of his supporters were the only exceptions. The views of Arius, aka Arianism, were declared heretical. The three dissenters were banished to Illyria on the east coast of the Adriatic Sea.[28]

The Jesus as fully equal to God precept would become established canon of the Catholic Church. This doctrine would be challenged by a deeply religious Puritan in the seventeenth century. He would see it as a violation of monotheism. As we shall see, his name was Isaac Newton.[29]

Heretics

What did Constantine do next? He proceeded to persecute Christians who did not hold to established Church dogma. Christians designated as heretics included Gnostics, Novatians, Marcionites, Valentinians, and Montanists of Phrygia. Also persecuted were followers of Paul of Samosata, who dared to propose Jesus was born a man and only later anointed by God.[30]

Fed by the anti-Semitism in the Gospels, Constantine also orchestrated the persecution of Jews — an estimated 5 to 7% of the empire's population. We humans need someone to hate.[31]

Constantine cemented Christian anti-Semitism in his letter to the Council of Nicaea. He called Jews "wicked people" and "murderers of our Lord." The four Canonical Gospels were edited under Constantine to further blame the Jews and exculpate Rome for the crucifixion of Jesus. All other gospel versions were destroyed, possession of which was a capital offense.[32]

Hatred of the Jewish people was now official policy of the Roman Empire. The malevolence would continue into post-Roman Europe, provoking some sixteen centuries of anti-Semitic persecutions and the horrific Holocaust of the 20th century.

For Good or Ill

Constantine's dedication to the moral principles of ethical monotheism was mixed, to say the least. He banned crucifixion, the "branding of certain criminals," and ostensibly, gladiator games.[33]

On the darker side, Constantine ordered the execution of the "twelve-year-old son of his sister," and later his eldest son Crispus, heir apparent to the throne. Constantine's reasons for these actions remain obscure.[34]

His mother Helena was now an openly devout Christian. Still, it seems she had Constantine order her rival, empress Fausta to be shut up in a hot bath where she was cooked to death. Perhaps to atone for her sins, Helena gave liberally to the poor, and "released prisoners and those condemned to work in the mines."[35]

Of course, court murders were typical of pagan rulers as well. Even the Hasmoneans — Jewish kings in Judea's brief period of independence in the first century BC — committed vile acts of familicide. It seems the high moral precepts of ethical monotheism are not enough to quell the human lust for power.

The adoption of Christianity by the Roman state changed both the empire and the religion. The Roman ideology of dominance would fuse with Christian themes of compassion and service.[36]

Lost in all this are stories of Christian institutions and everyday Christians that often do not make the history books — extraordinary acts of charity and kindness. They included the founding of orphanages and establishment of hospitals, public institutions to serve the needs of the sick and poor. These countless acts of compassion committed in the name of Jesus Christ go on to this day.[37]

A Christian City

Rome, the eternal city on the Tiber said to be founded by Romulus and Remus in 753 BC had not aged gracefully. Its spirit of republican government long gone, its citizen armies aroused to defend the city a thing of the past. Its streets were now overcrowded, its buildings decayed, its senate weak and corrupt, its economy on the decline. Its time had passed.

Constantine decided to move the capital of the empire to the east. It was in part a military decision, as it faced its "most formidable enemies" — Germanic tribes along the Danube and Persians in Anatolia.[38] The Christian emperor chose to locate the capital at the small Greek town of Byzantium — a "small trading town" on the Bosporus straight which straddled Asia and Europe.[39]

Byzantium was a judicious choice, with superb natural defenses and significant commercial advantages. Its trade routes linked three continents — North Africa to the south, Europe to the west, and Asia to the east — in a "network of caravan routes, rivers, seaways, and paved Roman roads."[40]

The site also featured a five-mile-long natural harbor, a nearly-landlocked location with exceptionally calm seas.[41]

Constantine had much of the old town destroyed. In its place, he ordered the construction of a new city of architectural wonders. Dedicated in 330 AD, it featured palaces, theatres aqueducts, porticos, public and private baths, and a Hippodrome enlarged to hold some 60,000 spectators. A regal highway ran through the center of the city, east from the gates of the imperial palace to the landward walls.[42]

Constantine turned the new capital into a Christian city. It was the "the religious center of the empire." He placed Christian crosses and items said to be holy relics throughout the city. He built a number of churches, including the Church of the Holy Peace and the Church of the Holy Apostles. He also began construction of the magnificent Hagia Sophia (Church of the Holy Wisdom) — now a Moslem Mosque.[43]

Constantine "kept some remnants" of the old Byzantium, including shines of pagan deities and the old acropolis (modern Topkapi), where the temples of Dionysus, Poseidon, Athena, Artemis, and Aphrodite stood. This was likely to ensure a more peaceful transition in the new capital. After all, the majority of the empire's population was still pagan.[44]

"After the defeat of Licinius, Constantine renamed the city Constantinople after himself." It is present-day Istanbul in Turkey.[45]

Constantine also gave particular attention to the city where Christ was crucified.

Jerusalem

Constantine rebuilt the Holy City around places where Jesus was said to have "suffered, was buried, and was raised." He had grand basilicas constructed in the city — including the Church of the Holy Sepulcher, built at the purported site of Jesus' crucifixion and tomb in Jerusalem. It is now reluctantly shared by Roman Catholic, Armenian Orthodox, Greek Orthodox, Egyptian Orthodox, and Syrian Orthodox sects. The site has "become the holiest place in Christendom," writes scholar of religions Shaye Cohen.[46]

Helena made a famous pilgrimage to the Holy Land in her seventies and founded two churches: The Church of the Nativity in Bethlehem at the supposed site of Christ's birth; and the Church of the Ascension at the supposed site of his rise to heaven on the Mount of Olives outside Jerusalem.[47]

Aftermath

Constantine has been called a scheming politician, a tyrant, a noble war hero, a warrior against heathens, and a model Christian prince. The wily Constantine had remained a supporter of pagan religion until he was forty years old, not openly supporting Christianity until the Battle of the Milvian Bridge. According to Christian tradition, he was baptized into Christianity on his deathbed*.[48]

Scholars continue to debate why Constantine chose to champion Christianity. His father's tolerance of Christians and mother's sympathies for the religion surely played a part. Perhaps he felt its ever-growing membership and empire-wide organizational structure was a way to unite the vast empire under one belief. Maybe the survival of Christianity in the face of persecution convinced him of the power of the Christian God. Perhaps he truly did have a religious epiphany. Or it was some combination of these things.

Religious epiphanies may be questioned, even seen as delusions by non-believers. They are typically perceived as very real to those who experience them. These events are often highly moving, unexpected and uncalled for, and life-changing. From a strictly scientific point of view, these visions bubble up from one's subconscious, e.g., an idea for an inventor, a story for a writer, a melody for a composer, a religious epiphany for a searcher. These revelations are often described as seeming to come from outside ourselves.

Yet the question remains. Was Constantine's conversion to Christianity sincere? There is "overwhelming evidence that he was a very real Christian during his reign," writes Ehrman. This includes his Edict of Milan, financing of numerous Christian churches across the empire, "beneficences on Christian clergy," establishment of Constantinople as a Christian city, establishment of holy sites in Jerusalem, and his baptism on his deathbed in 337 AD.[49]

> *Indeed my whole soul and whatever breath I draw, and whatever goes on in the depths of the mind; that, I am firmly convinced, is owed by us wholly to the greatest God.*
>
> — Constantine (*Life of Constantine* 2.29)[50]

As University of Toronto Professor of Classics Timothy Barnes puts it, "Constantine . . . was neither a saint nor a tyrant . . . After 312 [he] considered that his main duty as emperor was to inculcate virtue in his

---

*The delay of baptism until just before death was not uncommon at the time, as it assured a post-baptismal sinless soul thought necessary for entry into heaven. Source: Ehrman, *The Triumph of Christianity*, p. 35.

subjects and persuade them to worship God... with all his faults and despite an intense ambition for personal power, he nevertheless believed sincerely that God had given him a special mission to convert the Roman Empire to Christianity."[51]

Constantine remains "the most significant pagan convert in the history of Christendom."[52]

Later Emperors

After Constantine's death, his son Constantius II ascended to the throne. He ruled from 337–361 and observed a more strident Christianity than his father. Constantius II "ordered pagan temples closed and sacrificial practices stopped." Again, enforcement was generally weak. The practice of polytheism continued in most places.[53]

With the dynastic imperial tradition reestablished, all subsequent Roman emperors would be Christian — with one exception. Some fifty years after Constantine's demise, his nephew Emperor Julian (ruled 361–363) — though raised as a Christian — attempted to restore pagan worship throughout the empire. His successors reinstated Christianity and Constantine's policies.[54]

Emperor Theodosius I (ruled 379–395) — the last to rule over both the eastern and western empire — went even further. In 380, he effectively proclaimed Christianity as the imperial religion of the Roman Empire.[55] He again "made it illegal to perform pagan sacrifices." And to convert to Judaism, the religion of Jesus.[56]

A religious council was held under Theodosius in 381. It established the so-called Nicene Creed. The doctrine included the Holy Spirit with God the Father and Jesus the Son to form the *Trinity* — One God in three beings coeternal and of the same substance.[57]

Heretics, especially those who did not adhere to the correct theological understanding of the Trinity, such as Arians, were given a special warning:[58]

> *[To those who do not] embrace... the name... of Catholic Christians... whom we adjudge demented and insane, shall sustain the infamy of heretical dogmas... they shall be smitten first by divine vengeance and secondly by the retribution of our own initiative...*
>
> — Excerpt, Edict of Thessalonica by Theodosius I (380 AD)[59]

Though enforcement was spotty, "temples were leveled and sacred cult objects and art destroyed."⁶⁰ Among the destroyed was the magnificent Serapeum temple in Alexandria along with its breathtaking statues and striking artworks. The temple that "had stood for more than six centuries was demolished in 391." The ancient Greek sanctuary was replaced by a Christian martyr's shrine and church.⁶¹

Science and the Early Church

Persecution by Christians would reach "pagan" science itself. Hypatia of Alexandria — head of the Platonist school at Alexandria — was murdered by a mob of Christian fanatics in 415. Hypatia had studied under her father and assisted him in writing a new version of Euclid's *Elements*. She had also written commentaries on other great Greek geometer's works. At the time, she was the "the world's leading mathematician and astronomer," writes Australian researcher Michael Deakin, "the only woman for whom such claim can be made."⁶²

There is a long-standing argument as to the extent of early Christianity's effect on the demise of "pagan" science. A popular notion is that Church intolerance led to the suppression of the glorious discoveries of the ancient Greeks. Historian of science Thony Christie argues otherwise.

"Classical Greek learning began to decline in the ancient world from the middle of the second century (150's AD)," Christie writes. This was due to "a general socio-political and cultural decline. [It] had nothing to do with the rise of Christianity."⁶³

In addition, Emperor Constantine moved to the East in 330 AD. The Greek language was soon no longer spoken in the West. This exacerbated the decline.⁶⁴ In fact "the survival of the classical literature," historian of science James Hannam tells us, "was almost entirely due to the efforts of Christian monks laboriously copying out texts by hand."⁶⁵

Arguments for some early Christian suppression of science remain, e.g. Hypatia. Still, I lean towards Christie's point of view. There was no *universal* Christian suppression of pagan philosophy or natural science. Plato's metaphysics, Aristotle's cosmology, and Ptolemy's astronomy would become part of Christian theology, as we shall see.⁶⁶

With the power and financial backing of the emperors, by the end of the fourth century (300's) Christianity had grown to an estimated 20 to 25 million people — about half the empire's population.⁶⁷ By the end of the fifth century (400's), the great majority of the empire's population was now Christian.⁶⁸ Christianity's incongruous conquest of the Roman empire was for all intents and purposes complete.⁶⁹ It would become the dominant religion of Europe, a miraculous and wonderful transformation for Christians — and a nightmare for Jews.

Impact

Few events in the history of civilization have proved more transformative than the conversion of the emperor Constantine to Christianity in the year 312.

— Bart Ehrman[70]

John the Baptist had preached the imminent end of the world. Jesus had preached the same. Paul also said the Day of Judgement was near. It never came. What did come was a new religion of the God of Abraham which would subdue the Roman Empire — an altogether astounding historical phenomenon.

The lamb had conquered the eagle. The subjugation of the Roman Empire by a religion of ethical monotheism founded in a tiny country in the Levant, its divinity crucified by Rome for sedition, is nothing short of extraordinary. Secular scholars are at a loss to fully explain how this happened. Christians, of course, consider it an act of God.

In the centuries since, the Christian faith would come to be "the most important religious, social, cultural, political, and economic factor the Western world had ever seen," Ehrman writes. Christianity would have a profound impact on "art, music, literature, philosophy, ethics, economics, and law — and [arguably] prove to be the single greatest cultural transformation" on our planet.[71]

One small example is today's nearly worldwide dating system — BC for "Before Christ" and AD for the Latin *"Anno Domino"* or "Year of our Lord" — a testimony to the global influence of Christianity.[72]

Today, with some 2.3 billion followers representing 31% of the world's population, Christianity is the largest religion on Earth.[73]

The Next Chapter

The sciences of cosmology and astronomy would remain stagnant in the West for centuries.* Seeds of change would be planted some 700 years after

---

*One of the few exceptions was Martianus Capella of Carthage (360–428 AD). He put forward the notion that Mercury and Venus *orbit the Sun* rather than the Earth. This to account for the two planets always being near to where the Sun has risen or set in the sky (more on this later). Source: O'Connor & Robertson et al. "The Structure of the Solar System," st-andrews.ac.uk.

the birth of Christ by the third major religion of ethical monotheism: Islam. Sparked by Muslim astronomers, the cosmology of pagan Greece would be rediscovered and eventually come to be incorporated into Christian dogma.[74]

Medieval Europe would resurrect the conceptions of Plato, Aristotle and Ptolemy as tenets of Christianity. New ideas and discoveries of Renaissance natural philosophers and astronomers would come to challenge this Christian worldview and initiate the greatest conflict yet between science and religion.

This is the subject of Part IV of this book.

PART IV
Rebirth

Chapter 17

THE PROPHET

And it is He who ordained the stars for you that you may be guided thereby in the darkness of the land and the sea.

— Qur'an (Koran) 6:97

After the fall of Rome in 476 AD, Europe turned inward for nearly a millennium. The continent split into a number of separate empires, kingdoms, individual city-states, and fiefdoms.

During this so-called "Middle Ages," the Byzantine Empire came to rule in the East — with Eastern Orthodoxy as the state religion.[1]

With their great passion for learning, the Byzantines would produce notable religious and secular literature. It featured epic poems, stirring prose, insightful histories, and unmatched poetic hymns. Its contributions to the high arts would be characterized by art fresco masterworks, brilliant mosaics, embroidered tapestries, exquisite ivory carvings, fine enamels, "ecclesiastic-style architecture," and striking religious portraiture. The empire would thrive for over a thousand years — until the fall of Constantinople to the Ottoman Empire in 1453.[2]

The Roman Catholic Church was the unifying force in the divided West. The region would become for the most part "poor, rural, and largely illiterate."[3] Still, there were significant advances in agriculture — fostered by invention of the plough, whippletree, and three-field crop rotation. These farming innovations led to a population explosion. From 650 to the mid-thirteen-hundreds, Europe went from some 20 million to nearly 75 million people.[4]

The great science and mathematics of the Greeks was for the most part forgotten in the West, including the works of Euclid, Aristotle, and Ptolemy.[5]

Two epic events in particular helped end this isolation and restore the learning of the ancients to the western continent: the rise of Islam and the Crusades. Let's take a look,

Out of Arabia

If you indeed love Allah, follow me...

— Muḥammad (Qur'an 3:31)

<u>Note to Reader</u>: *Stories of the prophet's life were first written down some 120 to 170 years after his death. They are deeply influenced by religious belief and affiliations. Though modern scholars largely confirm the general history of the period, they cannot be sure of the details. Many accounts may have been made up later on.*[6]

His name was Abū al-Qāsim Muḥammad (c. 570–632) — an Arab merchant and trader from Mecca. He was known as al-Amin, "the Trustworthy" for his honesty and integrity. In 610 AD, the 40-year-old spoke of a revelation from God delivered to him by the Archangel Gabriel. Thus began the teachings of the Prophet Muhammad and the founding of the third major religion of ethical monotheism: Islam.[7]

Muhammad's visions over a period of some twenty-three years were collected after his death in the Muslim holy book, the Qur'an (Koran) — meaning recitation.[8] Widely regarded as "the finest work in classical Arabic," the Qur'an includes stories of Adam and Eve in the garden of Eden, Abraham and the near sacrifice of his son, Mary's birth and early years, and the virgin birth of Jesus — with modifications.

Considered by devout Muslims as "the final Work of God," the Qur'an lists 26 prophets. Among them are Adam; Noah; Abraham; his sons Ishmael and Isaac; Isaac's son Jacob; Joseph; Moses and his brother Aaron; David and his son Solomon; as well as John the Baptist and Jesus — Muhammad being the last.[9]

Islamic beliefs include "paradise and hell, angels and devils, life after death, and the last day of judgment." Sound familiar?[10]

Nonetheless, there are significant differences between Islam and the earlier religions of Abraham. There is no original sin in the Qur'an. Arabs are said to have descended from Abraham's son Ishmael. Jesus is not divine and was not crucified but raised to heaven by God. The Gospels have been corrupted by human alterations and additions.[11]

> *... the angel said, 'O Mary, indeed God gives you the good news of a word from Him, whose name will be the Messiah Jesus, the son of Mary, held in honor in this world and in the Hereafter ... '*
>
> — Qur'an 3:45-51[12]

How did Muhammad learn about Judaism and Christianity? Jews and Christians had been living in Arabia for centuries before Muhammad. The prophet likely first encountered their beliefs and stories in Mecca, a multi-religious town. He was also perhaps exposed to ethical monotheism in his travels as "an agent for wealthy merchants taking their goods on caravans throughout" the Arabic peninsula and into the Levant.[13]

Jews, Christians, and pagans generally tolerated each other in the towns of Arabia — unlike in the Christian Byzantine Empire and western Europe. In Mecca, there was a colony of Abyssinian Christians from Ethiopia across the Red Sea. In addition, some Arabs had converted to Christianity over the years.[14] Yathrib, which we will discuss later, had been settled by three

Jewish tribes in the 6th century BC. They still resided there in compounds outside the town in Muhammad's time.[15]

Muhammed's Mission

> *And recall when We made a covenant with the Children of Israel:*
> *Worship none but Allah. treat with kindness your parents and kindred,*
> *and orphans and those in need; speak fair to the people; be steadfast in prayer;*
> *and practice regular charity . . .*
>
> — Qur'an 2:83.

The last prophet of Allah began his preaching in his hometown of Mecca. Situated on the western edge of Arabia near the Red Sea, the major town was then on the "cross-roads of the lucrative caravan trade." Trade routes ran from Mecca through the desert south to Yemen, north to Alexandria and Damascus, and east to Iraq and Persia.[16]

Muhammad preached to the pagan people of Mecca that they must forego their gods and goddesses and worship the One God, whose name is Allah. His message of morality and charity appealed particularly to the poor and downtrodden.[17]

His own tribe, the *Quraysh*, rejected him. Muhammad's denunciation of polytheism threatened their prestige and wealth. As rulers of Mecca, they were guardians of the *Ka'bah*, the ancient pagan shrine in the center of the city. People from all over Arabia made pilgrimages to worship there. The *Quraysh* enjoyed a lucrative "commerce generated by the pilgrims."[18]

Muhammad and his Meccan followers suffered abuse and persecution. Some were stoned and beaten. One was even killed. Weary of increasing ill-treatment and fearful due to the demise of his uncle and clan protector Abu Talik, Muhammad decided to leave the city of his birth. In 622, he fled to Yathrib — a major oasis town some 250 miles (400 km) north of Mecca.[19]

When Muhammad arrived at the outskirts of Yathrib, he was met by followers whom he had sent earlier. It is said they rushed to greet him with a song. Muslims around the world still sing it today: *Tala 'al-Badru' Alaynā* or "The White Moon Rose over Us."*

---

*You can listen to this beautiful and haunting song in Arabic and English on YouTube: "Nasheed — The White Moon Rose Over Us. (Tala' al Badru 'Alayna)" https://www.youtube.com/watch?v=7mZaCwLdrwA.

Earlier, having heard Muhammad preach, a delegation from Yathrib had come to see him in Mecca. Knowing of his fair-mindedness, they had asked him to adjudicate an ongoing feud between two rival clans in their city.[20] Now in Yathrib, its leaders saw the arrival of the Meccan and his ever-growing group of followers as a way to strengthen the town against its rival, Mecca.

Muhammad soon "established an Islamic state" in Yathrib — its laws based on those "revealed in the Qur'an." He then began "to invite other tribes and nations to Islam." The city would become known as *Madinat al-Nabil*, "the city of the Prophet" or Medina, "the City" for short.[21]

Form Yathrib, Muhammad began to lead raids against Meccan caravans. This led to a series of battles between the two towns. The Battle of Badr in 624 was the first skirmish. With some 300 followers, Muhammad crushed a group of 1000 Meccans. This unexpected triumph over a much larger force was seen as a "sign of God's favor." The victory and its resultant booty drew more converts to Muhammad's side.[22]

The prophet gained an additional victory against the Meccans in the Battle of Uhud/Hamra al-Asad in 625. He then stopped a planned attack on Medina by a Meccan alliance in the Battle of Ahzab/Khandaq aka the Battle of the Trench in 627. Muhammad and his followers had become a force to be reckoned with. (More on these battles later.)[23]

In 630, a mere eight years after fleeing Mecca, Muhammad appeared before the gates of his hometown with an army of some 10,000 believers. With a number of Meccan leaders slain in prior battles and its people divided, the Meccans offered minimal resistance. In an unheard-of act of mercy, Muhammad spared nearly the entire town from the sword.[24]

Upon entering Mecca, Muhammad marched with his followers to the ancient *Ka'bah*.* Again, this was the pagan shrine in the center of the town. In those days, it was surrounded by 300 or so gods and goddesses. It held the "hallowed meteorite known as the black stone" within.[25] Entering the *Ka'bah*, he cried, "*laqad han alhaqiqat wazwal albatil*" or "Truth has come and falsehood has vanished."[26]

The pagans considered Allah their chief deity and creator of the universe. Muhammad declared that Allah was the only God and the God of Abraham — the same God worshipped by Jews and Christians. He then destroyed the pagan idols and images at the *Ka'bah*.[27]

---

* According to the Qur'an, Ibrahim (Abraham) and Ishmael had built the *Ka'bah* from instructions given to them by Allah. It had since been taken over by polytheism and idolatry. (Qur'an. 2:123-127.)

The Spread of Islam

With Muhammad now in control of Mecca and Medina, a number of Arabian tribes declared their allegiance to the Muslim state. His disciples began to spread the message of Allah, the One God of Righteousness, across Arabia. His troops drove further into Western Arabia — all the way to the Syrian border. Through conquest and conversion, the new religion spread across the desert sands, steppes, and wadis like wildfire.[28]

Muhammad and the Jewish Tribes

Muhammad envisioned that, in his advocacy of monotheism and the God of Abraham, Jews and Christians would become his allies. He referred to them as *Ahl al-Kitāb* or the "People of the Book." In Medina, he offered Jews "religious and cultural autonomy" in exchange for their political support.[29]

> *People of the Book! Come now to a word common between us and you, that we serve none but God . . .*
>
> — Qur'an 3:64.

The three Jewish tribes of Medina — the *Banu Qaynuqa*, *Banu Nadir*, and *Banu Qurayza* — remained aloof. Jewish scholars there ridiculed and rejected Muhammad's religious views. This soured Muhammad's attitude towards them.[30]

After the Battle of Badr (624 AD), he had laid siege to the fortress of the *Banu Qaynuqa* because they refused to convert. He showed mercy in victory — he let them live but banished them from Medina.[31] In 625, Muhammad also forced the *Banu Nadir* to leave Medina rather than kill them.

Then in 627 the *Banu Nadir*, along with the *Quraysh* of Mecca and a number of Bedouin tribes, "mounted a full-scale assault in Medina with some 10,000 troops," writes British journalist Desmond Stewart. Before the battle, Muhammad had a deep trench dug around the exposed part of the city. It rendered the enemy's cavalry ineffective. After some 40 days of vacillating, the coalition called off the siege. The so-called Battle of the Trench never took place.[32]

The *Banu Qurayza*, the last remaining Jewish tribe in Medina, had negotiated with the enemy to join their coalition. In revenge for their "planned treachery," Muhammad laid siege to their fortress. They surrendered. The prophet had the men tied up, brought to the main square in Medina, beheaded and thrown into ditches. Some six hundred were killed. Women and children were taken as slaves.[33]

Before Muhammad passed away, he is said to have "vowed to expel all Jews and Christians from Arabia." Umar ibn al-Khattab, the second Caliph (successor) of Muhammad, took up the prophet's promise and expelled remaining Jews and Christians to Iraq and Syria, where he granted them land in compensation.[34] "All Jewish and Christian communities in Arabia completely disappeared" — with the exception of Yemen.[35]

Demise and Aftermath

The Prophet Muhammad died in 633.[36] He was at once a holy man who inspired deep love and devotion from his followers, a wily political leader, and a masterful military strategist. He showed rare mercy in combat — along with cold brutality when he deemed it necessary. A highly intelligent man, he synthesized Judaism and Christianity with Arabic culture into a new religion. It would come to bring the high moral tenets of ethical monotheism to hundreds of millions of people across our planet.

Muhammad is recognized today as among the most influential individuals in human history. Islam, the religion he founded, now numbers some 1.8 billion people — nearly a quarter of the world's population.[37]

The Conquests

Within a decade after Muhammad's demise, Moslem forces subjugated Syria, Palestine, Egypt, Iraq, and Iran.[38] They then swept into North Africa, Iberia, and Central Asia. In a little over a century, they "dominated an area larger than the Roman Empire at its peak."[39] See Figure 17.1 for details.

The timing of their initial campaigns was fortuitous. The Christian Byzantine and Zoroastrian Sassanid empires had been weakened by decades of fighting each other. Repeated devastation from the Bubonic plague (The Plague of Justinian) further reduced their numbers. Under the Arab-Muslim invasions, the Byzantine Empire suffered great loss of territory. The Sassanid Empire collapsed.[40]

Muslims saw these conquests as an "act of altruism," as it opened the conquered population to the "light of Islam." One thinks of the ancient Hebrews who slaughtered "the wickedness of the Canaanites" in the name of *YHWH*. Or the Roman legions whose carnage brought "civilization to the barbarians." Or Constantine as "God's chosen instrument," crushing all who stood in his way. How we love to rationalize deadly aggression.[41]

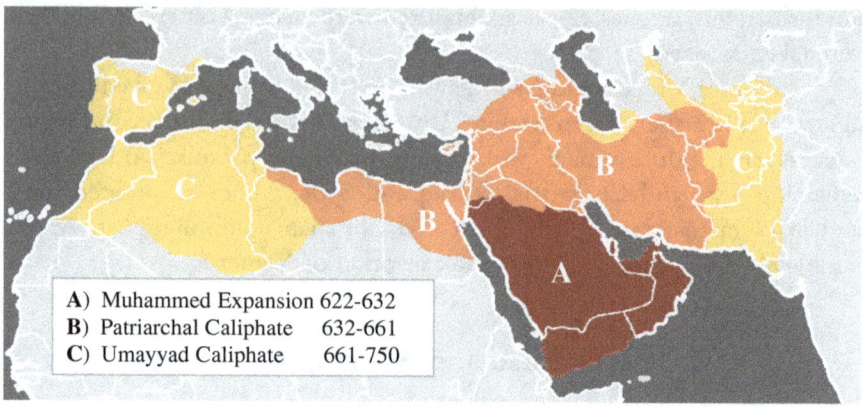

Figure 17.1. Muslim Conquests in the 7th and 8th Century. (A) Under Mohammed; (B) Additional Conquests under Patriarchal Caliphate (Abu Bakr, Umar, Uthman, Ali, and Hasan) — Eastern Arabia, Northern Egypt and Libya, Palestine, Lebanon, Syria, Iraq, Eastern Turkey, and Persia; and C) Additional Conquests under Umayyad Caliphate — Tunisia, Algeria, Morocco, Spain, Afghanistan, Pakistan, and southern Turkmenistan and Uzbekistan, eastern Kirgizstan, and Tajikistan. The great Islamic empire would "split up fairly early into competing caliphates" in the early ninth century (800s).[42]

Life under Muslim Rule

In the name of Allah, the compassionate, the merciful. [I] promise to give [the citizens of Damascus] security for their lives, property, and churches. Their city shall not be demolished, neither shall any Muslim be quartered in their houses . . . So long as they pay the poll tax, nothing but good shall befall them.

— Khalid ibn al-Walid, Arab commander who subjugated Damascus.[43]

Arab-Muslim conquerors generally ruled with tolerance. Conversion to Islam was typically voluntary.*[44] In most cities, "Muslim conquerors signed treaties which guaranteed the conquered people the freedom to practice their

---

*Christian Byzantine authorities persecuted Christians in Syria and Egypt who did not accept the Chalcedonian Creed of the Christian Orthodox Church. These "heretics" were more prone to accommodate Arab invaders—as they permitted freedom of worship as long as the tax was paid. Jews in Spain are said to have "welcomed Muslims as liberators from the oppression of Catholic Visigoth kings." Sources: Lapidus, Ira M. A History of Islamic Societies (Cambridge, UK: Cambridge University Press, 2014) p. 50.; Nicolle, p. 65.

religion," Shi'a scholar Sayyid Muhammad Rizvi tells us, "as long as they paid the required tribute to the caliph's treasury."[45]

Those conquered were treated as second class citizens. Anyone could escape inferior status — regardless of nationality, ethnicity, race, or prior religious adherence — simply by converting to Islam. Nonetheless, conversion to Islam was gradual. It took some two to three centuries before Muslims were the dominant population in conquered territories.[46]

Islam would also attract followers in places where no Muslim armies had set foot. This included China (as early as 7th century); sub-Saharan Africa (c. 900); Malaysia (1200–1300); and Indonesia (13th century). The latter is now the most populous Muslim country in the world.[47]

Early Islam was generally less dogmatic than Christianity regarding doctrines of faith (though there were exceptions). Muslim culture often promoted "rationality as the highest function given to human beings" by Allah. This helped foster an abiding interest in science and technology — an attitude which led to an age of discovery and advancement not seen since the ancient Greeks.[48]

The Golden Age of Islam

The unifying effects of Islam would come to incorporate the cultures of "ancient Semites, Classical Greece, and Medieval Indo-Persians." Out of this fusion, great advances were made in "science, astronomy, mathematics, algebra (itself an Arabic word), law, history, medicine, pharmacology, optics, agriculture, architecture, theology, and music," Desmond Stewart tells us.[49]

In the Islamic* Golden Age of the 8th to the 12th century, Muslim (as well as some Christian and Jewish) scholars translated the works of Plato, Archimedes, Euclid, Aristotle, and Ptolemy into Arabic. They also translated those of "celebrated Greek physicians Hippocrates, Dioscorides, and Galen — as well as major Persian and Indian scientific works." Latin translations of these works would later find their way to Western Europe, primarily through Muslim conquests of Sicily by the 11th century and Spain by the 12th.[50]

---

*We use "Islamic" as a general term here. In this era, Islamic scientists included "Arabs, Jews, Persians, Syriac Christians and a number of others." Source: Christie, "History of science on the Internet- the gift that keeps giving," thonywordpress.com. https://thonyc.wordpress.com/2019/04/03/history-of-science-on-the-internet-the-gift-that-keeps-giving/

Islamic scholars made their most lasting contributions in mathematics — particularly in geometry, trigonometry, and the algebra of ancient Babylonia and India. Translations into Arabic included Euclid's *Elements*, at least eleven works by Archimedes as well as later Greek authors, and a number of Indian mathematical works. The Latin translation of Persian mathematician Mūsā al-Khwārizmī's *Addition and Subtraction in Indian Arithmetic* introduced Hindu-Arabic numerals to the West.[51]

For a brief history of the development of Hindu-Arabic numerals, please see Appendix B.

In natural philosophy, the center of scientific discovery moved from Athens and Alexandria to Baghdad. Here in the 9th century, Caliph al-Ma'mūn — uneasy about applying pagan science to Allah's universe — reportedly "had a dream in which the ghost of Aristotle appeared and assured him there was no conflict" between reason and Islam.[52] From this he is said to have instituted the "House of Wisdom," an unrivaled center for research and learning. It had its own astronomical observatories and what was soon to be "the largest repository of books in the world."[53]

Islamic astronomers supported the religious needs of Islam. They studied the positions of stars to determine the location of Mecca for prayers in the vast areas under Muslim control. They studied the lunar orbital cycle, the basis for the Muslim calendar, and the "moment of sunrise and sunset for fasting during Ramadan."[54]

Ptolemy's work formed the basis of Islamic astronomy. Over time, Islamic astronomers improved the accuracy of his figures and tables. In c. 875, Yahya Ibn Abi Mansour conducted extensive astronomical observations. His book *Al-Zij al-Mumtahan*, "completely revised Ptolemy's *Almagest* values."[55]

Islamic scholars produced a number of ideas and findings which presaged modern science. In c. 800, Muhammad ibn Musa Ibn Shākir of Baghdad proposed that a "*force of attraction*" exists between heavenly bodies. Iraqi polymath Al-Kindi, Persian philosopher Algazel, and Jewish philosopher Saadia ben Joseph (c. 850) held that the universe has a *finite past* — rather than the infinite universe with no beginning of the ancient Greeks.[56]

In 968, Persian astronomer Al-Sufi generated the first detailed descriptions of what we now know are the Andromeda Galaxy and the Large Magellanic Cloud. In c. 1000, Al-Bīrūnī of Khwārezm (now in Uzbekistan) proposed that the Milky Way is a collection of numerous stars which *appear* to be "a continuous image due to the effects of refraction in the Earth's atmosphere." From Brahmagupta (7th century) and other Indian astronomers, Al-Bīrūnī also raised the possibility that Earth *goes around the Sun* and *rotates* on its axis.[57]

Figure 17.2. Ibn al-Haytham. Known as Alhazen in the West. Pioneer of the modern scientific method and father of modern optics.[58]

In c. 1021, Persian physicist extraordinaire Ibn al-Haytham (Fig. 17.2) proposed the concept of momentum and a "theory of attraction between bodies." He held that all heavenly bodies are "accountable to the laws of physics" — predating Isaac Newton by six centuries.[59]

Jewish convert to Islam Hibat Allah Abu 'l-Barakat al-Baghdadi (c. 1080–1165) correctly suggested that "a force applied continuously produces acceleration." Based on observation of comet locations, Persian polymath al-Tūsī (c. 1260) is said to have generated the first empirical evidence of the Earth's rotation.[60]

Muslim Conflicts with Science

We have indications of skepticism towards religion amongst Islamic scientists and philosophers. Based on his writings, Persian astronomer, mathematician, philosopher, and poet Omar Khayyám (1048–1131) appears to have been an atheist. The author of *The Rubáiyát* was called "a stinging serpent to the Sharī'ah*" after his death.

Around 1195, polymath Ibn Rušd, a follower of Aristotle, "was banished on suspicion of heresy," physicist Steven Weinberg tells us. In *The Tricks of the Prophets*, physician al-Razi wrote that "miracles are mere tricks." Yet theologian and astronomer al-Tūsī (1201-1274) was a devout Shiite. Astronomer al-Sufi's (903–986) name implies he may have been a Sufi mystic.[61]

---

* "Sharī'ah" is Islamic canonical law based on the Qur'ân.

The Islamic Golden Age petered out in the late 12th century. Scholars argue about why. It appears that growing religious antipathy towards science was a major factor — along with less tolerant Caliphs and conquests of Islamic lands by Mongols and Turks.

Theologian al-Ghazālī (c. 1058–1111) is said to have been "the most influential Muslim since Muhammad." In his *The Incoherence of the Philosophers*, al-Ghazālī wrote a scathing critique of Greek science. He labeled followers of "Socrates, Hippocrates, Plato, Aristotle, etc." as heretics. In another indication of growing intolerance, at the end of the 12th century "local religious scholars burned all medical and scientific books" in Islamic Córdoba (in what is now Spain).[62]

Nonetheless Islamic science lived on, feeding and inspiring scientific discoveries in what was to become the new center of world science: Western Europe.

The Crusades

From the 11th to the 13th century, Roman Catholic Europe waged a "holy war" to wrest Jerusalem and the "Holy Land" from Moslem control. Military campaigns were led primarily by the Franks of France and the Holy Roman Empire. Crusader armies also fought against non-Moslem peoples. This included Greek Orthodox Christians, Jews, "pagan Slavs, Russians, Mongols, other Eastern peoples, and political enemies of the popes," British historian Jonathan Riley-Smith tells us.[63]

The Crusades had unexpected effects. Medieval soldiers returned to Europe from the Levant with stories of the wonders and riches of the Moslem world. Trade in the Mediterranean grew by leaps and bounds. Soon Islamic advances in mathematics, astronomy, medicine, engineering, optics, chemistry, and architecture made their way to nascent European universities.[64]

> *The Crusades brought about results which the popes had never dreamed . . . (they) re-established traffic between East and West. Western knights . . . returned to their native land filled with novel ideas.*[65]

The conquest of the Byzantine Empire by the Ottoman Turks would further increase western exposure to ancient scientific knowledge. With the fall of Constantinople in 1453, Greek scholars fled to the West, bringing a huge number of ancient Greek texts with them.

The rediscovery of ancient Greek philosophy, mathematics and science along with exposure to Islamic mathematics and science would lead to an unprecedented revival of intellectual, artistic, and scientific innovation in Europe.

This is the subject of the next chapter.

Chapter 18

THE MEDIEVAL UNIVERSE

*The study of philosophy is not that we may know
what men have thought, but what the truth of things is.*

— Thomas Aquinas[1]

An unexpected series of events would come to further open up the West, this time to the wonders of Far East. It began with the rise of the Mongols.

Pax Mongolica

*The greatest happiness is to vanquish your enemies,
to chase them before you, to rob them of their wealth,
to see those dear to them bathed in tears, to clasp to your
bosom their wives and daughters.*

— Genghis Khan[2]

Genghis Khan
(c. 1162–1227)

In the 13th century (1200's), Mongol hordes led by the ruthless organizational and military genius Genghis Khan emerged from steppes of the Central Asian plateau. They possessed a force of arms and mobility the likes of which the world had never seen. The Mongols rode thick-necked, stocky horses, bred by their Mongolian masters for strength and endurance and the nature to "drive themselves to death if asked," writes historian John Mann. "With stirrups added to saddle, bridle, and bit, [Mongol] horsemen could out maneuver chariots (and) fire arrows, wield spears, or use the lasso while at full gallop."[3]

Their recurved bows were an ingenious composite of horn, wood, sinew, and glue which took up to a year to make. Arrows shot at close range from these formidable weapons possessed the penetrating power of bullets. At long range they outshot the English long-bow and were capable of reaching distances of nearly 500 yards (some 460 meters).[4]

The military strategy of Genghis Kahn and his generals featured rapid maneuver, surprise, and terror. Their armies exhibited remarkable agility for the time — traveling up to 100 miles in a single day. A pony-express method where messenger horse and rider were at the ready every 25 miles (40 kilometers) provided co-ordination of vast armies separated by huge distances.[5]

The Mongols used a simple yet most effective form of psychological warfare: surrender or be annihilated. If a city offered resistance, all within were exterminated. And they learned as they went, rapidly assimilating advanced siege technology.

An estimated 40 to 70 million people were annihilated by Genghis Kahn and his Mongol hordes. In what has been called the Muslim Holocaust, they obliterated the total populations of the Islamic cities of Otrar, Bukhara, Samarkand, Urgench, Merv, Nishapur, and Herat in central Asia.[6]

Although Mongol armies practiced brutal extermination during war, they governed the peace with a "general benevolence," writes Steve Dutch, Professor Emeritus of Natural and Applied Sciences at the University of Wisconsin." Cities (that surrendered) were "generally left under native governors," and "religious tolerance was practiced." In fact, Mongol "administration was generally more benign than pre-Mongol governments."[7]

At the height of its empire, the Mongol Khanate ruled from China in the East to Syria on the Mediterranean Sea — the "largest contiguous Empire in history."[8] (See Fig. 18.1) The so-called "Pax Mongolica" brought peace to a vast area and trade flourished. There was a "tremendous increase in Europe-Asia contact," writes Dutch, which opened the West to Eastern ideas and innovations.[9]

By the start of the sixteenth century, "most of the basic inventions of Asia were well established in the technology of Europe," historian David Larch tells us. Among the multitude of Chinese items which found their way

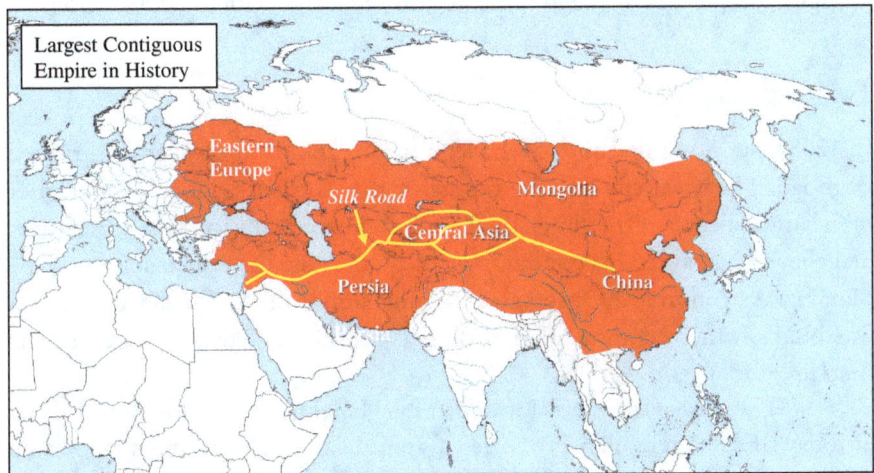

Figure 18.1. The Mongol Empire (1279). It was nearly five times the size of the Roman and Macedonian Empires. Trade Flourished along the "Silk Road".[10]

to the West were paper, the magnetic compass, gunpowder, charts, books, and paintings, as well as large ship-building technology.

In addition, "the suspension bridge, the fishing reel, the parachute, paper money, playing cards, the seismograph, brandy, rudders, cranks, matches, propellers, and biological pest control" also made their way to Europe from the East. Moreover, mechanical clocks, techniques for anatomical dissection, and movable type printing which enabled quicker and less expensive scholarly communication are suspected of having come to Europe from China as well.[11] Although spectacles, the blast furnace, the windmill, and farming equipment were invented independently in Medieval Europe.[12]

Chinese gunpowder in primitive firearms and cannons gave European kings the weapons to destroy baronial castles and unite fiefdoms into nation-states. The magnetic compass and key ship-building techniques adopted from the China would facilitate the fifteenth and sixteenth centuries oceanic voyages of Vasco da Gama, Christopher Columbus, Ferdinand Magellan, and many others.

Motivated to find new trade-routes to the East, the European Age of Exploration and Conquest had begun. It would prove horrific for indigenous populations, leading to near annihilation and enslavement in the Americas, Africa, Australia, the Pacific, and parts of Asia.[13]

Let us now return to the thirteenth century, where a particular Mongol conquest had a deep influence on the advancement of astronomy.

The Marāgha School

In 1258, the Mongol armies of Hulegu Khan, grandson of Genghis, sacked Baghdad. The great city was the capital of the Abbasid Caliphate and center of Islamic science at the time. To show his appreciation for astrologers who he believed had aided his conquests, Hulegu ordered the construction of the Marāgha Astronomical Observatory — the largest in the Medieval world. It was built on the recommendation of Naṣīr al-Dīn al-Ṭūsī, who became its first director.[14]

The observatory was located in the "heights west of Marāgha," a city in today's East Azerbaijan Province of Iran. Islamic astronomers associated with the observatory were referred to as the "Marāgha School." They initiated a profound shift in scientific thinking away from the Greek emphasis on philosophy to one based more on *empirical observations* and mathematics.[15]

Astronomers of the Marāgha School developed a comprehensive set of geometric mathematics to model a more accurate Ptolemaic Earth-centered universe. Some two centuries later, Copernicus would adopt this work to produce a Sun-centered solar system model, as we shall see.

The Awakening of the West

The first key to wisdom is assiduous and frequent questioning...

— Peter Abelard, 12th century French scholar[16]

The rediscovery of Greek and Roman science and mathematics, exposure to Islamic advances, and contact with the wonders of the East set off a firestorm of changes in Medieval Europe. Translations of Aristotle's works in particular would "form the basis for European natural science until the 17th century," writes science historian James Hannam."[17]

The West would soon become the dominant power on the world stage. The changes began slowly in the feudal society of the 10th century. French scholar Gerbert d'Aurillac (946-1003) — crowned Pope Sylvester II in 999 — was known as "the mathematical pope." He helped to bring the abacus, Hindu-Arabic numerals, and the decimal system from Islamic Spain to Christian Europe. Gerbert also introduced the spherical astrolabe, a mechanical model of the celestial sphere, to the West.[18]

Established in the 11th and 12th centuries, the first European universities enjoyed a measure of academic freedom and independence from royal meddling.* Theology was the most important subject. Though subservient to theology, natural science was encouraged as the study of God's creations.[19]

Hannam points out that the combination of "religious and cultural unity" within Christendom and the "political fragmentation" of many states and kingdoms was a boon to scholars. "No secular ruler could control them. And they enjoyed the overarching protection of the Church" — as long as they didn't contradict religious authorities.[20]

---

*The first university in the world is said to have been the University of Al-Qarawiyyin in Fez, Morocco. It was founded by Fatima al-Fihri, "daughter of a wealthy merchant" in 859 AD. Source: Stirone, Shannon. "How Islamic scholarship birthed modern astronomy" astronomy. com. Feb 14, 2017. http://www.astronomy.com/news/2017/02/muslim-contributions-to-astronomy. Retrieved April 10, 2019.

Rational thought was to become a greater subject of contemplation. In the early twelfth century, witty and fearless French scholar Peter Abélard (1079–1142) proposed that "the mysteries of Christianity were not to be taken for granted, but [to be] tested in the light of reason."

In *The Guide to the Perplexed*, Sephardic Jewish philosopher Moses Maimonides (1138–1204) sought to reconcile "the truth of the Holy Scriptures" with Aristotelian philosophy and science.[21] Islamic polymath Ibn Rušd (1126-1198) — Averroës in Latin — sought to reconcile Aristotle with Islam. Both were born in Córdoba, the glorious capital and center of learning of *al-Andalus* or Islamic Spain.

In *The Harmony of Religion and Philosophy*, Averroës tried to show that "philosophy and revelation do not contradict each other." Latin translation of his works led to the revival of the philosophy and science of Aristotle in the West. He has been called "the founding father of secular thought in Western Europe."[22]

In the 13th century, English philosopher and Franciscan friar Roger Bacon (1214–1292) advocated empiricism and the scientific method. These where seen as radical ideas at the time. Bacon criticized both Aristotle and the Church. He mastered Greek and Islamic texts, particularly al-Haytham's *Optics*. Bacon's work in mathematics, optics, the manufacture of gunpowder, and celestial bodies laid the foundations for the revival of European science.[23]

Bacon was a devout Christian. He studied nature in part as a tool to convert Muslims to Christianity before the coming of the Apocalypse, which he believed was imminent.[24] Referred to as "Doctor Mirabilis," he was an imaginative and prescient thinker.

> *It is possible that a car should be made that will move with inestimable speed... without the help of any living creature... it is possible that a device for flying shall be made such that a man sitting in the middle of it and turning a crank will cause artificial wings to beat the air in the manner of a bird's flight.*
>
> — Roger Bacon, c. 1240[25]

A contemporary of Bacon, the great Dominican priest, philosopher, and theologian Thomas Aquinas (1225–1274) had a profound influence on the Catholic Church and Western thought. He was inspired by the writings of both Ibn Rušd (Averroës) and Moses Maimonides. Aquinas held that truth is known through both faith *and* reason.[26]

Aquinas adopted Aristotle's philosophy and revived Ptolemy's astronomy. Through his efforts, "the study of Aristotle became the center of university education." The Greek geocentric model of the universe became Church doctrine as well, as we shall see.[27]

"Since the 13th-century the Catholic Church had integrated an uneasy synthesis of Aristotelian cosmology and Ptolemaic astronomy," writes historian of science Thony Christie, "largely created by Albertus Magnus and his pupil Thomas Aquinas, into their model of reality, one that seemed to fit the known empirical facts."[28]

As we learned, Plato and other ancient Greeks believed that celestial objects moved in perfect circles — the purest of geometric forms. The Church interpreted these perfect circles as a demonstration of God's divine plan: hence "divine circles."

Church doctrine also identified Aristotle's outermost Prime Mover with Christian Heaven — a great celestial sphere moved by the angels. At the center of the universe was the imperfect Earth, the abode of man created in the image of God.[29] Despite its inaccuracies, the ideas of Aristotle "were baptized into the Catholic Church and had assumed the power of religious dogma."[30]

In the early 14th century, French philosopher Jean Buridan (c. 1295-1358) further sowed the seeds of the upcoming scientific revolution with his concept of "impetus." His thinking foreshadowed the Newtonian concept of momentum. Buridan said scientific principles "are accepted because they have been *observed* to be true in many instances and to be false in none."[31]

Buridan based his theory of impetus on the work of Islamic scholars Ibn Sina (c. 980–1037), known in Latin as Avicenne, and Abu'l-Barakāt Hibat Allah ibn Malkā al-Baghdādī (c. 1080–1165). They in turn developed their theory based it on 6th century Christian theologian and natural philosopher John Philoponus (c. 490–c. 570 CE) of Byzantine Alexandria.[32]

Buridan also suggested the Earth may not be at rest. The rotation of the stars in the night sky could be an illusion caused by a rotating Earth. He felt this more elegant conception was a reflection of God's plan.*

---

*The great Indian astronomer Aryabhata (476–550) also proposed that Earth rotates about its axis daily, and that this diurnal rotation produces the apparent rotation of the stars we observe in the night sky. Source: Dutra, Amartya Kumar. "Aryabhata and Axial Rotation of Earth" Resonance May 2006 p. 58. https://www.ias.ac.in/article/fulltext/reso/011/05/0058-0072.

Recall that Ptolemy and others had argued there would be a great wind if our planet were really rotating through the atmosphere. Both Buridan and French philosopher Nicolas Oresme (c. 1325–1382) countered that that we would feel no wind. Why? Because our atmosphere would be carried along with the Earth's motion. This would be similar to Galileo's explanation some two centuries later. (More on this in Chapter 23.)[33]

In the 15th century, Nicholas of Cusa (1401–1464) made a series of speculations which turned out to be remarkably close to modern thinking. The German philosopher, theologian, and astronomer hypothesized that — to reflect God's glory — the universe is limitless. A century before Copernicus, he also suggested the Earth is not the center of the universe and must be moving. He also proposed that the sphere of fixed stars is not the outer border of the universe. Nicholas said Earth is just another star, though the most important. And perhaps most prescient, he suggested there may be "alien life elsewhere in the universe."[34]

Rebirth

From the 14th to the mid-15th century, Europe was beset by one catastrophe after another. They included The Hundred Years' War between France and England with an estimated death toll of 3.5 million; the schism of the Catholic Church with one pope in Rome and another in Avignon; and the Black Death, the most devastating plague in European history.[35]

The "Little Ice Age" came in the early 1300s. It was followed by widespread famine. And contact with the East brought more than new technologies and inventions, it brought the Black Death — bubonic and pneumonic plagues from the East. Peaking in mid-century, the plagues wiped out an estimated 50 to 70 million people, some one-third* of Europe's population.[36]

No one knew the source of this devastation. Some blamed the perennial scapegoat, the Jews — particularly in France and Germany. They were accused of "poisoning wells to kill Christians. Purges against Jews took place in 1349 and nearly every other year of plague outbreaks," writes L. Kip Wheeler, professor at Carson-Newman University."[37]

The resultant population vacuum brought political upheavals, religious turmoil, and breakdown of social order. This in turn opened up

---

*In some regions mortality rates were significantly higher, including many parts of England and France.

economic opportunities for survivors. Feudal structures were shattered. A new middle class emerged. Cities were in resurgence. Science in particular saw a shift from England and France to Germany and Italy.[38]

The maritime cities of Italy — Venice and Florence in particular — thrived and began to dominate the Mediterranean and eastern land trade-routes to East Asia. With their new-found wealth, Italian patrons funded intellectual endeavors and great works of art. A new humanist spirit arose which emphasized human self-worth and individual dignity.[39]

Then around 1450, German craftsman and publisher Johannes Gutenberg invented the moving-type printing press. The device would generate an information revolution. It enabled the mass production of books and gave people across Europe access to both sacred and secular texts for the first time.[40]

Science historians generally agree that this invention and the "arrival of the printed book in the middle of the fifteenth century was a major factor in the emergence of modern science in general and modern astronomy in particular," Thony Christie tells us. Works included reliable and consistent textbooks, astronomical charts, astrological calendars, inexpensive almanacs, serial printed globes, atlases, and maps.[41]

Also key to this scientific revolution was the base-ten counting system from India and the Islamic world — as well as the paper-making process which had originated in China. (See Appendix B for details.)

The continent experienced an explosion in new ideas.[42] It saw the rise of humanist philosophy blending Christian doctrine with classical Greek/Roman philosophy, science based on empiricism, a revolution in medicine and anatomy and the science of navigation, advances in trigonometry and algebra as a "central mathematical disciple," and the founding of geology and minerology. Also included were literature presented in the vernacular languages of the day rather than Latin, magnificent architecture in the ancient Greek style, masterpieces of neo-classical sculpture, superb realistic paintings of the human form, and modern church music.[43]

It was the time of Michelangelo, Da Vinci, Donatello, Cellini, Raphael, Titan, and Palestrina — an age of wonder and glory called the Renaissance or "rebirth" in French.

With the rediscovery of ancient texts, people of the Renaissance yearned for a return to the cultural glory of Rome, the intellectual climate of Athens. The ancient world was romanticized as "the pinnacle of human achievement, especially intellectual achievement." Still, Christian religious beliefs remained deeply held. The glory of God became the outlet for this yearning. Thus, the art of the Renaissance period depicts Biblical events yet emulates classical Greek and Roman forms.[44]

Explorers of the Universe

The conflict between ancient pagan high culture and Church-based beliefs influenced the works of Renaissance scientists as well — inspiring yet limiting the extent of their remarkable breakthroughs.[45]

It was in this climate of exploration that Renaissance scientists would challenge deeply-held religious beliefs, revolutionize man's worldview, and pave the way for Isaac Newton and Albert Einstein. Four in particular would gain great fame for their accomplishments. They were the visionary Nicolaus Copernicus, the meticulous Tycho Brahe, the brilliant Johannes Kepler, and the irascible Galileo Galilei. Their stories are told in the following chapters.[46]

Chapter 19

THE RELUCTANT REVOLUTIONARY

Yea, the world [Earth] is established, that it cannot be moved.

— Psalms 93:1[1]

The modern conflict between science and religion began with an unlikely source — a Polish Catholic cleric with a passion for astronomy. He would merge the ancient Greek idea of a Sun-centered universe with the mathematics of Islamic astronomers to produce a paradigm shift in humankind's world view. His conception would be later declared "suspected heresy" by the Roman Catholic Church.

His name was Mikolaj Kopernik. We know him as Nicolas Copernicus, the Latin name he chose for himself later in life.

The Path to Reality

<u>Note to Reader</u>: *A major source of this and the following four chapters is historian of science Thony Christie's blog: Renaissance Mathematics. It presents a scholarly history of the period's natural philosophers with an accuracy rarely found in popular accounts.*

Recall that Ptolemy considered his model of the universe a computation device. He did not consider it physically real. This led a number of scholars in late antiquity to separate the sciences. It was felt that physics, at the time restricted to earthly matters, described physical reality—while astronomy did not.[2]

Some felt that attempts to determine the motions of heavenly bodies were doomed to failure. In his *Guide for the Perplexed*, preeminent 12th century Sephardic Jewish philosopher Moses Maimonides (1135–1204) framed the argument in religious terms:

> *... regarding all that is in the heavens, man grasps nothing but a small measure of what is mathematical ... the deity alone fully knows the true reality, the nature, the substance, the form, the motions, and the causes of the heavens.*[3]

The belief that the essence and behavior of heavenly bodies was beyond the scope of human understanding became the prevailing view in Medieval Europe. Albeit mathematically clever, astronomical models could not hope to describe physical reality.

Islamic astronomers in particular rejected to the unphysical nature of Ptolemy's model and its geometric devices. Andalusian polymath Ibn Rushd aka Averroës (1126–1198) wrote:

> *Ptolemy was unable to set astronomy on its true foundations ... The epicycle and the eccentric are impossible ... what we have is something that fits calculation but does not agree with what is.*[4]

Rushd urged "new investigation concerning that genuine astronomy whose foundations are principles of physics." Islamic astronomers of the 13th and 14th century Marāgha school took up the challenge. They modified Ptolemy's construct to include what they felt were more physically acceptable geometries, as we shall see.

Nicholas Copernicus would adopt this Islamic work in the 16th century — and apply it to a *Sun-centered* model of the universe. Here the Earth is not at the center of the Universe. It and all the other known planets revolve around the Sun.

Fearing ridicule for this outlandish conception, he would hold off publication for most of his life.[5]

The Reluctant Revolutionary

At rest, however, in the middle of everything is the Sun.

— *Nicolaus Copernicus,*
De revolutionibus orbium coelestium

Nicolaus Copernicus
(1473–1543)

In his remarkable life, Nicolas Copernicus was a cleric, physician, administrator, military leader, currency reformer, cartographer, and, of course, astronomer.[6]

He was born in Toruń in Poland on February 19, 1473. The youngest of four children, Nicholas was ten years old when his father died.[7] His well-to-do Uncle Lucas Watzenrode became his guardian (along with becoming guardian of his older brother and two older sisters). Uncle Lucas was canon at the Roman Catholic Frauenburg (Frombork) Cathedral in "Warmia [Ermland in Germany], an autonomous self-governing Prince Bishopric." Here Copernicus would spend most of the rest of his life.[8]

Uncle Lucas funded Nicolas's education at the cathedral school of Wloclawek and then at the University of Kraków. Copernicus studied Latin, math, geography, astronomy, and philosophy.[9]

It was at Kraków that Copernicus found his life-long passion — astronomy. Here he began his extensive collection of books on astronomy and mathematics. One of the first books he acquired was Euclid's *Elements* — a prophetic step as Copernicus's astronomical theory would be rooted in geometry.

Uncle Lucas, now Bishop of Ermland, was not thrilled with Nicolas's new-found interest. To keep his uncle happy, Copernicus studied canon law at the University of Bologna from 1496 to 1501 in order to qualify for a position in the church. Bologna was Europe's oldest university. At his uncle's urging, he then studied medicine at the famous medical school in Padua, Italy — and obtained a doctorate in Canon Law at the University of Ferrara in 1503.[10]

Nicolas's passion for astronomy would not be denied. In Bologna he managed to room at the home of astronomer Domenico Maria da Novara (1454–1504), and helped him with observations. He also took as many astronomy courses as he could wherever he was enrolled.[11]

With his uncle's influence, Nicolas became a canon of Frauenburg Cathedral — "a lifetime appointment . . . with manageable duties and a comfortable pension," writes biographer Dava Sobel.[12] A career "under the auspices of the Catholic Church" was common for scholars of the period. On return to Poland, he became his uncle's doctor, private secretary, and personal advisor.

Copernicus also became a district administrator, organizer of the defenses of Allenstein Castle in war, peace negotiator, Commissar of Ermland, and currency reformer. In his spare time, he made astronomical observations and secretly worked on his heliocentric (Sun-centered) model of the universe.[13]

Like Albert Einstein some 400 years later, Copernicus was not a mainstream scientist. He worked alone. And like Einstein, Copernicus possessed the independence of thought to challenge deeply held assumptions of the day. Unlike Einstein, Copernicus was reluctant to make his radical views public.

Copernicus would later explain his hesitancy in the Preface to his masterwork, *De revolutionibus*:

> "... *the consensus of many centuries* ... *[is] that the earth remains at rest in the middle of the heavens* ... *[They] would* ... *regard it as an insane pronouncement if I made the opposite assertion that the earth moves.*
>
> *I debated with myself for a long time whether to publish the volume which I wrote to prove the earth's motion* ... *the scorn which I had reason to fear on account of the novelty and unconventionality of my opinion almost induced me to abandon completely the work which I had undertaken.*"[14]

The Polish cleric would disclose his revolutionary Sun-centered model for the first time around 1510 — anonymously.

The Copernican Model Revealed

At age thirty-seven, Copernicus penned a hand-written eight-chapter outline of his theory dubbed *Commentariolus (Little Commentary)*. No author was named on the title page. He distributed copies only to his friends. Nonetheless, it "seems to have circulated fairly widely."[15]

In the introduction, he summarizes the difficulties with the concentric spheres construct of Calippus/Eudoxus and Aristotle — particularly its failure to account for the observed increase in brightness of a planet when in retrograde. He also criticizes Ptolemy's eccentrics and epicycles model:

> *... the planetary theories of Ptolemy and most other astronomers ... were not adequate unless certain equants were also conceived. It then appeared that a planet moved with uniform velocity neither on its deferent nor about the center of its epicycle. Hence a system of this sort seemed neither sufficiently absolute nor sufficiently pleasing to the mind. (My underlines.)*[16]

Copernicus then gives the motivation for the development of his model:

> *I often considered whether there could perhaps be found a more reasonable arrangement of circles, from which every apparent irregularity would be derived while everything in itself would move uniformly, as is required by the rule of perfect motion ...*

The Model

In this "more reasonable arrangement of circles" of Copernicus, Earth is not the center of the universe, but rotates once a day and travels around the Sun once a year. (See Figure 19.1.)

More precisely, in the Copernican model:

(1) *Earth rotates on its axis* – the apparent daily motions of the Sun and stars are due to the daily rotation of the Earth;
(2) *the Moon orbits the Earth* as in Ptolemy's model; but

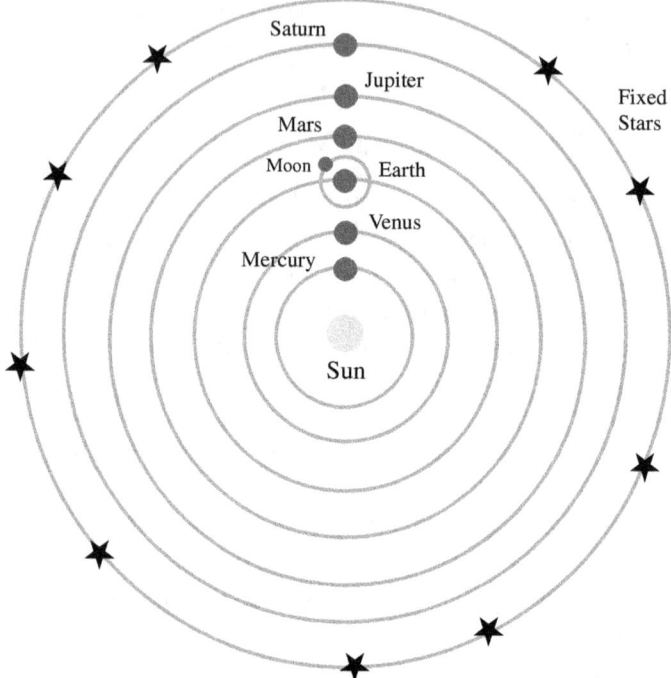

Figure 19.1. The Copernican Heliocentric (Sun-centered) Model of the Universe. The Earth orbits the Sun just like all the other planets. The planets as a solar system are given for the first time. Copernicus maintains Plato's circular orbits.

(3) *Earth and all the other five know planets orbit the Sun*; The order of the planets is Mercury, Venus, Earth, Mars, Jupiter and Saturn; and

(4) *Retrograde motion is an illusion.* The apparent periodic reversal of planetary motion as seen from Earth is due to planets orbiting the Sun at different distances and speeds (more on this later.)[17]

Others had proposed the idea of the Earth orbiting the Sun before Copernicus, as we have seen. This included Greek mathematician and astronomer Aristarchus of Samos in the 3rd century BC. We know that Copernicus was aware of Aristarchus from his own notes.[18]

Over the next thirty years or so, Copernicus continued to develop his heliocentric model. Despite painstaking analysis of observations and refinement of his planetary models, he remained hesitant to publish. The primary reason? He could offer no empirical physical evidence for a moving Earth.[19]

Near the end of his life, Copernicus finally consented to have his life's work published.

The First (openly published) Account

These phenomena, besides being ascribed to the planets, can be explained, as my teacher [Copernicus] shows, by a regular motion of the spherical earth; that is, by having the sun occupy the centre of the universe, while the earth revolves instead of the sun . . .[20]

— Georg Rheticus, Narratio Prima.

Johann Petreius (1497–1550) was a publisher of "astrological/astronomical and mathematical texts" from Nürnberg (now Nuremberg in Germany). Apparently, he had gotten wind of Copernicus's work. In 1539, he sent twenty-five-year-old mathematics professor Georg Rheticus (1514–1574) to Frauenburg to convince the Polish astronomer to publish.[21]

Copernicus was still hesitant. "To calm his fears," the young math professor wrote a booklet on the Copernican model called *Narratio Prima (The First Account)*.[22]

Published in 1540, "*Narratio Prima* is strong on rhetoric and polemic but rather weak on its scientific content," writes Thony Christie. "There are no diagrams and Rheticus tends to rely on philosophical arguments rather than mathematical ones."

Nonetheless, the booklet received a positive reception. This and the "lack of negative reactions seem to have finally convinced Copernicus" to let his life's work be published. He titled it *De revolutionibus orbium coelestium* (*On the Revolutions of the Heavenly Spheres*).[23]

That same year — some 30 years after his anonymous *Little Commentary* — 67-year-old Nicholas Copernicus gave *De revolutionibus* to Rheticus for publication by Petreius.[24]

De Revolutionibus

This contains, as it were, the general structure of the universe.

— Copernicus, Preface to De revolutionibus[25]

De revolutionibus, Copernicus's six-book masterpiece (Figure 19.2) was a "heavy duty, large-scale mathematical text," Christie tells us. It was "the first major, extensive work of mathematical astronomy [in Europe] since Ptolemy's *Almagest* in the middle of the second century." Only a "comparatively small group of mathematical astronomers [were] capable of understanding it."[26]

268   *Cosmic Roots*

Figure 19.2.   De revolutionibus orbium coelestium (On the Revolutions of the Heavenly Spheres). Nicholas Copernicus's magnum opus published in Latin in 1543 by Protestant printer Johannes Petreius in Nuremberg.

The book's most radical proposition was heliocentricity. Where did Copernicus get his inspiration for a moving, rotating Earth? From the ancients Greeks. He writes in *De revolutionibus*:

> . . . *I found in Cicero that Hicetas supposed the earth to move. Later I also discovered in Plutarch that certain others were of that opinion . . . from these sources, I too began to consider the mobility of the earth.*[27]

As noted in Chapter 5, Heraclides of Pontus had proposed that the Earth *rotates* on its axis in the 4th century BC.[28] Copernicus mentions this and other classical sources in *De revolutionibus*:

> *This opinion [that the Earth rotates] was indeed maintained by Heraclides and Ecphantus, the Pythagoreans, and by Hicetas of Syracuse, according to Cicero . . . they ascribed the setting of the stars to the earth's interposition, and their rising to its withdrawal.*[29]

If certain ancient Greeks, as well as more recent astronomers, had proposed that Earth rotates and orbits the Sun, what was so special about Copernicus? He was the first to provide both a Sun-centered model *and* the detailed mathematics to support it. And perhaps most significant, it was Copernicus who first treated the Sun and all the planets as a *solar system*, revealing the order of the planets for the first time.[30]

The notion that Earth is not at the center of the universe threw a monkey wrench into Aristotelian physics and its later modifications. In Book 1 of *De revolutionibus*, Copernicus proposed new physics to accommodate his heliocentric theory. Let's take a look.

Copernican Physics

<u>Note to Reader</u>: *This section is based on Professor of Renaissance studies at University College London, Dilwyn Knox's scholarly evaluation of Copernican physics.*[31]

As noted in Chapter 6, Aristotle held that all things move according to their nature. In his theory, the natural motion of the elements earth and water is downward in a *straight line* towards the center of the Earth — i.e., the center of the universe. The natural motion of air and fire is upwards away from the center. The heavens are made of a fifth element: imperishable aether. The natural motion of heavenly objects is *circular*.

Copernicus argued that the natural motion of elements earth and water is not a straight line, but *circular*. This to explain the diurnal (daily) rotation of Earth. He proposed that objects fall towards the center of Earth by a *combination* of motions:

(1) the natural straight-line motion of gravity, coupled with
(2) an unnatural acceleration due to "the impetus of their weight."[32]

The Polish astronomer's heliocentric model "disrupted Aristotle's division between sublunar and supralunar worlds. To get around this problem Copernicus suggested that each planet had its own gravity, like Earth."[33]

He wrote in De revolutionibus:[34]

> *For my part I believe that gravity is nothing but a certain natural desire, which the divine providence of the Creator of all things has implanted in parts, to gather as a unity and a whole by combining in the form of a globe.*

> *This impulse is present, we may suppose, also in the sun, the moon, and the other brilliant planets, so that through its operation they remain in that <u>spherical shape</u> which they display.* (My underlines.)

Copernicus also rejected the Aristotelian notion of aether as the substance of the heavens, as Earth is now part of the heavens. In addition, he eliminated Aristotle's "sublunary air and fire spheres."[35]

The physics of Copernicus was quite muddled and incorrect — a mix of natural motion and natural desire ala Aristotle. Still, "during the late 16th and early 17th centuries," Dilwyn Knox tells us, "Copernicus's doctrine was commonly proposed as an alternative to Aristotle's account of gravity, levity and natural elemental motion."

The first definitive answer on how motion and gravity actually behave would come a century and a half later with the brilliant work of Isaac Newton, as we shall see. Copernicus had asserted that gravity produces the spherical shape of Earth, as well as that of the Sun, Moon, and other planets. This inferred that the *same physics* applies to Earth and the heavens. This conception would become a core principle of Newton's Theory of Universal Gravitation.

<u>Heliocentricity</u>

In the Introduction to his magnum opus, Copernicus paid tribute to Ptolemy:

> *To be sure, Ptolemy of Alexandria, who far excels the rest by his wonderful skill and industry brought the entire art to perfection with the help of observations extending over a period of more than four hundred years, so that there no longer seemed to be any gap which he has not closed . . .*

Then he later argued for his new paradigm and its advantages over Earth-centered models:

Rotation of Earth — The Polish astronomer explained why a rotating Earth and stationary starry vault made more sense. "Indeed, a rotation in twenty-four hours of the enormously vast universe" he wrote, "should astonish us more than a rotation of its least part, which is the earth."[36]

Mercury and Venus — The Copernican model explained why Mercury and Venus always appear in the sky close to where the Sun had risen or set, thus "only in the hours after sunset and before sunrise."[37] Copernicus again gives credit to classical sources:

> *In my judgment, therefore, we should not in the least disregard what was familiar to Martianus Capella [fl. 410–420 CE.], the author of an encyclopedia, and to certain other Latin writers. For according to them, Venus and Mercury revolve around the sun as their center. This is the reason, in their opinion, why these planets diverge no farther from the sun than is permitted by the curvature of their revolutions.*[38]

Ptolemy's Earth-centered model included this curious link of these two planets to the Sun — with no physical explanation. Copernicus's simple explanation: Mercury and Venus are the first two planets which orbit the Sun. So, from Earth, the third planet, they always appear close to the Sun.

Orbital Periods — Copernicus' model allowed astronomers to calculate *how long* it took each planet to complete one revolution around the Sun (its orbital period). It also allowed them to "compute relative distances between the planets and the Sun for the first time."[39]

Planetary Distances — Copernicus showed that the relative "distances of the planets from the sun bore a direct relationship to the size of their orbits."[40] The more distant planets have longer orbital periods, and move more slowly. For example, Jupiter, much further away from the Sun than Earth, takes some twelve earth-years to complete one revolution around the Sun. Still further out Saturn takes over 29 earth-years.[41]

This was a major reason why Copernicus was drawn to a Sun-centered system. "In no other way do we find a sure bond of harmony between the movement and magnitude of the orbital circles," he wrote in *De revolutionibus*.[42]

Planetary Positions — Copernicus's Sun-centered model made it much easier to calculate the predicted positions of the planets.[43]

Planetary Motion — The greatest simplification in the Copernican model was on the observed motion of the planets. His construct explained for the first time *why* orbiting planets sometimes appear to stop, turn in the opposite direction, and then return to their original direction. Let's take a look.

Retrograde Explained

The irregular patterns of the planets seen in the sky are produced simply because we view their orbits around the Sun *from Earth* — also a planet orbiting the Sun.[44]

To understand this, let's look at a diagram of the Earth and Mars orbiting the Sun. (Figure 19.3.) Here we see the Earth and Mars orbiting the

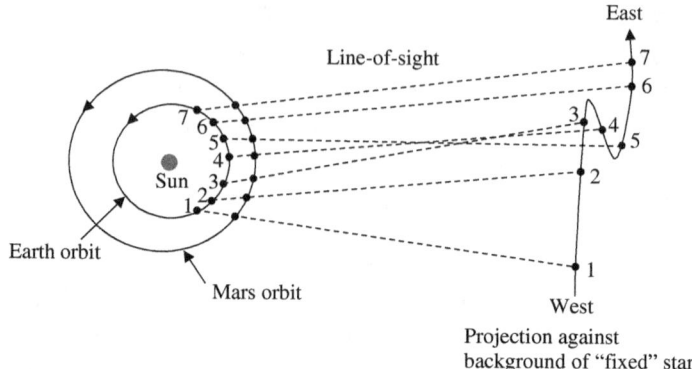

Figure 19.3. Copernicus Explains Retrograde Motion — Earth/Mars Example. Earth is closer to the Sun so it orbits faster than Mars. As seen from Earth: Steps 1–3 — Mars appears to be moving to the East. Step 4 — As Earth catches up to Mars, the red planet appears to be moving backwards (to the West). Steps 5–7 — Mars again appears to be moving to the East.[45]

Sun according to Copernicus. The figure shows the line-of-sight from Earth to Mars (dashed lines), and its projection onto the background of "fixed" stars.

Mars initially appears to be moving eastward, as seen from the Earth (Steps 1, 2, and 3). Since Earth is closer to the Sun, it moves in a faster orbit than Mars. As a result, from time to time the Earth catches up with Mars and overtakes it (Step 4). At this point Earth's "faster motion makes Mars appear to be going backwards" (westward).*

Copernicus's Sun-centered model gave the first *physical* explanation for the observed retrograde motion of planets.[46] Earth occupies the third position from the Sun. It has the faster inside track with respect to Mars, which holds the fourth position from the Sun. Earth thus overtakes "the slower, more distant Mars every couple of years."[47] To us on Earth, it appears that Mars is moving in the opposite direction during that time.

Copernicus still assumed the planets travelled in *circular* orbits in *uniform circular motion*. The "divine circles" of the Catholic church and Platonic/Aristotelian geometry remained sacrosanct. Because of this,

---

* For an excellent depiction of retrograde motion, please see: James O'Donoghue. "An animation to explain the (apparent) retrograde motion using actual 2020 planet positions" youtube.com. https://www.youtube.com/watch?v=hOjrPcD6Iuc. Retrieved Jul 27, 2021.

Copernicus had to incorporate epicycles and eccentrics in his model to better match observations

Islamic Roots

As noted, Islamic astronomers had objected to the unphysical nature of Ptolemy's model. Some European astronomers also objected to Ptolemy's equant point as a violation of Plato's principle of uniform circular planetary motion. Marāgha school astronomers of the thirteenth and fourteenth century modified Ptolemy's construct to include geometries which eliminated the equant.

Nicholas Copernicus would adopt this Islamic work in the 16th century. "The removal of the equant point for exactly this reason was the starting point of [Copernicus'] reform efforts," Thony Christie tells us.[48]

I had read that Copernican mathematics was similar to that of Islamic astronomers. On further research, the extent of the duplication shocked me.

> ... *the relationship between the models of [his Islamic predecessors] is so close that independent invention by Copernicus is all but impossible.*
>
> — Esteemed science historian Noel Swerdlow.[49]

Copernicus's detailed mathematical models for the orbits of the known planets, the orbit of the Moon, and the obliquity of the ecliptic* are nearly identical to the works of four astronomers of the Marāgha School. These Islamic astronomers are Naṣīr al-*Dīn al-Ṭūsī* (1201–1274); Mu'ayyad al-Din *al-'Urdi* (ca. 1200–1266); Qutb al-Din *al-Shirazi* (1236–1311); and Abu al-Ḥasan Ibn *al-Shatir* (1304–1375).

The evidence is overwhelming. To summarize:

Copernicus uses the same geometric device invented by *al-Ṭūsī* to eliminate equants — the so-called "*Ṭūsī* couple." He makes the same mistakes in copying Arabic letters in an illustration of *Al-Ṭūsī's*.[50]

---

*The obliquity of the ecliptic is "the angle between the plane of the ecliptic (the plane of Earth's orbit) and the plane of Earth's equator," i.e., the tilt of Earth's axis. Its current value is about 23.4 degrees. Source: "Obliquity of the Ecliptic", *American Meteorology Society*. ametsoc. org. http://glossary.ametsoc.org/wiki/Obliquity_of_the_ecliptic. Retrieved Jan 19, 2020.

The Polish astronomer's arrangement of Mars, Jupiter, and Saturn is identical to that of *Al-'Urdi* and *al-Shirazi*. Copernicus uses "*Urdi's* lemma" for the same purposes in his model.[51]

Copernicus bases his planetary models for longitude on *al-Shatir's*. For Mercury and Venus, he converts the smaller epicycle into an equivalent rotating eccentricity — an adaption of *al-Shatir's* model. He even makes the same mistake for Mercury as *al-Shatir*.

In addition, Copernicus' lunar model is identical to *al-Shatir's*. His illustration shows a striking similarity to that of *al-Shatir's*.[52]

The Center of the Universe

> *Near the sun is the center of the universe.*
>
> — Nicholas Copernicus, *De revolutionibus*

Once he had revised the Ptolemaic model in the same way Marāgha school astronomers had done, all Copernicus had to do was "reverse the direction of the last vector connecting the Earth to the Sun," science writer Dick Teres points out.[53] Thus, the heliocentric model of Copernicus was born.

Mathematically, he moved the center of the coordinate system used by astronomers to a different point. Rather than at the middle of the Earth, the center of the Copernican system was now at the center of Earth's *circular orbit* — a location near to but not exactly at the Sun.[54]

In effect, Copernicus moved the origin of the universe from the center of the Earth to the center of Earth's circular orbit around the Sun. This is shown in Figure 19.4. The co-ordinates for the geocentric (Earth-centered) model of Ptolemy and Islamic astronomers is on the left. The Copernican heliocentric (Sun-centered) version is on the right.

Plagiarism?

Copernicus mentioned five Islamic astronomers in *De revolutionibus*: Ibn Qurra (836–901), al-Battani (c. 850–929), al-Zarqali (1029–1100), Ibn Rushd or Averroes (1126–1198), and al-Bitruji (c. 1150–1200).[55]

Yet there is no mention in *De revolutionibus* or the earlier *Commentariolus* of al-Ṭūsī, al-'Urdi, al-Shirazi, or al-Shatir. Why?

Perhaps Copernicus learned of the Marāgha school models but was unaware of who created them. Perhaps he knew who did, but choose not to mention their names.

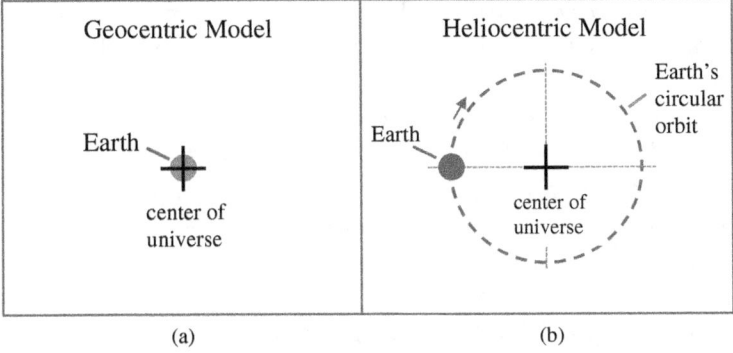

Figure 19.4. Copernicus Moves Center of Universe to Center of Earth's Circular Orbit. (A) <u>Geocentric model of Ptolemy and Islamic Astronomers</u>. — Center of universe at center of Earth. (B) <u>Heliocentric model of Copernicus</u> — Center of universe close to but not at the Sun. (Other planets, Sun, Moon not shown. Not to Scale.)

Some historians suggest the models may have come to Copernicus when he was in Italy, where Islamic science and philosophy was seriously studied. Still, no one has (yet) identified a direct path from Marāgha school astronomers to Copernicus.[56]

How Copernicus learned of their works, and whether or not he knew their original sources remains a mystery.[57]

The notion of a Sun-centered universe with a moving Earth was received with great skepticism — as is often the case with radical new ideas. Especially those unsupported by empirical evidence (at the time).

Scientific Objections

> *... as for the quietness [stationary nature] of the earth, I need not to spend any time in proving of it, since that opinion is so firmly fixed in most men's heads, that they accompt it mere madness to bring the question in doubt . . . . it is as much folly to travail to prove that which no man denieth . . .*
>
> — Welsh mathematicus Robert Recorde (1556)[58]

The two most common objections of natural philosophers to Copernican heliocentricity were similar to those made by Aristotle and others some nineteen centuries earlier (as discussed in Chapter 6.)

Earth's motion — The age-old question was again raised: If the Earth is moving, why don't we feel it? Copernicus gave his own explanation:

For when a ship is floating calmly along, the sailors see its motion mirrored in everything outside, while on the other hand they suppose that they are stationary, together with everything on board. In the same way, the motion of the earth can unquestionably produce the impression that the entire universe is rotating.[59]

So far so good, but then Copernicus adds:

As a quality, moreover, immobility is deemed nobler and more divine than change and instability, which are therefore better suited to the earth than to the universe.[60]

The latter semi-religious explanation is based on his Aristotelean-based physics of motion, which again is incorrect. Galileo would clear up the issue of a moving Earth a century later with his now famous dictum on uniform motion (as we shall discuss in Chapter 23).

Stellar Parallax — Recall that Aristotle had argued that if the Earth orbited the Sun, stars would appear to change position with respect to each other over the months. No such effect was observed.

As noted earlier and as suggested by Copernicus, the stars are way too far away to detect stellar parallax with the naked eye.

A third issue dealt with a foundational principle of Aristotelean physics:

The Center of the Universe — Aristotle held that "heavy bodies fall to their natural place, the Earth, which is the center of the universe." How else could one explain why all (heavy) things fall to the Earth?[61] Again, the answer would come a century and a half after Copernicus with Isaac Newton's Law of Universal Gravitation (see Chapter 25).

Religious Reaction

. . . and he said in the sight of Israel:
Sun, stand thou still upon Gibeon;
and thou, Moon, in the valley of Aijalon.
And the Sun stood still, and the moon stayed,
Until the nation had avenged themselves of their enemy.

— Joshua 10:12-13

In this Hebrew Bible passage, Joshua commands the Sun and Moon to stand still. This reflects the belief they both normally go around a stationary Earth. Copernicus of course knew of this as well as other Bible passages which indicate a geo-centric (Earth-centered) universe.

Nonetheless, several high-placed Church authorities encouraged Copernicus to publish. They included

(1) Danticus, Prince-Bishop of Warmia;
(2) Copernicus's best friend, Tiedmann Giese, Bishop of Kulm, and
(3) Nicholas Schönberg, Cardinal of Capua.[62]

Still, Copernicus was concerned about religious reaction to his heliocentric theory. So he dedicated *De revolutionibus* to Pope Paul III. And he placed a letter of support from Cardinal Schönberg in the front material of his magnum opus.

A third action by his editor helped the most to stave of potential theological objections. It has been called "a most outrageous act."[63]

Osiander's Preface

Unlike Ptolemy, Copernicus believed his new model of the universe represented physical reality — that the Earth indeed rotates daily and orbits the stationary Sun yearly.

Andreas Osiander (1498–1552) was a Lutheran preacher in Nürnberg. He also worked as an editor for Petreius. He took it upon himself to replace Copernicus's original Preface (following the title page) with an unsigned statement:[64]

> *To the Reader Concerning the Hypotheses of This Work:*
> *There have already been widespread reports about the novel hypotheses of this work, which declares that the earth moves whereas the sun is at rest at the center of the universe. Hence certain scholars, I have no doubt, are deeply offended...*
>
> *[The astronomer] must conceive and devise the causes of these [celestial] motions or hypotheses about them. Since he cannot in any way attain to the true causes, he will adopt whatever suppositions enable the motions to be computed correctly from the principles of geometry for the future as well as the past...*

> *For these hypotheses [of Copernicus] need not be true nor even probable. On the contrary, if they provide a calculus consistent with the observations, that alone is enough . . .*
>
> *So as far as hypotheses are concerned, let no one expect anything certain from astronomy, which cannot furnish it, lest he . . . depart from the study a greater fool than when he entered it. Farewell.*[65]

Was Copernicus aware of this "outrageous act?" Surviving fragments of a "correspondence between Osiander and Copernicus make it clear that Osiander discussed the stratagem . . . with him," Christie tells us, "We don't know how or even if Copernicus reacted to this suggestion."[66]

The three actions discussed above did not wholly prevent religious objections — though criticism of *De revolutionibus* was initially rare.[67] Dominican theologian and mathematician Giovanni Maria Tolosani declared Copernican heliocentricity "contrary to human reason and also opposes holy writ." Protestant leader Philip Melanchthon "rejected heliocentric hypotheses proposed by some ancient and 'recent' authors."[68]

Nonetheless, "in the first seven decades following its publication," Christie writes, "there was almost no [other] rejection of heliocentricity on religious grounds." There was, however, "very little acceptance by astronomers." This was due to perceived issues with the physics of a moving Earth.[69]

As we shall see, in 1616, seventy-three years after its publication, Roman Catholic Church authorities would place *De revolutionibus* on the Index of Forbidden Books — so, as they put it, Copernicus' "opinion may not creep any further to the prejudice of Catholic truth."

Sidebar: Martin Luther and the Protestant Reformation

A Roman Catholic contemporary of Copernicus would mirror his scientific revolution with a religious upheaval that would change Christianity forever. Martin Luther (1483–1546) was born on November 10, 1483 in the German town of Eisleben some 100 miles (170 km) southwest of present-day Berlin. It was then in the Electorate of Saxony, part of the Holy Roman Empire.

As a child, as Luther tells it, he suffered severe physical abuse from both his father and mother (Fig. 19.5). "It was this harshness and severity of the life I led with them," he later wrote, "that forced me subsequently to run away to a monastery and become a monk."[70]

(*Continued*)

Figure 19.5. Martin Luther's Parents, Hans and Margarethe. Portraits by Lucas Cranash the Elder.

The adult Luther was subject to deep melancholy, spiritual conflict, and self-loathing alternated with fits of blind rage. Friends and foes alike found him stubborn, combative, and imperious—as well as a man of "dauntless courage and sweeping eloquence." As his fame grew, he "recognized no superior and tolerated no rival."[71]

On October 31, 1517, the thirty-four-year-old Augustine monk sent a letter to his superiors titled *Disputations on the Power of Indulgences*, aka *Ninety-Five Theses*. His brazen document railed against the corruption and wealth of the Catholic Church. It protested the sale of indulgences — sold to believers by the Catholic Church in exchange for forgiveness of sins.

Of particular concern was a new set of indulgences announced by Pope Leo X to fund construction of St. Peters Basilica in Rome. In Thesis 86, Luther wrote:[72]

> *Why does not the Pope build St. Peter's Minster with his own money — since his riches are now more ample than those of Crassus — rather than with the money of poor Christians?*[73]

Luther sent his *Ninety-Five* Theses to his archbishop, who sent it on to Rome. The Pope was outraged. He ordered Luther to "retract his claims immediately." Luther refused. The Augustine monk was declared a heretic and excommunicated from the Catholic Church. Only the protection of Frederick the Wise, ruler of Saxony, saved him from "trial and likely execution."[74]

In his writings, Martin Luther proved a great theologian and profound thinker. His Pauline canon provided Protestant believers with a set of spiritual tenets without the need for Rome. He argued that the true foundations of Christianity are to be found in the gospels, not the Church; that the Bible is the only source of divine knowledge. And that it can be interpreted by any individual true believer, not just Church authorities.

Luther declared that the grace of God — the "free unearned, and unmerited favor of God" — cannot be won by righteous acts, such as giving money to the Church. It can only be gotten by faith alone. Salvation is granted freely, despite one's sins, through the love and mercy of Jesus Christ.[75]

Martin Luther's incendiary words of 1517 and subsequent theological writings, preaching, and debates set off a religious, social, and political revolution that changed the map of Europe.

The movement Luther founded became known as the Protestant Reformation. German princes saw it as liberation from the papacy and a chance to seize church lands and property. Long-suffering small land owners and peasants sought release from onerous church taxes. Luther's words also awakened "latent national aspirations" in Germany and other regions across the continent.[76]

Dozens of new Christian denominations and thousands of new churches would form — all independent of Papal authority. The Roman Catholic Church's nearly twelve-hundred-year hold on most of Europe would come to an end. Northern Europe would become Protestant, while southern Europe remained Catholic, and France became a split of the two.[77]

Luther's translation of the New Testament from Latin into vernacular German is considered perhaps his greatest literary contribution to the Reformation. "Its linguistic merits were indisputable," wrote twentieth century Roman Catholic historian Henry George Ganns, "... It unfolded the affluence, clarity, and vigour of the German tongue ... with a result that stands almost without a parallel in the history of German literature."[78]

Near the end of his life, Martin Luther suffered from kidney disease, gall stones, gout, vertigo, and other ailments. His days were marked with "increasing irritation, passionate outbreaks, and hounding suspicions." In this agitated state, given license by the gospels, he wrote two vicious anti-Semitic polemics: "On the Jews and their Lies" (1543) and "Of the Unknowable Name and the Generations of Christ" (1543).

Luther called for the destruction of Jewish synagogues and homes, confiscation of their money, and restrictions on their liberty. His writings were later cited by conservative Protestant churches in Germany in support of the NAZI's and the Holocaust.[79]

His last publication was yet another ugly rant against the pope; "Against the Papacy established by the Devil" (1545).[80] Martin Luther died a year later on February 18, 1546. His body is buried in the All Saints' Church, Wittenberg, Saxony, where it is said he first posted his Ninety-Five Theses.

Today some 37 percent of the world's Christian population identifies as Protestant, numbering over 8 million people.[81]

Scientific Considerations

There is need of art and more exacting toil in order to investigate the motion of the stars, to determine their assigned stations, to measure their intervals, to note their properties.

—— John Calvin, Protestant theologian and reformer.[82]

Scholars argue as to the impact of the Protestant Reformation on the Scientific Revolution.* Some say the Reformation was key. Others argue advances in science would have happened without the Reformation.[83] I agree with the latter.

Great discoveries in science were made by both Catholic and Protestant natural philosophers and astronomers during this period. This was made possible in no small part by the movable type print press, invented by Johannes Gutenberg in c. 1440. As noted, it allowed the mass production of pamphlets and books for the first time in human history.[84]

Martin Luther's friends translated his *Ninety-Five Theses* (1517) from Latin to German and made multiple copies of the document via the printing press. "Suddenly everyone was reading it across Europe"[85] His other writings, especially his German translation of the New Testament, enjoyed similar exposure.

Scientific treatises, such as the Copernican *De revolutionibus* (1543) also benefitted from the ability to mass produce pamphlets and books. As we shall see, radical new ideas in science would find their way to scholars and lay people alike — ideas that could not be suppressed by religious authorities.[86]

---

*The so-called "Scientific Revolution" is generally said to have begun with Copernicus' publication of *De revolutionibus* in 1543 and ended with Isaac Newton's publication of the *Principia* in 1687.

A New Vision of the Cosmos

Copernicus not only paved the way to modern astronomy, [he] also helped to bring about a decisive change in man's attitude towards the cosmos . . . the illusions of the central significance of man himself became untenable.

— A. Einstein[87]

Nicholas Copernicus handed the finished manuscript of his masterwork, *De revolutionibus orbium coelestium* to Rheticus in September 1541.[88] The document arrived at Petreius' printing office in Nürnberg in 1542 — followed by Rheticus who would see it through the press."[89]

In December of that year, Copernicus suffered a cerebral hemorrhage. It left him partially paralyzed. The reluctant revolutionary astronomer died some six months later on May 24, 1543 — "having circled the Sun 70 times," as science historian Michael Crowe put it.[90]

The story goes that on the day of his death, a friend reportedly placed the first printed copy of *De revolutionibus* in his hands. Copernicus is said to have "awakened from (his) stroke-induced coma, looked at his book, and died peacefully."[91]

A Hollywood ending for sure. Whether it really happened or not, it still makes me smile.

His Legacy

One might argue that Copernicus's extensive use of the mathematics and models of Islamic astronomers in *De revolutionibus* without acknowledgement was akin to thievery. Yet he did not seek fame. He reluctantly consented to the publishing of his masterpiece at age sixty-seven. Although plagiarism was quite common in this era, this does not seem to me to have been the behavior of a plagiarist.

Whatever his motivation, the heliocentric universe he left humanity was at once conservative and radical. It kept the perfect circles, uniform motion, and epicycles of Ptolemy yet placed the Earth in orbit around the Sun.[92] Copernican physics was new but still founded in Aristotelian thinking. Like the Renaissance itself, Copernicus' work was a bridge between medieval beliefs and our modern understanding of the world.[93]

The so-called Sun-centered solar system of Copernicus stands amongst the greatest scientific discoveries of all time. It gave mankind a new worldview, a new perspective on the universe and our place in it. Like a child who grows to find the world does not revolve around him, mankind in a sense lost its innocence. Such was the Copernican revolution — for the globe we inhabit is no longer the center of the universe, but merely one of a number of planets orbiting the Sun.

Though it made little impact at first, the heliocentric worldview of astronomer Nicholas Copernicus would spark a seminal clash between science and religion.[94] An astronomer obsessed with the accuracy of observations, along with several of his peers, would unwittingly take the next step. This is the subject of the next chapter.

Chapter 20

THE OBSERVER

*Man hath weav'd out a net, and this net throwne
Upon the Heavens, and now they are his owne*

—John Donne[1]

Historically, most advances in science were achieved by individual effort. In this and the next chapter we explore great discoveries borne out of a partnership between two strikingly different men.

One was robust and wealthy; the other weakly and poor. One a meticulous observer of the heavens; the other with the poorest of eye sights. The first unexceptional in mathematics; the second a brilliant mathematician and geometer. What did they have in common? An all-consuming passion for understanding the heavens.

It was most fortuitous that they found each other; two great men who complemented each other's strengths and weaknesses. We begin with the man who has been called "the finest pre-telescopic astronomical observer of all time."[2] (We introduce his fellow collaborator in the next chapter.)

Tycho Brahe

Those who study the stars have God for a teacher.

— Tycho Brahe[3]

Tycho Brahe
(1546–1601)

At first glance, Tycho Brahe seems an unlikely candidate for the role he would play in the scientific revolution. He was flambouyant, brash, quick-tempered, and hard-nosed — literally. A fellow student had lopped off the bridge of his nose in duel — over a disagreement in mathematics, of all things. From then on, he sported replacement noses he fabricated out of brass.[4]

One wouldn't necessarily think this high-born eccentric would possess the patience, seriousness, and disciple required to study and record the stars night after long night. Yet he was a diligent practitioner of the observational arts with scrupulous attention to detail and an unsurpassed dedication and passion for astronomy.[5]

Royal Beginnings

The great observer was born on December 14, 1546 — three years after the death of Copernicus. The blessed event took place at Knudstorp Castle on the Brahe ancestral family estate in the Danish peninsula of Scania (now part of southern Sweden). Baptized as Tyge Ottesen Brahe, he would later "go by the Latinized name Tycho."[6]

Tycho was from a powerful Danish noble family. His father Otto served on the *Riksrådet*, the Council of the Realm and State of King Christian III of Denmark-Norway. His mother Beate was Chief Lady-in-Waiting to Queen Dorothea.[7]

At age two, Tycho, the eldest son, was taken from his parents and raised by his Uncle Jørgen, admiral of the Danish navy and his wife Aunt Inger.[8] They had no children of their own. It appears Tycho's parents did not dispute this action — a not uncommon practice at the time. Perhaps this separation contributed to Tycho's fiery personality.[9]

Denmark had become a Lutheran kingdom under King Christian III in 1536, ten years before Tycho Brahe's birth. Uncle Jørgen and later Tycho became leading patrons of the Philippists — after Philipp Melanchthon, an inseparable friend and devotee of Martin Luther.[10] Philippists encouraged reliance on human reason and the use of astronomy and mathematics to reveal God's mysteries.

Tycho the Astronomer

Tycho became enchanted with astronomy at a young age. His first experience was the total eclipse of August 21, 1560. The 13-year-old was impressed that it had been predicted yet disturbed that the prediction was off by a day. "This was the beginning of his realization that more accurate observational data was needed," Thony Christie tells us.[11]

Tycho's foster father, uncle Jørgen was not happy with Tycho's passion for astronomy. Similar to Copernicus, he pushed him to be a lawyer. While studying for the law at the University of Leipzig in 1562, Tycho began making nightly observations of the sky in secret.[12]

There he recorded a conjunction of Jupiter and Saturn (both planets near each other in the sky) in August. "Neither tables based on Copernicus nor Ptolemy gave the correct date for the conjunction." Ptolemy was off by almost a month, Copernicus by days. With the confidence of youth, the seventeen-year-old Tycho thought he could do better.[13]

Tycho's Uncle Jørgen died in 1565. His biological father Otto passed away six years later. Soon after, Tycho turned to his maternal uncle, Steen Bille for financial support for his dream. Uncle Steen consented to fund the construction of an observatory and laboratory on his estate at Herrevad

Abbey, a former Roman Catholic monastery complex. Here Brahe made his first great discovery.[14]

A New Star

> *(Stella Nova) is a wonderful prodigy of God...*
> *now shown to the eventide world.*
>
> — Tycho Brahe[15]

On November 11, 1572, twenty-six-year-old Tycho was among those who observed a new star in the constellation Cassiopeia. It was christened *Stella Nova*. We now know it was a type 1a supernova explosion of a previously existing star. Some 13,000 light-years* from Earth, it was too far away to be seen by the naked eye before it exploded.[16]

Stella Nova is now known as "Tycho's supernova." A remnant can be seen by telescopes to this day. (See Fig. 20.1.)[17]

At the time, *Stella Nova* was brighter than Venus. It was visible in daylight for a few weeks and seen in the night sky for some 15 months. Tycho Brahe took meticulous measurements of the new star during the

Figure 20.1. **Remnant of *Stella Nova* as Seen Today — Tycho's Supernova SN 1572.** The remnant is the red circle (see arrow). Star forming nebula S175 is in the lower center right. Image by NASA's Wide-field Infrared Survey Explorer (WISE).[18]

---

*A light-year is a measure of *distance*. It is how far light travels in a year. One light-year is about 6 trillion miles or about 9.5 trillion km.

entire period. A number of astronomers had measured the location of the new star. Tycho Brahe's observations were likely the most accurate.[19]

Brahe was among several astronomers who detected *no parallax* in observations of *Stella Nova*. This is diurnal parallax: the apparent shift in a stellar object's position due to the daily rotation of the Earth — or conversely, rotation of everything else around a stationary Earth. No observed parallax indicated that the new star must be "far above the sphere of the Moon* in the very heavens," as Brahe put it.[20]

This challenged Aristotelian and Medieval Church dogma that the heavens are permanent and unchanged beyond the orbit of the Moon. Brahe reported his findings in a small book of 1573, *De Nova Stella* (*On the New Star*). With the aid of the printing press, it made him famous.[21]

It was time for Brahe to look for a more advanced facility. It would be the observatory of his dreams.

The Search

Nobleman and astronomer Wilhelm IV had built the first permanent European observatory in his castle in 1560-61. With custom-built instruments, the Count of Hesse-Kassel in what is now northern Germany had achieved the best star position accuracies known to date.[22]

Searching for possible sites for an observatory on his own land, Tycho paid Wilhelm IV a visit in 1575. "Wilhelm was so impressed" by the twenty-nine-year-old, "that he sent a message to now Danish king Frederick II, son of Christian III urging him to support this genius."[23]

After checking him out, Frederick II gave the Danish astronomer the nearly 2000-acre island of Hven in Copenhagen Sound (Øresund) — along with lavish funds to construct a state-of-the-art observatory there. Built from c. 1576 to c. 1580, it would be dubbed a "celestial palace."[24]

---

*Brahe's state-of-the-art instruments could "measure parallax down to 2 arc minutes (1/30 degree)." Their inability to detect any parallax implied that Stella Nova was at least "700 times further away from Earth than Saturn"—the furthest planet detectable by the naked eye. This computes to a distance of at least some 500 billion miles (800 billion km) from Earth. The Moon is only some 240 thousand miles (385 thousand km) from Earth. Source: "Tycho Brahe's Observations and Instruments." hao.ucar.edu.

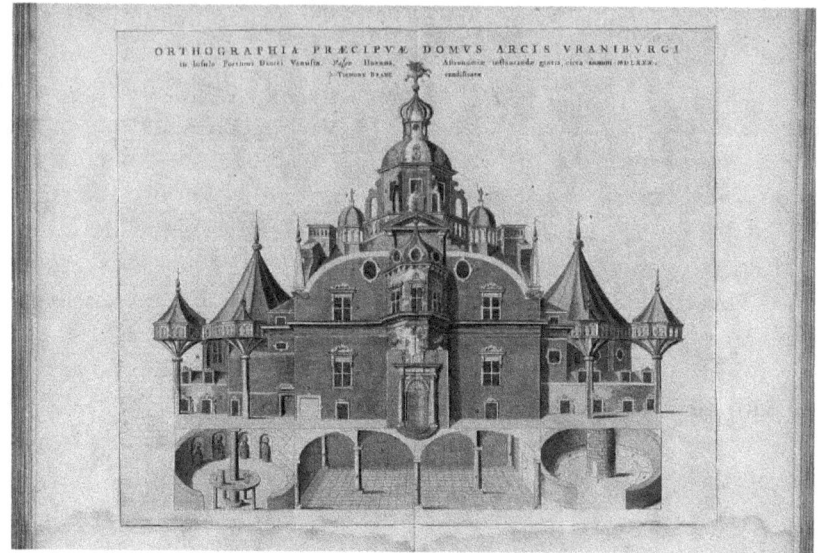

Figure 20.2. Uraniborg — Tycho Brahe's "Celestial Palace." From Blaeu's Atlas Maior (1663). Main building shown here. Its "conical roofs mark Uraniborg's four observing rooms." On the balconies of these outer towers were "a large array of self-designed and constructed instruments." The complex also contained a library, an alchemical laboratory, facilities for designing and building instruments, as well as paper mills and a printing press to publish his findings.[26]

Uraniborg

On the island of Hven, Tycho Brahe designed and built the most advanced astronomical observatory in human history to that date. He named his remarkable building complex and extensive gardens "Uraniborg" — after the Greek Muse of astronomy, Urania. (See Fig. 20.2.)[25]

For the next twenty years, Brahe conducted a series of observations of stars and planets at Uraniborg which were unprecedented in scope and accuracy. Ever methodical, Tycho erected a smaller observatory called Stjerneborg on the island to verify findings through simultaneous observations.[27]

Along with a "small army of servants and assistants, over a hundred students and astronomers" worked and studied at Uraniborg. Together with Brahe, they filled book after book with ever more precise observational data — producing a library of unparalleled astronomical information.[28]

The isolated island was infamous for its wild and lavish parties. Led by the heavy drinking Tycho Brahe, they would become legendary.[29]

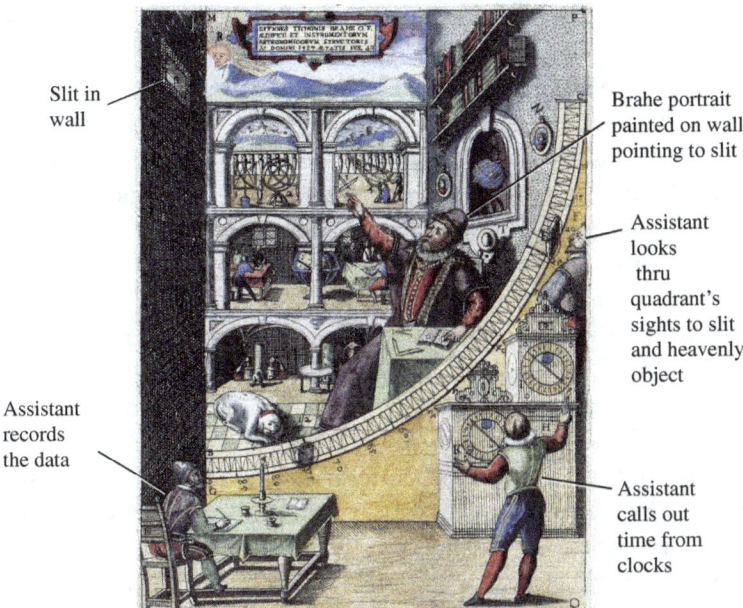

Figure 20.3. Tycho Brahe's 19-foot (5.8 meter) Mural Quadrant. Mounted on a wall for stability and oriented to lie on the meridian, it measured angles of celestial objects from 0 to 90 degrees. A portrait of Brahe (painted onto wall) points to a "slit in the wall through which heavenly bodies were viewed." An observer looks through the quadrant's sights as one assistant calls out the time from clocks and another records the data."[30]

The key to Uraniborg's observational accuracy was its exquisite astronomical instruments. Conceived by Brahe and based on astronomer Wilhelm IV's custom-built instruments, they were masterful in design, manufacture, and implementation.

Brahe's Instruments

Tycho's measuring apparatus included a number of custom-made quadrants, sextants, equatorial armillary, and other precision instruments. Most impressive was a nineteen-foot (5.8 meter) high mural quadrant attached to a north-south wall. "One of the largest astronomical instruments of its time," it measured the altitude of a star or planet "as it passed through the meridian." (See Fig. 20.3 above.)[31]

For a more detailed description of Brahe's instruments, please see endnote.[32]

The Danish astronomer described his instruments and their installation at Uraniborg in his 1598 book *Astronomiae Instauratae Mechanica* (*Instruments for the Restoration of Astronomy*). The volume included Brahe's observational methodology and "detailed specifications of his instruments complete with illustrations." This work would set the standards for future astronomical instrumentation and observation.[33]

Brahe bragged that the King "had spent a ton of gold supporting his observatory." Estimates put it at some one percent of Denmark's national treasury.[34]

The great Danish observer continued to conduct detailed observations of stars and planets for the next two decades. He completed tens of thousands of observations and catalogued over a thousand stars — with "an accuracy several factors higher" than previous astronomers.

Unlike prior observers, Brahe "observed the planets throughout their entire orbits."[35] This detailed data on the motion of planets would facilitate a new paradigm for planetary orbits, as we shall see in the next chapter.[36]

While Uraniborg was still under construction, Brahe and others made another discovery which challenged Aristotle and Church dogma — the observation of a "fiery messenger."[37]

The Comet of 1577

Recall Aristotle had taught that comets were confined to below the moon — the so-called "sublunary sphere" which contained the changing, impure abode of man.[38] The comet of interest made its appearance in November 1577 and departed in January of the following year. With his usual diligence, Brahe tracked it for the entire period.

Observations by several astronomers including Brahe revealed no measurable parallax. Thus, Brahe wrote, the comet was at least "six times farther from Earth than the moon." Like *Stella Nova*, it was in the supposedly unchanging realm of the heavens — yet another challenge to Aristotle and Christian dogma.[39]

Astronomers saw an additional conflict with Aristotle's cosmic model. By all indications, the comet had *shot through* the celestial realm. Do you see the issue?

Such a comet would crash into at least one of Aristotle's celestial spheres — the 56 invisible concentric spheres that allegedly surround the Earth and give motion to the attached Sun, Moon, and planets.

Recall Aristotle claimed his crystalline spheres were physically real. Observations of the comet contradicted this notion. This verified a critical new insight: There are no crystalline spheres. The Sun, Moon, and planets float freely in the heavens.[40]

The observation of a comet well beyond the Earthy realm threw confirmed Aristotelians and "Ptolemaic astronomers into a dither," writes Michael Crowe. Beginning with the 1570's, more and more young astronomers began to accept the cosmology of Copernicus.[41]

The heliocentric universe of Copernicus had a competitor beyond the geocentricity of Aristotle/ Ptolemy — a strange compromise between the two advocated by the great observer himself, Tycho Brahe.

Brahe's Geo-heliocentric Model

I will endeavor to adapt my restorations [of a stationary Earth] in the course of all the planets to my own hypothesis, not one already invented . . . God willing.

— Tycho Brahe[42]

Tycho Brahe criticized Ptolemy's model for its use of "the equant and lack of elegance." In particular, it failed to explain why planets exhibit retrograde motion. The Danish astronomer also found a number of issues, both scientific and religious, with "that newly introduced innovation of the great Copernicus," as he put it.[43]

Brahe's Scientific Objections to Copernicanism

Brahe commended Copernicus for his model's "conceptual simplicity" — and for avoiding the "mathematical absurdity of the equant." Apparently, he did not realize that Copernicus's mathematics came from Islamic Marāgha School astronomers.[44]

Brahe objected to Copernicus' heliocentric system regarding the usual issues:

Moving Earth — Steeped in Aristotelean physics, Brahe saw Earth as a "lazy, sluggish body . . . unfit for motion." How could such a ponderous Earth move at such great speed around the Sun. (Some 67,000 miles per hour or about 110,000 kilometers per hour in today's measurements.)[45]

As to diurnal rotation of the Earth, Brahe wavered. He could see the advantages of a spinning Earth rather than the daily rotation of the entire universe. Nonetheless, he ended up arguing for a non-spinning Earth.[46]

> *It is likely nonetheless that such a fast motion [the daily rotation of Earth] could not belong to Earth, a body very heavy and dense and opaque, but rather belongs to the sky itself whose form and subtle and constant matter are better suited to perpetual motion, however fast.*
>
> — Tycho Brahe

Recall Aristotelean physics held that all heavenly bodies, i.e., the Sun, Moon, planets, and stars, are made of a fifth substance: aether. They are not subject to the "heaviness" properties of the elements earth and water which, along with fire and air, were said to make up our planet.

Other objections to a moving Earth included the familiar notion that if one dropped a lead ball from a tower, it would land beyond the bottom of the tower due to Earth's motion. Based on the (incorrect) understanding of the time, Brahe declared that a moving Earth "violated physical principles."[47]

No Parallax — The Danish astronomer also pointed out that, if the Earth were truly orbiting the Sun, one should detect a change in the positions of stars as viewed from a moving Earth — an argument that went all the way back to Aristotle.

As most everyone before him, Brahe had grossly underestimated the size of the universe. At the great distances to stars, stellar parallax is way too small an effect to be detected by Tycho's naked eye instruments. As noted in Chapter 6, our nearest star, Proxima Centauri. is some 25 trillion miles or 40 trillion kilometers away.[48]

The huge distances to the stars led to another issue: their apparent size.

Enormous Star Sizes — Brahe was struck by the sheer sizes of the far away stars, as seen in his naked eye observations. Copernicus' moving Earth model and no measurable parallax implied that they were at least billions of miles away. Thus, if Copernicus was right, their actual sizes had to be of astounding proportions — a great many times that of the Sun.[49]

What Brahe did not know was whenever you put light through an opening — like the iris of your eye — the light spreads out. This phenomenon is now known as diffraction.* Turbulence from the atmosphere (due to air temperature differences) and eye defects (none are perfect) also tend to spread out light from a star. All these effects make stars look bigger than they are. Without these effects, stars would appear point-like to us.[50]

Brahe's Religious Objections

The Danish astronomer held the conviction that "many truths of physics are contained in holy scripture." Thus, he argued, a moving Earth is "opposed to . . . the authority of Holy Writ."[51,52]

A second religious objection involved the implications of Copernicus' model and no observed stellar parallax. If the stars are truly such tremendous distances beyond the furthest planet, there would be a "vast and useless vacant space between Saturn and the starry vault." It seemed absurd to Brahe that a purposeful God should construct such a universe.[53]

It is tempting to scoff at Brahe's objections to a moving Earth, both scientific and religious. Let us not be too smug. Whether correct or not, we are all influenced by the thinking of our time.

In the mid 1580's, Tycho Brahe developed his own model of the solar system — a "perfect compromise" between Ptolemy's Earth-centered model and the Sun-centered system of Copernicus.

The Tychonic System

In the so-called Tychonic system, the Earth is *at rest*. The Moon and Sun orbit this stationary Earth. What about the other five planets? They orbit the Sun. (See Fig. 20.4.)[54]

Brahe's model is geocentric since the Sun and Moon orbit Earth — which is at the center of the universe. It is a heliocentric in that the other planets go around the Sun.[55]

It had two major advantages:

---

*Diffraction can be attributed to the wave nature of light. It increases as the size of the opening or aperture gets *smaller*. We also see diffraction effects with sound waves and water waves. The phenomenon of light diffraction was not understood until the 19th century. Source: Haqq-Misra, bostonglobe.com.

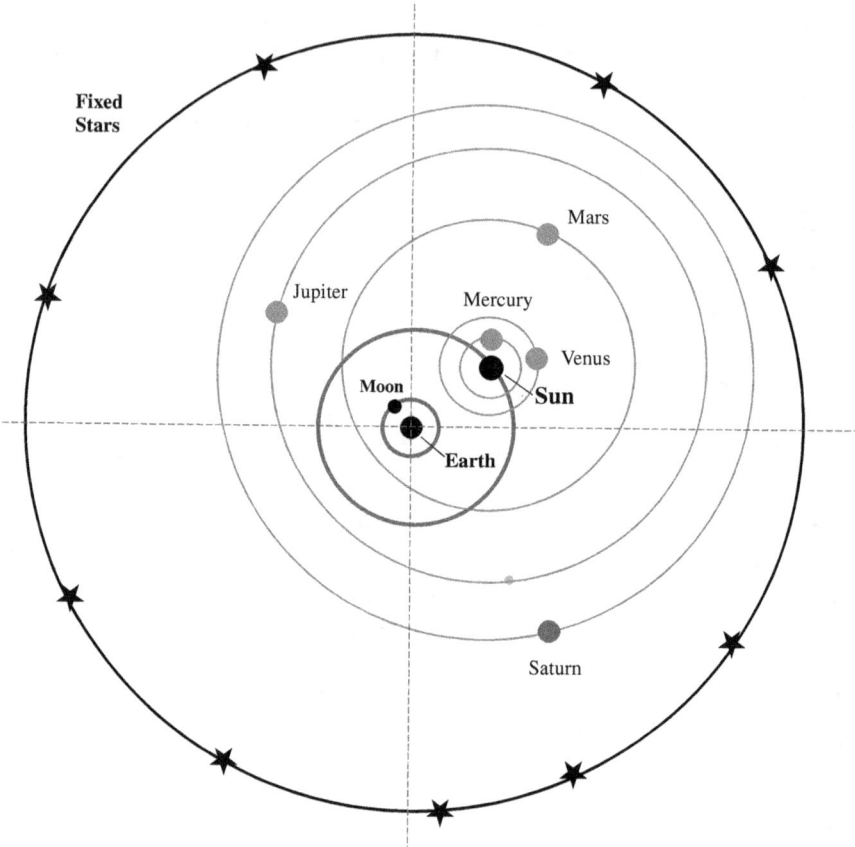

Figure 20.4. The Tychonic System — Tycho's Brahe's Model of the Universe. (1) In black: Earth at rest at the center of the universe. The Moon, Sun, and fixed stars circle the Earth. (2) In grey: Mercury, Venus, Mars, Jupiter, and Saturn circle the Sun. Not to scale.

(1) Like Aristotle and Ptolemy's models, it kept the Earth stationary, which satisfied physical and religious objections; and
(2) Like Copernicus' model, it had Venus and Mercury orbited the Sun. This explained why the two planets always appear close to the Sun in the sky, the so-called "bounded elongation."

Tycho saw his construct as real. "Nearly all astronomers following Copernicus now believed in the reality of their models," Thony Christie tells us.[56]

Brahe published an outline of his Tychonic model in 1588 in *De mundi aitherei recentioribus phaenomenis* (*Recent Phenomena in the Celestial World*) — along with his work on the Comet of 1577.[57] In *De mundi*, Brahe said he would flesh out the details of his model in a future work, including epicycles, etc. He fervently hoped his model would eventually win the day — and he would be remembered as the greatest astronomer ever.[58]

Tycho was not the first to publish a geo-helio model. German astronomer Nicolaus Reimers Baer (1551–1600) aka Ursus published his version earlier the same year. Except for diurnal rotation, it was identical to Brahe's construct.* Furious that Ursus had published "his system" before he did, Tycho "accused him of plagiarism . . . citing an earlier visit to Hven" by Ursus.

Due to his greater fame, Brahe is typically credited with the invention.[59]

The Danish astronomer's compromise model had great appeal in the "last decade of the sixteenth century and early decades of the seventeenth century." As we shall see, the Roman Catholic Church came to adopt the Tychonic system and its stationary Earth as its "official astronomical conception of the universe."[60]

Legacy

Tycho Brahe set a new standard for naked eye observation astronomy. Previous observations were typically accurate to "perhaps 15 arc minutes," historian of astronomy Albert Van Helden tells us. Brahe generally achieved accuracies of "perhaps 2 arc minutes." His finest observations were accurate "to about half an arc minute."[61]

In one example, the Danish astronomer determined the duration of an earth year to a "difference of less than one second."[62] The accuracy of Brahe' measurements "would not be surpassed for almost three centuries."[63] In addition, his remarkable measurement devices "revolutionized astronomical instrumentation." He established the astronomical standard for "physical measurement and data analysis." He pioneered methods for "error percentages and ranges" for the calculation of measurement precision. He

---

*Recall Martianus Capella (360–428 AD) had also hypothesized that Mercury and Venus *orbit the Sun* rather than the Earth. Unlike Brahe and Ursus, in Capella's model they are the only planets which orbit the Sun. "The Capellan system was well known in Europe during the Middle Ages," Source: Christie, "https://thonyc.wordpress.com/2020/01/08/the-emergence-of-modern-astronomy-a-complex-mosaic-part-xxvii/" The emergence of modern astronomy – a complex mosaic: Part XXXI," thonyc.wordpress.com.

was also the first to correct for stars near the horizon appearing to have greater altitude due to atmospheric refraction.[64]

Brahe's Tychonic system eliminated Aristotle's crystalline spheres said to carry the Sun, Moon, planets, and stars. Although his planetary model would later come to be discredited, his "astronomical observations were an essential contribution to the scientific revolution."[65]

A New Paradigm for Planetary Orbits

What did the ancient Greeks Aristotle, Hipparchus, and Ptolemy; Islamic astronomers al-Ṭūsī, al-'Urdi, al-Shirazi, and Al-Shatir; and European astronomers Copernicus and Brahe all have in common? They held Plato's *circular* planetary orbits or combinations thereof as sacrosanct. It would take a new astronomer working with Tycho Brahe's remarkable observational data to break with this 2000-year-old paradigm. He would provide the next revolutionary step in separating science from religion.

This is the subject of the next chapter.

Chapter 21

THE VISIONARY

"The Laws of Nature are but the mathematical thoughts of God."

Johannes Kepler
(1571–1630)

After a falling out with the new king of Denmark, Christian IV over a lack of funding, Tycho Brahe moved his family and instruments to Prague in 1599. The Protestant astronomer erected a new observatory in a castle in Benátky nad Jizerou, a town about 30 miles (50 kilometers) northeast of the capital city — under the patronage of Holy Roman Emperor Rudolf II, a Catholic. Here he continued his precision observations of the heavens.[1]

Brahe was now Imperial Mathematicus and Astrologer to the emperor. Ironically, mathematics was not his strong suite. To help with the subject, he took on an assistant — a small, frail, nearsighted man "plagued by fevers and stomach ailments" named Johannes Kepler.[2]

The Pious Geometer

Johannes was born in 1571 in the German town of Weil der Stadt on the edge of the Black Forest in Baden-Württemberg, then part of the Holy Roman Empire of German States.[3] His early years were "beset by smallpox, headaches, boils, rashes, worms, piles, and the mange," writes Michael Crowe. He had "double vision in one eye and myopia in both eyes." What he lacked in physical advantages he more than made up for with high intelligence, dogged persistence, and fierce independence of thought.[4]

Family troubles haunted his early life. In 1574, when Johannes was three years old, his father Heinrich disgraced his Lutheran family by signing up with a group of mercenaries to fight the Protestant uprising in Holland. Heinrich returned in 1576, only to join Belgium's military. He abandoned his family forever in 1588.[5]

Johannes later described his father as "criminally inclined and quarrelsome;" his mother Katharina as "thin, garrulous, and bad-tempered."[6]

At age six, Johannes "was taken by his mother to a high place," he later recalled, to look at the great Comet of 1577.[7] What he did not know was that his future mentor, Tyco Brahe, was observing the same event from Uraniburg.[8]

Despite being from a poor family, Kepler's exceptional intelligence earned him a college scholarship to nearby University of Tübingen, "the leading Lutheran center of higher learning."[9] Here he was taught the standard Earth-centered model of Ptolemy. His renowned astronomy professor, Michael Mästlin also taught Kepler, one of his select students, the new Sun-centered cosmology of Copernicus.[10]

The Tübingen senate said Kepler was of "such a superior and magnificent mind that something special may be expected of him."[11] Still, he was denied a university position upon graduation. The profoundly religious Kepler's thinking was deemed too close to Calvinism for the Lutherans of Tübingen.

Though he had studied to be a Lutheran minister, the "first [job] to offer itself was an astronomical position," Kepler later wrote. "I was driven to take on the task by the authority of my teachers."[12]

Kepler came to believe his becoming an astronomer was part of a divine plan. He saw his work as "fulfillment of his Christian duty to understand the works of God," writes historian J. V. Field. The young astronomer felt "that God made the Universe according to a mathematical plan, and it was his duty to uncover this cosmic plan."[13]

Kepler favored the Sun-centered cosmology of Copernicus. He saw the Sun as the Holy Father, the stellar sphere that held the stars as the Son, and the intervening space as the Holy Ghost.[14]

Graz and Mysterium Cosmographicum (1594–1600)

> *I wanted to become a theologian . . . Now, however, behold how through my effort God is being celebrated in astronomy.*
>
> — Johannes Kepler in letter to Michael Mästlin, 1595.[15]

In 1594, Kepler moved to Graz in Styria in southern Austria, a deeply Catholic area. He went there to take up his new position teaching at the Lutheran high school. "He was also appointed district mathematicus."

Two years later, the twenty-five-year-old wrote his first major astronomical work, *Mysterium Cosmographicum* (*Cosmic Mystery*). His former professor, Michael Mästlin, called it "wonderfully imaginative and learned."[16] In *Mysterium*, Kepler made the first published defense of Copernicus's Sun-centered solar system. "I myself agree with Copernicus and allow that the earth is one of the planets," Kepler wrote.[17]

In the Sun-centered system of Copernicus, there were six planets — the five visible in the sky plus Earth. "Why six?" Kepler asked. Surely the number of planets and their distances to the Sun were part of God's plan.

Recall from Chapter 5 that Plato proposed all matter is composed of five different elements — earth, air, fire, water, and the cosmic aether.

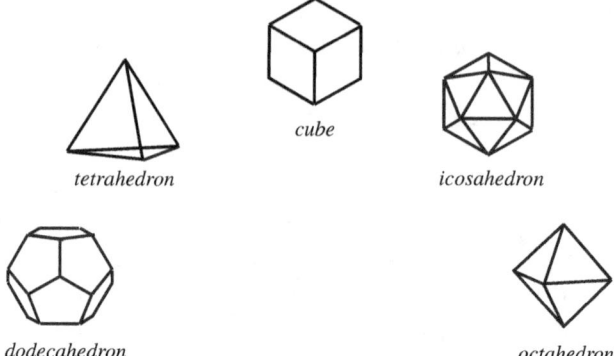

Figure 21.1. The Five Platonic Solids. Kepler would use the five solids (regular polygons) centered on the Sun along with six spheres holding the planets to model the Solar System. The five solids would fill the gaps between the spheres.

They in turn are made of different geometric atoms: cubes, octahedrons, tetrahedrons, icosahedrons, and dodecahedrons, respectively. (See Fig. 21.1.)

This geometric atom theory remained the prevailing belief at the time of Kepler. The young astronomer hoped these unique regular convex "Platonic Solids" — the fundamental atoms of the cosmos — would also reveal God's plan for our solar system.[18]

Starting with the orbit of the Earth, he constructed a six-planet solar system model. Six spheres held the six planets. They were nested with the five Platonic solids — all this centered on the Sun.[19]

To understanding Kepler's approach, please consider a simplified nesting example in 2-D. (See Fig. 21.2.) Here we start with

(1) a *small circle*;
(2) a *triangle* circumscribes (just surrounds) the small circle;
(3) a *larger circle* circumscribes the triangle;
(4) a *hexagon* circumscribes the larger circle: and
(5) an *even larger circle* circumscribes the hexagon.

Do you see it? Given the size of the initial small circle, the size of the triangle, hexagon, and two larger circles are fixed. We could also start with the size of the outmost circle and work our way inwards. Either way, *unique* sizes follow in the nesting of these geometric figures.

For his Solar System model, Kepler conducted a similar nesting in 3-D. He placed the five Platonic solids in a certain order all centered on the Sun. He

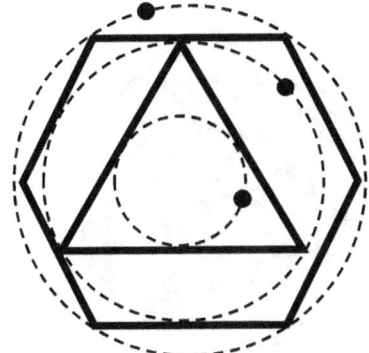

Figure 21.2. Simplified Nesting Example in 2-D. Two shapes: triangle and hexagon. Dashed lines: three inscribed/circumscribed circles. The circles represent planetary orbits. The dots represent planets. Not to scale.

then inscribed and circumscribed six spheres, each containing one of the six planets on their respective surfaces — Mercury, Venus, Earth, Mars, Jupiter, and Saturn.[20] (Recall that in Aristotle's model, each planet is attached to a rotating celestial sphere.) Thus their orbits are "separated by the ratios of the volumes of the five Platonic solids," as Thony Christie points out.[21]

In *Mysterium*, Kepler drew two 3-D pictures of his model. The outer planets are shown in Figure 21.3. The Earth and inner planets are shown in Figure 21.4. I have added labels for clarity.

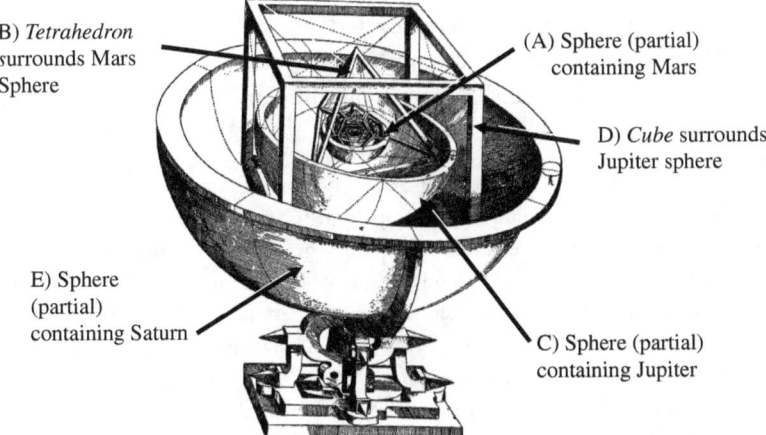

Figure 21.3. Kepler's 3-D Model — the Outer Planets. (A) Mars sphere; (B) Tetrahedron; (C) Jupiter sphere; (D) Cube; (E) Saturn sphere.

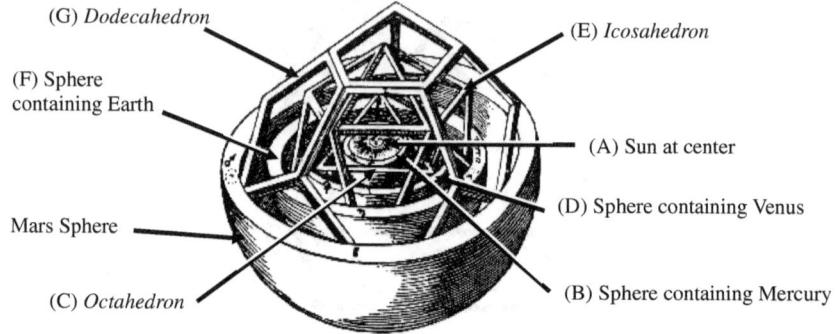

Figure 21.4. Kepler's 3-D Model — Earth and the Inner Planets (Blown up). (A) Sun at center; (B) Mercury Sphere; (C) Octahedron; (D) Venus sphere; (E) Icosahedron; (F) Earth sphere; (G) Dodecahedron. (Mars sphere repeated for reference.)

Geometry is one and eternal shining in the mind of God

— Johannes Kepler.[22]

In Kepler's 3-D nesting model, the sizes of the spheres *roughly* matched the *relative* sizes of the six planetary orbits — as "given by observations of the day."[23] In fact, most of his model agreed with Brahe's observations to within 5%. The one exception was Jupiter. In defense of this problem, the German astronomer wrote, "no one will wonder, considering such a great distance." Nice try, but Kepler's convoluted geometric construct was wrong.*

Here we see a noble attempt by Kepler to find the *cause* for the number of planets and their distances from the Sun. Astronomer and science

---

*We of course now know there are eight planets in our solar system (sorry Pluto). There is no particular reason for that number nor any pattern for their distances from the Sun. It is just how our Solar System happened to form — in the gravitational compressions and chaotic collisions from a spinning flattened cloud of interstellar gas and dust. Exoplanets, planets which orbit stars other than our Sun, show a wide variety of orbital shapes and distances to their respective stars. Source: "Our Solar System" nasa.gov. Dec. 19, 2019. https://solarsystem.nasa.gov/solar-system/our-solar-system/in-depth/#:~:text=Our%20solar%20system%20formed%20about,spinning%2C%20swirling%20disk%20of%20material. Retrieved Aug 7, 2020.

historian Owen Gingerich points out that this was a striking change from medieval Aristotelian thinking. The latter "considered the 'naturalness' of the universe sufficient reason."[24]

Solar Influence — Again, in the Copernican model, the *further away* a planet is from the Sun, the *longer it takes* to complete its orbit. Kepler derived a formula for this relationship which was a reasonable approximation at the time.[25] He saw this as evidence that the *Sun influences the motion of planets* — yet another break with the medieval mindset.

The Assistant

In 1597, Kepler sent a copy of *Mysterium* to Tycho Brahe "asking his opinion of my little book." Brahe replied and mentioned his own observations. This "ignited in me [Kepler] an overwhelming desire to see them," Kepler later wrote.[26] Why this deep interest in Brahe's more accurate planetary data sets? Perhaps, Kepler thought, they would enable him to fine-tune his Platonic Solids model.[27]

Impressed by Kepler's astronomical knowledge and mathematical skill, Tycho invited the budding astronomer to visit. The Roman Catholic Counter Reformation would soon trigger Kepler's response.[28]

In late September of 1598, "Archduke Ferdinand, who had become ruler of Styria in 1596, expelled all Protestant teachers and pastors from the province," Christie tells us. "Kepler was initially granted an exception because he had proved his worth as district mathematicus."[29]

Conditions remained tense. In January of 1600, Kepler decided to test the waters with Brahe.[30] He arrived at Benátky Castle in February for a trial period.[31]

Working for Brahe was not easy. Known for his hot temper and overbearing manner, Tycho treated Kepler like a beginner. He gave him "little opportunity to participate except at meals." The two great astronomers soon had a falling out. After some three months in Bohemia, Kepler returned to Graz.[32]

In August of 1600, the Catholic city now ordered the expulsion of all Protestants and Jews. Kepler refused to convert and was banished. "Depressed and in poor health," he returned to Prague with his family. Tycho's chief assistant, Christian Longomontanus, had just resigned. The great observer "gladly took Kepler back."[33]

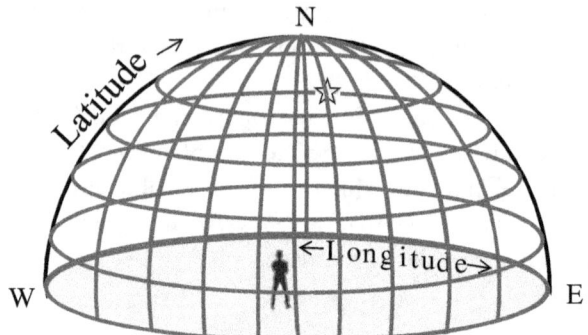

Figure 21.5. Longitude and Latitude. Longitude is the East-West location or azimuth of a stellar object. Latitude is its elevation.

The Martian Catastrophe

Knowing how bright Kepler was, Brahe feared his young assistant would upstage him with a new theory of the solar system. So he restricted Kepler's access to a small part of his voluminous data base. Brahe gave Kepler a particularly tough assignment: understanding the orbit of the planet Mars.[34]

Astronomers use *longitude* and *latitude* to record the location of a planet (or star) in the sky. Put simply, longitude is the planet's left-right or East-West angular location. Its latitude is its up-down or North-South location. Both are referenced to the ecliptic.* (See Fig. 21.5 above.) Prior astronomers had used different mathematics to predict longitude and to predict latitude. Not Kepler. He demanded a single physically acceptable model for both.

Back in August of 1593, Tycho had examined Mars in *opposition*. That is when it and the Sun were in opposite sides of the sky as viewed from Earth. (See Fig. 21.6.) To Tycho's great surprise, his observations of the celestial longitude of Mars showed a **4 to 5-degree** error with predictions — well above his previously recorded maximum error of 2 degrees.[35]

Kepler tackled this so-called "Martian catastrophe" and soon found the problem. It was not with Mars, but with Earth. Recall that Copernicus had designated the center of the solar system as the center of Earth's circular orbit. He also had the Earth travel in uniform motion around the Sun, unlike all the other planets. "Why should Earth be different?" Kepler asked.[36]

---

*More formally, longitude is "the position of a planet on the ecliptic, measured from the spring equinox point. Its latitude is its angular distance north or south of the ecliptic. The ecliptic is the projection of the Earth's orbit onto the celestial sphere." Source: Donahue, p. 106.

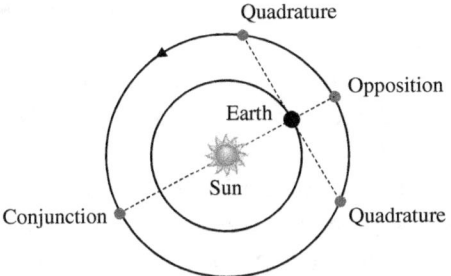

Figure 21.6. Opposition, Conjunction, and Quadrature. The four grey discs represent Mars in various orbital positions. Mars is in opposition when, as seen from Earth, it is in the opposite part of the sky from the Sun. Conjunction is when Mars is the same place as the Sun in the sky, as seen from Earth.

The young German astronomer placed the center of the solar system at the *Sun's true location*. Then he changed the eccentricity (offset) in Earth's circular orbit so it too traveled in non-uniform motion.[37]

The results were astonishing. As noted, Tycho's model predicted a longitude location of Mars which disagreed with his observations by 4 to 5 degrees. Kepler's model with a true Sun center and non-uniform motion Earth reduced the error to **half a degree**.[38]

Still Kepler was not satisfied. In October of 1600, he took on a much more subtle problem — the orbit of Mars itself. He tried the standard circular orbit for Mars with an *arbitrary location* for the offset point or equant — but still along a line joining the Sun and the center of its orbit. By Spring of 1601, Kepler's predicted longitudes for Mars were in much greater agreement with Brahe's observations. But Mars latitudes were still way off.[39]

Cosmic Reformation

Tycho possess the best observations . . . He only lacks the architect who would put this all to use.

— Johannes Kepler[40]

Tycho Brahe died under "unclear circumstances" on October 24, 1601. He was fifty-four years old. Kepler had worked with the great Danish observer for a mere eleven months.[41] Two days after Brahe's passing, Kepler was named

Brahe's replacement as Imperial Mathematicus to Emperor Rudolph II — at one-sixth Brahe's salary.[42]

Kepler's primary duty was to provide astrological advice to the emperor. Kepler considered conventional astrology "evil-smelling dung." Still, he felt there was an "occasional . . . pearl or gold nugget" in the works of conscientious scientific astrologers (like himself).[43]

On his deathbed, Brahe had asked Kepler to collect his two decades worth of astronomical data and publish the *Rudolphine Planetary Tables* — named in honor of Rudolph II.[44] The great observer added his hope that Kepler would frame the data according to his Tychonic solar system hypothesis.[45]

Now with access to all of Brahe's remarkable data, Kepler compared the astronomical models of Ptolemy, Copernicus, and Brahe. He found all three models *made identical predictions*, as long as one used the same observational data as input. "The three opinions (models) are for all practical purposes equivalent to a hair's breadth, and produce the same results," Kepler later wrote.[46]

Kepler then returned to the Mars data that Brahe had so meticulously collected. He continued to iterate on his circular orbit with arbitrary equant location. After some 70 arduous trials, he was able to find a Mars model which gave longitudes to an accuracy of 2 arc-minutes. But predicted latitudes gave erroneous distances.[47]

By Spring of 1602, Kepler was still in trouble. To obtain correct distances, he repositioned the orbit of Mars so its center would be midway between the equant and the Sun — a so-called "bisected eccentricity." The error in longitude jumped up to 8 arc-minutes. Rather than despair, Kepler saw this as Divine Providence.

> . . . *it is only right that we should accept God's gift with a grateful mind . . .*
> *Because these 8 (arc-minutes) could not be ignored, they have led*
> *to a total reformation of astronomy.*
>
> — Johannes Kepler[48]

He turned from mathematics to the physical side of the problem. Kepler updated his speculation that a rotating, magnetic Sun is the cause of planetary motion. (More on this later.) He developed his so-called distance law, where a planet's orbital velocity is inversely proportional to its distance to the Sun.[49]

From this he came up with his famous time-area law — a radius line joining a planet to the Sun sweeps out equal areas in equal time intervals. (More on this later as well.) This worked for Earth, but not for Mars. The 8-minute discrepancy was still there.

Upon further examination, he saw something earlier, less accurate observations had not revealed. Using an ingenious triangulation method on Brahe's data, he determined three locations in the orbit of Mars. This confirmed Kepler's hunch — the orbit of Mars was close to but not *a perfect circle*. It "bowed in from a circle." It was difficult to determine the exact amount.[50]

> *My first error was to suppose that the path of the planet is a perfect circle . . .*
> *a thief of time . . . endowed with the authority of all philosophers . . .*
> *It appears, then, from the most reliable observations, that the course*
> *of a planet through the aethereal air is not a circle, but an oval figure . . .*
>
> — Johannes Kepler[51]

Kepler knew that Mercury's epicycle produced an ovoid curve. So he added a small epicycle to the orbit of Mars. Thus led, as he put it, "to a labyrinth of calculation." Results now showed a difference of 4 arc-minutes between the distance law and time-area longitudes. He was still way off.[52]

"If you are wearied by this tedious procedure," he later implored his readers, "take pity on me who carried out at least seventy trials . . . I was almost driven to madness in considering and calculating the matter."[53]

Interlude

From September of 1602 to the end of 1603, Kepler took a break from the Mars problem. He worked primarily on his book, *Astronomiae pars optica* (The Optical Part of Astronomy). It has since been recognized as a masterpiece in the science of optics. "Kepler was my principal teacher in optics," the great French philosopher, scientist, and mathematician René Descartes later wrote, ". . . he knew more about this subject than all who preceded him."[54]

It seems this interlude served to help clear Kepler's mind. When he returned to astronomy in 1604, he began to write down his thoughts in what was to become his greatest work: *Astronomia Nova (New Astronomy)* or

Commentarius de stella martis (*Commentary on Mars*). "Its subtitle emphasized its repeated theme: *Aitiologitos, seu physica coelestis* or 'Based on Causes, or Celestial Physics.'"[55] By early 1605, he had written 51 chapters.[56]

In parallel, Kepler renewed his attack on Mars. He took on "an operation of great subtlety," as he later put it, "the procedure mechanical and tedious."[57] He tried various oval shapes — including "ovoids, lemon-shaped, and puffy-cheeked curves."[58] Using a revised triangulation to the orbit of Mars, he coupled Tycho's Earth-centered observations of Martian position angles with his own Sun-centered model's predicted Martian longitudes. This showed that the orbit of Mars lay "midway between an oval and a circle."[59]

At first he thought there must be an error. After several weeks of doubt, Kepler realized his area law "demanded an oval orbit," science historian William H. Donahue tells us. "Only then did he trust the observational data."[60]

Kepler's rigorous analysis had paid off.[61] "By most laborious proofs and by computations on a very large number of observations," Kepler wrote, "I discovered that the course of a planet in the heavens is . . . perfectly *elliptical.*" (My italics.)[62]

Not quite. Kepler is being a bit sanguine here. Any of the variety of oval curves had tried "could have fit the observations equally as well as the ellipse," Owen Gingerich points out. Kepler saw that "one focus of an approximating ellipse coincided with the Sun." This link to a physical explanation is what drew him to choose an elliptical orbit.[63]

Isaac Newton would later write: "Kepler knew the Orb to be not circular but oval and guest it to be Elliptical." If so, it was a prescient guess.[64]

Regardless of how Kepler had arrived at his insight, the change from circular to elliptical orbits proved remarkable.

New Predictions

Kepler compared predictions from his new elliptical model with Brahe's observations. The error in the location of Mars now decreased from a half-degree or 30 arc-minutes to a mere **2 arc-minutes!**

It had been over four long years since Brahe had given the "Mars catastrophe" to his assistant. Now Kepler had what he called his first law: Planets go around the Sun in elliptical orbits.

> *. . . it was as if I were awakened from sleep to see a new light.*
>
> — Johannes Kepler [65]

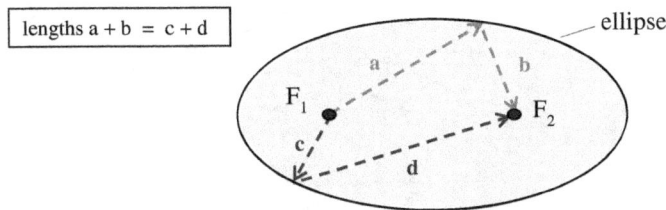

Figure 21.7. Ellipse. All lines from one focus to any point on the edge of the ellipse and on to the second focus are of equal length. F_1 and F_2 are the foci. (Not to scale.)

The Ellipse — What is an ellipse? It is a projection of a circle. Look at the circular top of a cup or drinking glass head-on. Then tip the cup or glass at an angle. The shape you see is an ellipse.

More formally, an ellipse (Fig. 21.7) is a closed curve containing two central points called foci. If you run a line from one focus to anywhere on the edge of the ellipse and then to the second focus, the total distance is always *the same*. For example, in the figure the total length of lines a plus b is the same as the total length of lines c plus d.

Kepler later proposed that *all* planets travel in elliptical orbits. His elliptical geometry with the Sun at one focus (and nothing at the other) proved the most accurate model yet for predicting the motions of planets. And it explained why ancient Greek and later models introduced offsets, obliquities, and equants to their circular orbit models to better match observations. The fact that the Sun is at one of the foci of a planet's elliptical orbit further appealed to Kepler's desire to find physical connections to geometric representations of the heavens.[67]

Those of us who are not astronomers perhaps do not appreciate how remarkable Kepler's discovery was. We look at elliptical orbits (as depicted in Figure 21.8) and perhaps ask why it was so hard for astronomers of the time to see that they are not circles. For one thing, Plato's "perfect" circular orbits were venerated ancient Greek knowledge. They had since been enshrined as Christian religious dogma, as noted.

In addition, the elliptical orbits in the figure are not to scale. They are, in fact, very close to circles and only slightly elliptical — including Mercury and Mars. The minor axis of Mercury is only 2% shorter than its major axis. The minor axis of Mars only 0.4% shorter. The ellipticity of all the other planetary orbits was undetectable with Brahe's data sets.[68]

312 *Cosmic Roots*

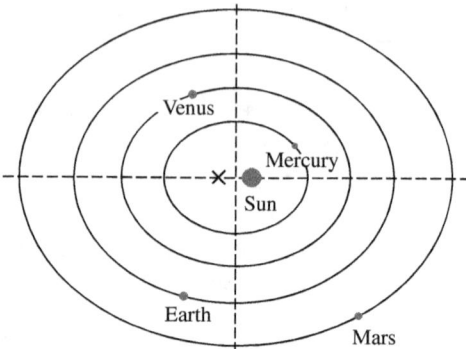

Figure 21.8. Kepler's Elliptical Planetary Orbits — the Inner Planets. The Sun is at one focus of ellipse (other focus marked with x). The Sun also has "a little orbit of its own" (not shown), which, as Newton later explained, is due to the gravitational influence of all the planets on it.[66] (Ellipses exaggerated. In reality, they are very close to circles and not all the same ellipticity. Not to scale.)

And Figure 21.8 above shows the (inner) solar system from a "bird's eye" view high above the Sun. Kepler and others did not have this advantage. They had to rely on naked eye observations from Earth — as it spins on its axis and orbits the Sun. All this while the other planets spin and orbit the Sun at varying speeds and different distances relative to us. To add to the confusion, as seen from Earth the other planets periodically exhibit an apparent looping motion we call retrograde.

To determine the actual shape of planetary orbits from Earth's vantage point was a pain-staking process which required highly accurate observations. Thank you, Tycho Brahe. It also necessitated ingenious geometric manipulations, sophisticated mathematical analysis — and intuition. Thank you, Johannes Kepler.

Kepler published his remarkable discoveries in *Astronomia Nova* in 1609. The full title of his work in English is: *A New Astronomy Based on Causation or a Celestial Physics Derived from Investigations of the Motions of Mars Founded on the Observations of the Noble Tycho Brahe.*[69]

Let's take a closer look.

Astronomia Nova

. . . most honest and grateful mention is made, and recognition given [to Tycho Brahe], since I built this entire structure from the bottom up upon his work . . .

— Johannes Kepler, *Astronomia Nova*[70]

Kepler saw Tycho's assignment of the Mars data to him as no accident. "I . . . think it to have happened by divine arrangement that I arrived at the same time in which he [Brahe] was intent upon Mars," Kepler wrote in *Astronomia Nova*, "whose motions provide the only possible access to the hidden secrets of astronomy . . ."[71]

Kepler was correct. As noted, Mercury and Mars are the only planets whose orbits are elliptical enough to be seen as such by Brahe's instruments. Mercury's closeness to the Sun limited observational data — leaving Mars as the key to Kepler's discovery.

True to its 32-year cycle, Mars returned to opposition in 1625. Predictions using Kepler's Sun-centered solar system with non-uniform Earth motion agreed with observations to within a half-degree or **30-arc-minutes**. The addition of Kepler's *elliptical* orbit for Mars brought maximum prediction errors down to **2 arc-minutes**.[72]

The New Astronomy and Holy Scriptures

> *The [Bible's] message is a moral one, concerning something self-evident and seen by all eyes but seldom pondered.*
>
> — Johannes Kepler, *Astronomia Nova*[73]

Kepler argued in *Astronomia Nova* that Biblical references to a stationary Earth should not be taken too literally. "While in theology it is authority that carries the most weight, in philosophy [physical science] it is reason," he wrote. "The Inquisition nowadays is pious, which . . . denies [Earth's] motion. To me, however, the truth is more pious still, and, with all due respect for the Doctors of the Church, I prove philosophically . . . not only is it contemptibly small, but also that it is carried along among the stars."[74]

He went on to say, ". . . to teach mankind about ordinary things is not the purpose of Holy Writ, which speaks to people about these matters in a human way in order to be understood by them and uses popular concepts . . . you are not listening to any teaching in physics [but rather] moral admonitions."

Kepler's defense of Copernicanism and support of science over Church authorities was seen as scandalous. Yet of all of his writings, his "arguments on interpretation of scripture were . . . the most widely read," historian of science William H. Donahue tells us.[75]

Perhaps Kepler's greatest accomplishment — often unrecognized by the general public — was his attempt to find a *physical cause* for the motions of the planets.

Magnetic Gravity?

> *For it may appear that there lies hidden*
> *in the body of the sun a sort of divinity.*
>
> — Johannes Kepler, *Astronomia Nova*[76]

The cosmic models of Ptolemy, Copernicus (for the most part), and Brahe were purely mathematical. Kepler attempted to determine *why* planets move — to ascertain the physics behind the mathematics.

Kepler knew that each planet speeds up as it approaches the Sun and slows down as it retreats. "It has been demonstrated that the increases and decreases of the velocity [of Mars] are governed by its approaching towards and receding from the sun," he wrote in *Astronomia Nova*, "And in fact the same happens with the rest of the planets."[77]

From this, Kepler made a grand leap. He proposed that "the cause of this intensification and weakening resides in . . . the center of the world [the Sun]."[78]

Based on William Gilbert's *De Magnete* (*On the Magnet*) of 1600, the "first modern investigation of magnetism," Kepler hypothesized that the Sun's *rotation* produces a magnet-like force. And this force holds the planets in their orbits.[79]

"By demonstration of the Englishman William Gilbert," Kepler wrote, "the earth itself is a big magnet . . ."[80] Kepler theorized that the Sun too is magnetic. And that it rotates on its axis like Earth. From this he proposed "the source of the five planets' motion [and the Earth's] is the sun itself."[81]

In Kepler's theory, the Sun's rotation causes the planets to orbit around it — like the spinning blades at the center of a food processor cause the liquid to rotate. To Kepler, the celestial spheres of Aristotle do not exist. (They didn't work with elliptical orbits anyway.) He proposed that ". . . every detail of the celestial motions is caused and regulated by facilities of a purely corporeal nature, that is, magnetic . . ."[82]

And since the Earth is magnetic, it in turn causes the Moon to orbit our home planet. As he put it in *Astronomia Nova*, "the monthly motion of the moon arises from the diurnal [daily] rotation of the earth."[83]

Of course, Kepler's magnetic theory is incorrect. He himself questioned the idea: "I will be satisfied if the magnetic example demonstrates the general possibility of the proposed mechanism. Concerning its details, however, I have doubts."[84]

Even though Kepler's magnetic, spinning Sun as the source of planetary motion was wrong, it was still a milestone in science. Aristotelian physics held that motion is a natural internal property of each body, and "not from an outside source." Here was the first attempt to explain an *external* physical cause behind the motions of the planets.[85]

Kepler's Laws

As noted, Kepler had figured out that each planet orbiting the Sun sweeps out an equal area over an equal time. With his prodigious mathematical talent, he later determined the mathematical relationship between a planet's period (the time for a planet to complete a single orbit) and its distance from the Sun.

Collectively, these relationships are known as Kepler's Laws of Planetary Motion. They are:

(1) all planets revolve around the Sun in elliptical orbits,
(2) a radius line joining any planet to the Sun sweeps out equal areas in equal time intervals, and
(3) the square of the period of revolution of a planet is proportional to the cube of its mean distance from the Sun.

The second law, the so-called time-area law is depicted in Figure 21.9.[86] Kepler developed the second law before the first. He used it as his "primary tool to determine the orbit of Mars." Both laws were first presented in *Astronomia Nova*.[87]

Kepler discovered his third law thirteen years later when he was working on *Harmonice mundi (The Harmony of the World)*. It "gives us is a direct mathematical relationship between the size of the orbits of the planets and their duration," writes Christie. "[It] only works in a heliocentric system . . . nobody really noticed its true significance until Isaac Newton at the end of the seventeenth century."[88]

Kepler published his Laws of Planetary Motion from 1609 to 1621. His elliptical orbits, Sun at one focus, and associated laws provided the most exact astronomical tables (yet) known. Fulfilling his promise to Tycho Brahe, Kepler published the *Rudolphine Tables* in 1627. They were not based on Brahe's geo-heliocentric model, as the great observer had wished. They were founded rather on Kepler's modifications to the Copernican heliocentric system.

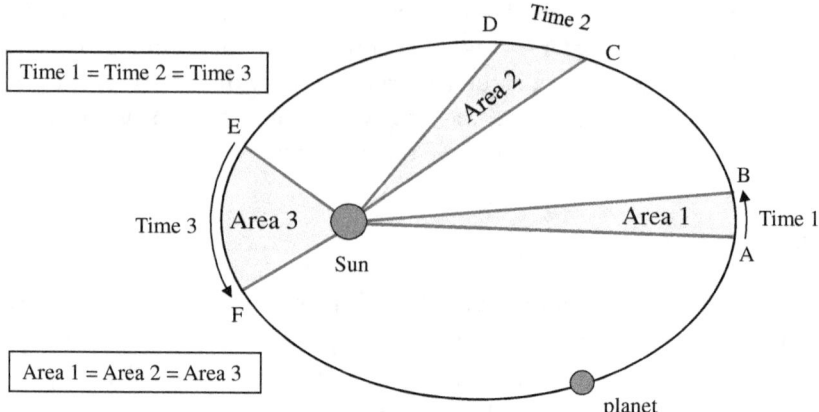

Figure 21.9. Kepler's Time-Area Law. A radius line joining a planet to the Sun sweeps out equal areas in equal time intervals along its orbit. All three times depicted are equal, as are the three areas they sweep out, respectively. The figure indicates that a planet orbits the Sun at a faster speed closer to the Sun. For example, look at path E to F (Time 3) on the left of the figure. It is longer than path A to B (Time 1) on the right. But their two times are equal. Thus the planet must travel faster on the left to cover the longer path, and slower on the right to travel the shorter path in the same amount of time. (Sun location, elliptical orbit exaggerated. Not to scale.)

Kepler's Rudolphine Tables based on Brahe's "vast collection of observational data" and his heliocentric elliptical orbit model were in such excellent agreement with observations that they began to sway the tide towards a Sun-centered solar system.[89]

A famous example: From his tables, Kepler predicted that Mercury would transit over the disk of the Sun — "a phenomenon never before observed." Parisian Astronomer Pierre Gassendi witnessed the transit on November 7, 1631, within hours of Kepler's prediction.[90] Later examination showed that "the tables of Ptolemy, Copernicus, and Longomontanus each erred by about five degrees in predicting this transit," Owen Gingerich writes, "whereas Kepler missed by less than 10 minutes of arc!"[91]

Kepler published four major works between 1617 and 1627: *De cometis* (1619), *Epitome Astronomiae Copernicanae* Vol I (1618), Vol II (1620) and Vol III (1621). These writings would lay "the foundations for the eventual triumph of heliocentricity."[92]

De cometis would become "a standard reference work" on comets in the 1680s. Nonetheless, Kepler mistakenly thought that the "trajectory of comets was rectilinear," and only appeared curved because of the movement of the earth," Christie tells us.[93]

Religious Conflicts

Religious authorities harassed Kepler throughout his life.

In 1612, Emperor Rudolf II was deposed by his younger brother Archduke Matthias. Forced to leave Prague, Kepler moved to Linz in Austria to be its District Mathematicus.[94]

Suspected of harboring Calvinist views since his days in Tübingen, he was questioned by the local Lutheran church in Linz and found wanting — particularly his beliefs regarding the presence of Jesus in the Eucharist ceremony.[95]

Lutheran doctrine at the time held that the "body and blood of Christ are truly and substantially present" in the transformed bread and wine. Kepler argued this was impossible. It would require Christ to be physically present everywhere at all times. (Calvanists also denied Christ's descent from heaven during the Eucharist ritual.)[96]

For this heresy, the Lutherans of Linz excommunicated Kepler. So much for Martin Luther's advocacy of the right to interpret Scriptures based on one's own conscience.

Not that Catholic authorities left Kepler in peace. Three of his books — *Astronomia Nova (New Astronomy)*, *Astronomiae Copernicanae (Epitome of Copernican Astronomy)*, and *Harmonices Mundi (Harmony of the World)* — were banned by the Roman Catholic Church.[97]

The Thirty Years War

While Kepler was publishing his Laws of Planetary Motion, the Thirty Years War (1618–1648) broke out in Central Europe.

Fighting between Catholics and Protestants within the Holy Roman Empire began this religious-political conflict. It would come to involve Austria, the Czech Republic, Sweden, Denmark-Norway, Transylvania, Poland, and Spain, as well as troops from Scotland and France.

It "remains one of the longest and most brutal wars in human history." Over eight million died in "military battles and related famine and disease." One-third to two-thirds of the population of middle Europe is said to have perished.[98]

The Catholic army of Duke Maximilian of Bavaria occupied Linz in 1620. Despite having been excommunicated by the Lutherans, Kepler still resided there. At the time, he was "fighting for the freedom of his mother, who had been accused of witchcraft." He won the court case. She died shortly after.[99]

The Counter-Reformation reached Linz five years later. Protestants were once again persecuted. As Imperial Mathematicus, Kepler was allowed to stay. Still, his "library was confiscated," Christie tells us, "making it almost impossible for him to work." So he left Linz.[100]

"After two years of homeless wandering," in 1628 Kepler settled in Sagen in Silesia He would never find "peace or stability again."

My question: Catholics and Protestants both worshipped Jesus Christ, the non-violent Prince of Peace who advocated "turn the other cheek." How could these "Christians" kill each other? For thirty years!

We humans form groups, then split and fight. We saw this in the Hebrews of Israel and Judah; we see it in modern times with Catholics and Protestants in Ireland; with Sunni and Shiite Muslims in the Near East; and in numerous other sad examples throughout human history. It is not just about religion. It happens with tribal, political, and national groups as well. There have been nearly sixty civil wars worldwide in the twentieth century alone — and seventeen already in the twenty-first, as of this writing. (c. 2021).[101]

All this recalls the words of Albert Einstein at the outset of World War I: "At such times as this, one realizes what a sorry species of animal one belongs to."[102]

The Great Visionary

. . . the planets complete their courses in the pure aether, just like birds in the air.

— Johannes Kepler[103]

Kepler, in his way, was as much a revolutionary as Copernicus. His model of the solar system eliminated "divine" circles and crystalline spheres, effectively cutting ties to the ancient Greek astronomy of Aristotle and Ptolemy. His elliptical orbits replaced the mistaken Platonic belief held sacred for some 2000 years that stellar objects orbit in circular motion. The need for epicycles, equants, eccentrics and other geometric devices was finally eliminated.

Kepler was also first to state that the other planets were material bodies like the Earth.[104]

> *Then burn thy Epicycles, foolish man;*
> *Burn all thy spheres, and save thy head.*
> *Faith need no staffe of flesh, but stoutly can.*
> *To heav'n alone both go, and leade.*
>
> — George Herbert, 17th century
> Welsh poet and Anglican priest.[105]

And perhaps most significant, Kepler assigned a *physical cause*, rather than a religious one, to the motions of the planets around the Sun — and of the Moon around the Earth. In his various writings, he suggested "gravity" was a physical force, a mass dependent mutual attraction between celestial entities.

Still, he proposed incorrectly that this force drops off with the distance, r between them, or $1/r$. And he was murky on the details.[106]

> *[My goal is] to show that the celestial machine is not so much a divine organism but rather a clockwork . . .*
>
> — Johannes Kepler[107]

Although wrong as to the source of this force and how it weakened with distance, Kepler's' ideas would presage the theory of universal gravitation of Isaac Newton some 50 years later.[108] For this, Carl Sagan called Kepler "the first astrophysicist."[109]

Other Keplerian Innovations

In his remarkable life, Johannes Kepler "wrote and published eighty-three books and pamphlets." Among these was what is said to be the first work of science fiction — the novel *Somnium (Dream)*.[110]

He also proposed correctly that the Moon and Sun causes Earth's tides. His calculations of areas and volumes by infinitesimal techniques has been called "one of the significant works in the prehistory of the calculus."[111]

Kepler also explained for the first time how a telescope produces an image, how the human eye works through refraction, and how to improve eyeglasses to correct for far-sighted and near-sightedness.[112]

But it is for Kepler's work in astronomy that he is most remembered.

The Next Chapter

I used to measure the skies,
Now I shall measure the shadows of the earth,
Skybound was the mind,
Earthbound the body resides.

— Kepler's self-written epitaph.[113]

Of modest demeanor, scrupulous honesty, deep piety, and brilliance of mind, Johannes Kepler died in 1630. He was 59 years old. His grave was later destroyed during the horrific Thirty Years War.[114]

A contemporary of Kepler would also defend the Sun-centered system of Copernicus and have his books banned. His mark on astronomy would not be so much a theoretical advance like Kepler's but an empirical one. His name is Galileo Galilei.

Chapter 22

THE MICHELANGELO OF SCIENCE

Philosophy is written in this grand book, the universe . . . It is written in the language of geometry, and its characters are triangles, circles, and other geometric figures.

— Galileo, *Il Saggiatore (The Assayer)*[1]

Galileo Galilei
(1564–1642)

Looking back, it may be hard for us to understand why the contemporaries of Copernicus and Kepler had such a difficult time accepting their ideas. Even given religious considerations, why couldn't the people of the Renaissance see what to us seems so obvious: The Earth spins on its axis daily and revolves around the Sun annually?

Let us not be too hard on our forebears. At the time, there was no evidence for a rotating, orbiting Earth — and no explanation as to why we don't feel the effects of its motion. Thus, it was hard for scientists as well as laypeople to accept that Earth actually moves.

What was needed was some indication from the heavens that would finally put the cosmology of Aristotle and Ptolemy to rest. The first evidence to reject their Earth-centered models came from observations by the great Renaissance natural philosopher Galileo Galilei — along with several other seventeenth century astronomers.

The fiery Galileo spoke and wrote in a flamboyant, humorous, and sometimes mocking style. His books were captivating, his lectures electrifying. Staged debates with the brilliant scientist were often caustic, leaving challengers embittered. He delighted in ridiculing his fellow philosophers, "skewing them with his keen sarcasm." Galileo's arrogance and "suffer no fools" approach produced a number of admirers and a number of enemies.[1]

In physics, Galileo was the first to express the concept of *inertia* mathematically, and to introduce the notion of frictional force. This and his studies of the motions of bodies along with the work of Kepler and Descartes would become the foundation for Newton's laws of motion. Galileo also formulated the basic law of *falling* bodies; deriving a law of how far a uniformly accelerated object will travel before it falls to the Earth. This so-called *time-squared* law is still taught in high school physics classes today.[2] In addition, he determined that the trajectory of uniformly accelerated artillery projectiles marks out a parabola (neglecting air effects).[3] The latter works became the foundation for Newton's Law of Universal Gravitation (as we shall discuss in chapter 25.)[4]

Galileo was a brilliant natural philosopher and astronomer — with a deep craving for recognition. This lust for fame would ultimately lead to his downfall.

The Early Years

Galileo Galilei was born in Pisa, then part of the Duchy of Florence in Italy, on February 15, 1564 — three days before the death of Michelangelo.

Galileo's family "were impoverished descendants of a leading Florentine family," writes physicist Hans Ohanian.[5] His proud father, a composer, lutenist, and music teacher, taught his eldest son to question authority and think independently.[6]

> *I . . . wish to be allowed freely to question . . . without any sort of adulation [of authority], as well becomes those who are in search of the truth.*
>
> — Vincenzo Galilei, father of Galileo[7]

At age thirteen, Galileo's father sent him to the Camaldolese Monastery to learn Latin, Greek, and logic. Young Galileo took his religious training quite seriously. He joined the Benedictine order as a novice — with hopes of becoming a monk. Vincenzo wouldn't hear of it. He removed his son from the Monastery. He wanted his firstborn to secure a well-paying vocation to help provide dowries for his four sisters and help support his two younger brothers.[8]

To please his father, seventeen-year-old Galileo enrolled in medical school at the University of Pisa in 1581. When he encountered the geometry of Euclid, he switched his emphasis to mathematics.[9] After four years of study, the now twenty-one-year-old quit the University without earning a degree to start his career in mathematics.

Galileo began writing "proofs and papers" in geometry, tutored students in mathematics, and gave occasional public lectures. A year later, he became a math teacher in Sienna.[10] He developed "ingenious theorems on centers of gravity." He also gave two invited lectures at the prestigious Academy in Florence on the dimensions and location of Hell in Dante's *Inferno*.[11]

Pisa

With his reputation growing, Galileo secured the chair of mathematics* at his alma mater, the University of Pisa in 1589. The argumentative Tuscan "quickly developed a reputation for obstinately disputing the doctrines . . . of

---

*"In the 15th century, humanist universities of Northern Italy and Poland had introduced dedicated chairs for mathematics, whose principal purpose was the teaching of astrology to medical students. The chairs for mathematics that Galileo would occupy at the end of the 16th century in Pisa (and later in Padua) were two such astrology chairs." Source: Christie, "The emergence of modern astronomy — a complex mosaic: Part III," thonyc.wordpress.com.

Aristotle and [2nd century] Greek physician Galen," writes biographer Dava Sobel.[12] Nothing gave the twenty-five-year-old Galileo more pleasure than refuting the authority of the exalted Aristotle. He was his father's son.

Galileo further irritated his colleagues and superiors by refusing "to wear the regulation academic regalia at all times," Sobel tells us. He deemed the attire "a pretentious nuisance."[13] In a ribald rhyming spoof, Galileo argued that official doctoral dress prevented "men's and women's frank appraisals of each other's attributes ... [while] the dignity of the professor's gowns barred him from the brothels ... resigning him to the equally sinful solace of his own hands."[14]

And yes, it was here that, according to tradition, Galileo carried out his famous Leaning Tower of Pisa experiment. The story goes that he dropped two balls of unequal weight from the roughly 180-foot (55-meter) high tower to show they both fall or accelerate downward at the same rate (neglecting air resistance). This refuted Aristotle's claim that the heavier ball would fall faster than the lighter ball by the ratio of their weights and reach the ground first.[15]

Did Galileo actually conduct this test? Science historians generally feel it was more likely a "thought experiment." And he was not the first to make this argument against Aristotle. Galileo had tapped into "an extensive stream of previous work on the subject," Thony Christie tells us. See endnote for details.[16]

Modern experiments on Earth, in space, and on the Moon have since confirmed that objects do fall at the *same* rate (in a vacuum) independent of their mass or composition.[17]

Padua

Seventy-year-old Vincenzo Galilei passed away in 1591. Galileo was now head of the family, responsible for the support of his sisters and brothers. With an ever-growing reputation, he secured a position as chair of mathematics at the University of Padua in the "Serene Republic of Venice" — at three times his former salary. He was twenty-seven years old. The Tuscan mathematician would remain at Padua for eighteen years. He would later describe it as the happiest time of his life.[18]

At Padua, Galileo taught the standard Earth-centered astronomy of Aristotle and Ptolemy. He gradually convinced himself that the Sun-centered universe of Copernicus was much more probable.[19] In three public lectures, Galileo argued against Aristotle's natural philosophy and astronomy.[20]

The Letters

Galileo received a copy of Johannes Kepler's *Mysterium* in 1597. Along with Tycho Brahe, the German astronomer had sent copies to a number of professors. What was Galileo's reaction? He did not like the "neo-platonic mysticism" of Kepler's solids.[21] He did approve of Kepler's impassioned defense of Copernicanism.

Galileo wrote a letter to Kepler a year later, in which he confided his hesitance to publish his own support of Copernicanism:

> *Like you, I accepted the Copernican position several years ago . . . I have written up many of my reasons and refutations on the subject, but I have not dared until now to bring them into the open, being warned by the fortunes of Copernicus himself, our master, who . . . stepped down among the great crowd (for the foolish are numerous), only to be derided and dishonored. I would dare publish my thoughts if there were many like you; but, since there are not, I shall forebear . . .*[22]

Kepler's letter in response encouraged Galileo to be more forthcoming:

> *. . . I could only have wished that you, who have so profound an insight, would choose another way . . . after a tremendous task has been begun in our time, first by Copernicus and then by many very learned mathematicians . . . . would it not be much better to pull the wagon to its goal by our joint efforts . . . and gradually, with powerful voices, to shout down the common herd . . . Thus perhaps by cleverness we may bring it to a knowledge of the truth. . . Be of good cheer, Galileo, and come out publicly. . . If Italy seems a less favorable place for your publication, and if you look for difficulties there, perhaps Germany will allow us this freedom.*[23]

It is instructive that, in these deeply troubled religious times, Kepler, a devout Lutheran, and Galileo, a loyal Catholic, had no trouble communicating with respect and mutual admirations. Perhaps due to the encouragement of Johannes Kepler, Galileo's confidence seemed to grow. He would soon have an opportunity to reveal his radical views. A chance encounter would lead him to great fame and great troubles.

The Starry Vault

About ten months ago a report reached my ears that a certain Fleming had constructed a spyglass by means of which visible objects, though very distant from the eye of the observer, were distinctly seen as nearby.

— Galileo[24]

Despite what you may have read, Galileo did not invent the telescope. The invention is generally credited to Dutch eyeglass maker Hans Lippershey of Middelburg in Zeeland in September of 1608.[25]

Soon after, Galileo received a letter from his friend, the eminent Venetian scholar and priest Paolo Sarpi. He told Galileo "about a spyglass that a Dutchman had shown in Venice."[26] Composed of a convex objective lens and concave eyepiece, the original Dutch telescopes had a magnification of about 3×. This means the image in the telescope appears three times nearer and larger than the object appears to the naked eye.[27]

Within a few days of hearing the news and likely after seeing one, Galileo made his own telescope — as did a number of astronomers. Produced with available lenses, it had a magnification of perhaps about 6×.[28]

Not to miss an opportunity for fame when he saw one, Galileo marched off to Venice to present his new device to the Venetian Senate in August of 1609. (See Fig. 22.1.) In response, the Senate doubled his salary and "set him up for life as a lecturer at the University of Padua," according to biographer Stillman Drake.[29]

Not satisfied with his first telescope, Galileo taught himself to grind and polish lenses. He would later produce a telescope of some 20× magnification.* It was with this telescope that he would make most of his celestial discoveries.[30]

We can imagine the ever-curious scientist setting his new instrument in the garden behind his house, carefully aiming it across the sky, scanning here and there, sighting it on one celestial object after another. Most everywhere he looked, he saw a brilliant array of glowing stars. There were many more than can be seen with the naked eye, a dizzying array of tiny lights scattered against the blackness of space.[31]

Galileo later reported "innumerable stars grouped together in clusters. . ." in the Milky Way. They were *individual stars*, just as Muslim polymath Abū Rayḥān al-Bīrūnī had proposed in c. 1000. "Many are rather

---

*Galileo would also "build a 30X magnification telescope... His competitors used very similar telescopes." Source: Christie, "Galileo's the 12th most influential person in Western History — Really?" thonyc.wordpress.com.

Figure 22.1. Galileo Demonstrates his Telescope to the Venetian Senate. Man sitting behind telescope is Leonardo Donato, 90th Doge (chief magistrate and leader) of the Republic of Venice. Painting by H.J. Detouche c. 1900.[32]

large and quite bright," Galileo wrote, "while the number of smaller ones is quite beyond calculation." In the constellation Orion alone, he counted "more than five hundred new stars distributed among the old ones within limits of one or two degrees of arc."[33]

This observation was not new. It was first made by Lippershey, Christie tells us, "and those who had taken part in Lippershey's first ever public demonstration of the telescope in Den Haag in September 1608" — and later by others who looked at the night sky with a telescope.[34]

Lunar Discovery

A number of natural philosophers in antiquity had proposed that the "markings on the Moon" indicated a "mountainous landscape."* In late 1609,

---

* "That the moon was earth-like and for some that the markings on the moon were a mountainous landscape was a view held by Thales, Orpheus, Anaxagoras, Democritus, Pythagoras, Philolaus, Plutarch and Lucian." Christie tells us, Source: Christie, "The emergence of modern astronomy — a complex mosaic: Part XXI." thonyc.wordpress.com.

Galileo was the first to verify the existence of mountains and craters on the Moon with a telescope.[35]

> ... the surface of the moon is not smooth, uniform, and precisely spherical as a great number of philosophers believe it (and the other heavenly bodies) to be, but is uneven, rough, and full of cavities and prominences, being not unlike the face of the earth, relieved by chains of mountains and deep valleys.[36]

The Moon Galileo observed was certainly not the *perfect* sphere in the heavens of Aristotelian philosophy and Christian dogma.[37] This was the first verification that the "pure" heavens were not perfect and more like the "impure" Earth.

The Moons of Jupiter

Galileo then focuses his instrument on a "wandering star," the planet Jupiter. It was the night of January 7, 1610. Johannes Kepler's *Astronomia Nova* had been published five months earlier. What Galileo saw must have surely set his heart to pounding. He later described observing "three fixed stars, all within a short distance of Jupiter and *lying on a straight line through* it"[38] This was a unique sight never before seen by man: a linear pattern* within a sea of random stars.

When I first saw the moons of Jupiter through a friend's homemade telescope, almost perfectly lined up with the great planet, a chill ran through my body. I brought my young son over to look at the sight. "This is what Galileo saw!" I told him as he gazed into the eyepiece. He had that bored look of a teenager indulging his geeky father once again.

Galileo observed this sight with *his* homemade telescope — a "spyglass" with two lenses at opposite ends of a tube.[39] As noted, Galileo improved telescope now had a magnification of some 20×.[40]

He would return to Jupiter's strange companions on following nights. They had moved! On January 10, he observed that one of the small "stars" had disappeared entirely. What was going on?

Galileo realized the missing stellar object was now hidden *behind* Jupiter. He concluded that these were not tiny stars, but three *moons* orbiting

---

*The moons of Jupiter all orbit the planet in nearly the *same plane*. So they appear lined up nearly in a straight line as viewed from Earth.

Jupiter "like a planetary system in miniature."[41] These were the moons we now call Io, Europa, and Callisto.*

On January 15 he saw a fourth moon, Ganymede.[42] He later wrote:

Our sense of sight presents to us four satellites circling about Jupiter, like the Moon about the Earth, while the whole system travels over a mighty orbit about the Sun . . .

— Galileo, "The Starry Messenger."[43]

By March 2, Galileo had completed sixty-four sightings of Jupiter's moons.[45] He would later describe what he called his "most important" discovery (see Fig. 22.2):

The disclosure of four PLANETS never before seen from the creation of the world up to our time . . .

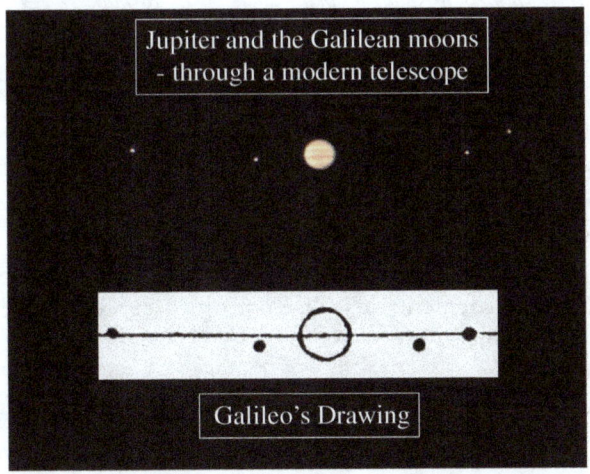

Figure 22.2. Jupiter and its Four Brightest Moons. Top: Modern image through a 10 in. (25 cm) Meade LX200 telescope, courtesy of Jan Sandberg. Bottom: Galileo's drawing from observations using his telescope.[44]

---

*Astronomer Simon Marius observed three moons of Jupiter on January 8, 1610, a day after Galileo. The Tuscan scientist would publish on March 12, 1610. Marius didn't publish until four years later. Source: Christie, "The telescope — claims and counterclaims." thony.wordpress.com.

Galileo astutely named these moons of Jupiter the four *Medicean planets* in honor of his patron, Cosimo II de' Medici. The nineteen-year-old had become Grand Duke of Florence after the death of his father Ferdinando a year earlier. This is the same Cosimo that Galileo had tutored in previous summers.[46]

These largest four moons of Jupiter are now referred to as the Galilean satellites in Galileo's honor.[47]

Initially "many astronomers and philosophers refused to believe that Galileo could have discovered such a thing," writes Dava Sobel.[48] Why such doubts? Because Galileo's observations ran against Aristotelian dogma that *all* celestial bodies revolve around a stationary Earth.

It didn't help that a number of astronomers had great difficulty finding the moons of Jupiter in their less acute telescopes.[49] In addition, early telescopes suffering from poor lens quality with "imperfections and distortions." Galileo was remarkably adept at discerning features through a telescope, despite optical artifacts. It appears that others were much less so.[50]

> *I render infinite thanks to God for being so kind as to make me alone the first observer of marvels kept hidden in obscurity for all previous centuries.*
>
> — Galileo[51]

Johannes Kepler wrote an "enthusiastic reply" to the news of Galileo's discovery of four moons orbiting Jupiter.[52] Galileo wrote back: "You are the first and almost the only person who, even after but a cursory investigation, has, such is your openness of mind and lofty genius, given entire credit to my statements…" He went on to express his frustration to the German astronomer:

> *I think, my Kepler, we will laugh at the extraordinary stupidity of the multitude . . . What do you say to the leading philosophers of the faculty here, to whom I have offered a thousand times of my own accord to show my studies, but who with the lazy obstinacy of a serpent who has eaten his fill have never consented to look at planets, nor moon, nor telescope?*[53]

Starry Messenger

Galileo rushed to report his telescopic findings to make sure he got the credit. He published a short pamphlet called *Sidereus Nuncius (Starry Messenger)* in March of 1610. It was the first published work on astronomical observations through a telescope. It described the mountains and valleys of the Moon; the "countless" stars in the Milky Way; and the "real sensation," the four moons of Jupiter.[54]

The little pamphlet — "more a press release than a scientific paper" — was a hit.[55] "It sold out within a week of publication," writes Sobel, "while news of its contents quickly spread worldwide."[56]

As noted, other astronomers had observed the multitude of stars in the Milky Way, the mountains of the Moon, and moons of Jupiter with their telescopes. They included "English polymath Thomas Harriot, Franconian court mathematicus Simon Marius, and Jesuit astronomers Odo van Maelcote, Giovanni Paolo Lembo and Christoph Grienberger," Christie tells us. Why is Galileo so often given sole credit for these discoveries? He published first.[57]

Galileo dedicated *Starry Messenger* to Cosimo II and his family. In June 1610, three months after its publication, Cosimo appointed Galileo as Royal Professor of Mathematics and Philosophy to the Grand Duke — with a substantial salary. "As Galileo's reputation spread though Italy and Europe," historian of science Albert Van Helden writes, "his Medici patrons bestowed celebrity and protection. In return, Galileo insured his discoveries were unveiled at the court of the Medici as a form of court entertainment."[58]

It was a glorious time for Galileo. When the Tuscan scientist arrived in Rome in March, 1611, he was received as a "leading celebrity." The Jesuits honored him at a grand banquet at the *Collegio Roman* (Roman College). The prestigious *Accademia dei Lincei*, founded eight years earlier, made him their sixth member.[59]

Galileo and rival astronomers continued their examinations of the night sky for the next three years. This included observations of Venus, Saturn, and the Sun itself. Arguably the most important observations were when they trained their telescopes on the planet Venus.

Venus Revealed

With absolute necessity we shall conclude, in agreement with the theories of the Pythagoreans and of Copernicus, that Venus revolves about the sun just as do all the planets.

— Galileo Galilei[60]

Recall that Venus is always seen close to the Sun in the sky. This has implications for its *phases* as viewed from Earth.

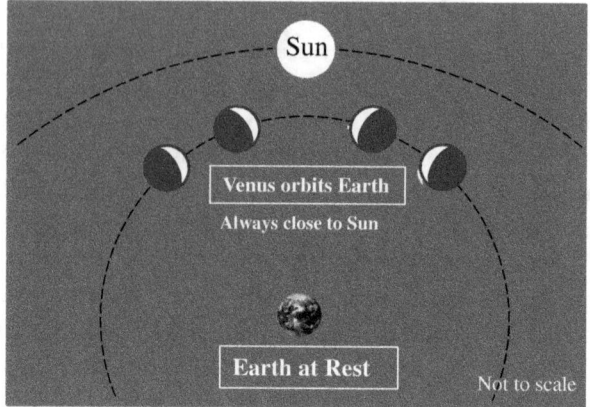

Figure 22.3. <u>Earth-centered</u> Solar System ala Aristotle/Ptolemy. Venus is always seen near the Sun's location in the sky — towards the East before dawn and towards the West after dusk. In an Earth-centered system, only partial phases of Venus would be observed from our planet. Why? Because the Sun is always behind Venus.

In the *Earth-centered* models of Aristotle and Ptolemy, the Sun is always *behind* nearby Venus. As seen from Earth, this would produce only *two* phases of Venus — a crescent and new phase (as shown in Fig. 22.3).

Now let us consider the *Sun-centered* model of Copernicus. Here Venus *orbits the Sun* as does the Earth. As seen from Earth, Venus is in front of the Sun at times and behind the Sun at other times. Thus, you would expect to see a *full set of phases* of the planet Venus, similar to the phases of the Moon. (See Fig. 22.4.)

What did Galileo and other seventeenth century astronomers observe in their telescopes? Over time they saw a *full set of phases* of Venus[*] from thin crescent to half-shaded to the equivalent of a full Moon.[61]

Telescopic observations of the full phases of Venus represented the first definitive evidence that the Earth-centered cosmos of Aristotle/Ptolemy, revered for nearly two thousand years, was wrong.

Yet, as Galileo and others fully understood, it was not proof of Copernicus' or Kepler's Sun-centered universe. Why? Because full phases of Venus also support Tycho Brahe's compromise geo-heliocentric system (and other geo-helio models). They too have Venus (and Mercury) orbiting the Sun.[62]

---

[*] At the time, at least four observers independently discovered the full phases of Venus. They were "Thomas Harriot, Simon Marius, Galileo and Jesuit astronomer Paolo Lembo . . . dating the discoveries is almost impossible." Source: Christie, "Galileo's the 12th most influential person in Western History — Really?" thonyc.wordpress.com.

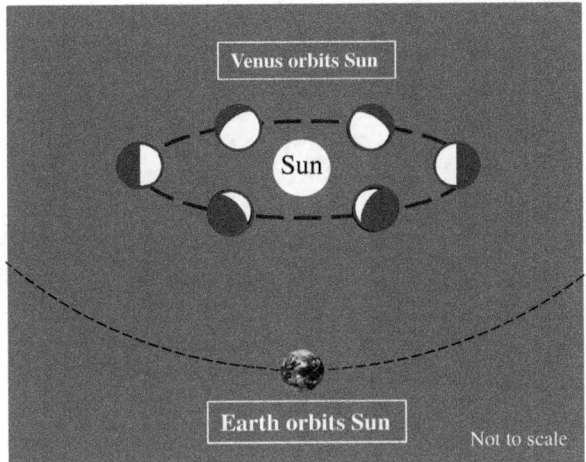

Figure 22.4. <u>Sun-Centered</u> System of Copernicus. <u>Full phases</u> of Venus are seen because the planet orbits the Sun <u>inside</u> the orbit of Earth. So we see Venus at times in front of, to the side, or behind the Sun. Note that full phases of Venus would also be observed in the Geo-helio system of Brahe. Although Earth is at rest in the Tychonic model, Venus still orbits in front of, on the sides, and behind the Sun, as seen from Earth. (As was shown in Chapter 20, Fig. 20.4.)

It would be "almost three decades before anybody succeeded in observing the full phases of Mercury."[63]

More Discoveries

With their telescopes, Galileo and other astronomers also discovered that: 1) planets are disks, not points of light; 2) Saturn has "ears" (rings of Saturn); and 3) the Sun* has sunspots.[64]

The last two observations in particular gave additional evidence against Aristotle and the Church's "unchanging perfection of the heavens."[65]

In tracking the movement of sunspots on the Sun's surface, Galileo determined that the Sun *spins*, just as Kepler had proposed. From his observations,

---

* "In a dispute with Christoph Scheiner, Galileo kept changing the date he first observed sunspots to establish priority," Christie tells us. "Both Galileo and Scheiner were unaware that Harriot had observed sunspots before both of them and Johannes Fabricius had already published on them." In addition, "Chinese and Korean astronomers had been recording naked-eye observations of sunspots since the first millennium BC. There are also scattered observations of sunspots beginning with the ancient Greeks and down through the Middle Ages in Europe." Source: Christie, "The emergence of modern astronomy — a complex mosaic: Part XXII," thonycwordpress.com.

Galileo calculated that the Sun makes a complete rotation in about 27 days.[66] The modern value for this so-called *synodic* rotation is 26.24 days.[67]

The Trial

Galileo Galilei's support of the Copernican model and its spinning Earth orbiting the Sun would be challenged by Church authorities. This would lead to arguably the most famous trial in Western history. The story is told in the next chapter.

Chapter 23

THE TWO CHIEF WORLD SYSTEMS

I do not believe the same God who has endowed us with senses, reason, and intellect has intended us to forego their use. He would not require us to deny sense and reason.

— Galileo[1]

The year 1611 was a heady time for Galileo Galilei. As noted, Jesuit astronomers honored him at a grand banquet at the *Collegio Romano* in March. The following month they certified his celestial discoveries. He was also inducted into the prestigious Lincean Academy.[2] It seems the musician's son from Pisa had it all. Recognition from his peers, a lucrative position as Royal Professor to the Medici Grand Duke of Tuscany — and most desired: great fame.

Perhaps it wasn't the currently favored Father, Son, and Holy Ghost, but the ancient gods who took umbrage at his hubris. For he would soon witness his undoing, instigated by his own words.

The Scientist and the Pope

Not everyone was thrilled with Galileo's publication of his discoveries and his unorthodox ideas. Professors steeped in Aristotelian teachings felt threatened. Some tried to get the Catholic Church to ban his works.[3] Dominican friar Tommaso Caccini, for one, took up the charge.

In 1614, Caccini denounced Galileo's heliocentric views as "dangerous and close to heresy." Galileo was surprised by this accusation of heresy — a crime which he considered "more abhorrent than death itself," Sobel tells us.[4] On the other hand, the number of those who supported Galileo was growing. This included more than a few Jesuits.[5]

> *The intention of the Holy Ghost is to teach us how one goes to heaven, not how heaven goes.*
>
> — Galileo quote of Vatican librarian Cesare Cardinal Baronio.[6]

In the summer of 1615, Galileo defended his views in a famous letter to Grand Duchess Madama Cristina of Lorraine, the mother of his Medici patron Grand Duke Cosimo II. She had objected to his Copernican stance at an earlier lecture.

Galileo wrote: "These men [Aristotelian professors] . . . resolved to fabricate a shield for their fallacies out of the mantle of pretended religion and authority of the Bible."[7] Galileo also quoted Cardinal Baronio on the Holy Ghost (see above) and, like Kepler, argued for a non-literal interpretation of Holy Scripture.

In the letter, he sealed his fate with these words[8]:

I hold the sun to be situated motionless in the center of the revolution of the celestial orbs while the <u>earth revolves about the sun</u> . . . (My underline.)

No longer the cautious university professor of his first letter to Kepler, emboldened by his mounting fame, Galileo declared that for him the Copernican model was more than a mathematical calculating tool. It was physical reality.[9]

Galileo's words made their way to Pope Paul V himself. The Pontiff ordered Cardinal Roberto Bellarmino, the Church's leading theologian to convene a Sacred Congregation of Index. The Cardinal assigned a commission of eleven theologians to investigate the complaint against Galileo.[10]

The commission's report condemned the Copernican system. It declared the notion of a sun-centered system with a moving Earth as ". . . foolish and absurd in philosophy [science]; and formally heretical, since it explicitly contradicts in many places the sense of Holy Scripture."[11]

The commission and inquisitors that followed were "well aware of the *scientific* objections to Copernican cosmology," writes research scientist Haqq-Misra. This included the lack of evidence for a moving Earth — and, if it did move, the stupendous distances to the stars and colossal star sizes implied by no measurable parallax (as discussed in Chapter 20).

They also knew that Tycho Brahe's geo-heliocentric model with its stationary Earth did not suffer from these issues. "The best science of the 17th century tended to favor a universe with Earth at rest at its center," Haqq-Misra points out. Thus, the Church had valid religious and scientific reasons to support the Tychonic system and reject Copernicanism — given the understanding of the times.[12]

The Cardinals of the Inquisition met on February 25, 1616. Based on the commission's findings, they condemned the teachings of Copernicus. And, as noted, they suspended the Polish astronomer's magnum opus, *De revolutionibus* until corrected by "removing claims* that heliocentricity represented reality."[13]

---

*The Congregation of the Index specified corrections to Copernicus's book in 1620. The "ten emendations were designed to make Copernicus' book appear hypothetical and not the description of a real physical work." (See source for examples.) Once implemented, the book became available for reading. Source: Alvarez, Pablo. "Collection Highlights, Copernicus, De Revolutionibus..." University of Rochester, Rush Rees Library. https://rbscp.lib.rochester.edu/3338. Retrieved Jul 7, 2021.

Galileo was summoned to Cardinal Bellarmino's residence. He was ordered by injunction of the Congregation of the Index to "abandon completely . . . the opinion that the Sun stands still at the center of the world and the earth moves, and henceforth not to hold, teach, or defend it in any way whatever, either orally or in writing."[14]

At Galileo's request, Bellarmino later wrote a letter stating that he, Galileo, had not been summoned to appear before the Inquisition for heresy, nor did he receive any penance. The Cardinal added that Galileo had been notified that "the doctrine attributed to Copernicus . . . cannot be defended or held."[15]

Note that the injunction of 1616 stated "not to hold, teach, or defend it [Copernicanism] *in any way whatever, either orally or in writing.*" (My italics.) The subsequent letter from Bellarmino to Galileo stated only that the doctrine attributed to Copernicus "cannot be defended or held." This difference would come up in the later trial of Galileo in 1632.

Bellarmino's letter seems to have made Galileo even more confident. Some would say more arrogant. This attitude would play a part in his later more serious run-in with the next pope, as we shall see.[16]

O Lord my God, Thou art great indeed . . .
Thou fixed the Earth upon its foundations, not to be moved forever.

— Psalm 104:5

Religious authorities used the passage above (and others) in the Hebrew bible to argue against the Copernican theory that the Earth moves. What about the phrase *"upon its foundations"*? Why would a spherical Earth have or need foundations? This is a reference to the pillars which support the flat, rectangular Earth of Sumerian cosmology — a heritage long since forgotten.

Against advice of friends, the stubborn Galileo had traveled to Rome to defend himself. He felt more than his honor was at stake. The prescient Galileo was concerned that should the Church continue to take a stand against Copernicanism, it could be proven wrong by more advanced instruments of future astronomers. This would cause irreparable damage to Catholicism.[17]

This was not a good time to challenge Church leaders. The Roman Catholic Church at that time was a religious and political entity, with its own territory and an army defending the Italian peninsula and Papal States.

The Church was feeling ever increasing pressure in its fight against the now century-old Protestant Reformation.[18]

Again, Protestant founder Martin Luther had stated that he (and any individual of faith in Jesus) had "the right to a personal reading of the Bible." In response, the Roman Catholic Council of Trent in 1564 declared biblical interpretations as the exclusive prerogative of the Catholic Church. The Council stated that "no one, relying on his own judgment . . . shall dare to interpret [the Sacred Scriptures]." Galileo had taken it upon himself in his writings *to publicly interpret Scriptures.*[19]

Galileo had every reason to fear the Church. Astronomer and mathematician Giordano Bruno was a proponent of heliocentricity and "fighter for freedom of thought and expression," Christie tells us. He was an avowed Pantheist — one who did not "believe in a personal god" but (similar to Einstein) saw God in nature and the grand workings of the universe. In the late 1500s, Bruno had read texts forbidden by the Church and publicly argued against Aristotle.[20]

Despite warnings from Church authorities, the headstrong Bruno continued to speak and write about his beliefs. By 1600, the Church had had enough. Refusing to recant, Bruno was "bound and his tongue [placed] in a gag" at Piazza Campo de' Fiori in Rome and burned alive for heresy.[21]

Pope Paul V died in 1621. He was succeeded by Gregory XV, who passed a mere two years later. Telltale white smoke rose once more into the skies above the Vatican.

A certain Tuscan scientist saw his chance.

Dialogo

The crowd of fools who know nothing, Sarsi, are infinite. Those who know very little of philosophy [science] are numerous. Few indeed are they who really know some part of it, and only One knows all.

— Galileo, Il Saggiatore (The Assayer).[22]

In 1623, the new pope in Rome was Urban VIII, aka Maffeo Barberini (1568–1644). He was a fellow Tuscan and long-time admirer of Galileo. Pope Urban had granted Galileo "a couple of private audiences, an unusual honour for a mere mathematicus," Christie writes.[23]

Seeing his opportunity, the great scientist asked for permission from the Pontiff to write a book on two competing theories; Ptolemaic and Copernican. Urban consented.*

Galileo was given two conditions:

(1) do *not* take sides, and
(2) conclude that "man could not determine how the world works in any case because God could bring about the same effects in ways unimagined by man."

Galileo was given permission to write a book whose conclusion was already ordained.[24]

After nearly a decade of off-and-on effort, Galileo finally received the required approval. Travel complications due to the plague led to authorization to publish by two Church censors: Niccolò Riccardi, Master of the Sacred Palace in Rome and Clemente Egidi, Inquisitor of Florence. Galileo also neglected to tell them about the injunction of 1616 against Copernicanism.[25]

Galileo published his *Dialogo sopra i due massimi sistemi del mondo: Tolemaico e Copernicano (Dialogue on the Two Chief World Systems: Ptolemaic and Copernican)* in 1632.[26] (See Fig. 23.1.) The book was modeled after Plato's dialogues. With brilliant analogies, clever deductions, and artistic portrayals it has been called "a masterpiece of Italian literature."[27]

Appended to the Latin version of *Dialogo* and translated into the vernacular was Johannes Kepler's *Astronomia Nova*. The great visionary had died two years earlier.[28]

Dialogo's "Two Chief World Systems" — Ptolemaic and Copernican — were not favored by astronomers at this time, as Thony Christie points out. Instead, stargazers favored Kepler's elliptical heliocentric model or Brahe's geo-heliocentric model with diurnal rotation of Earth. Galileo knew this. Why did he choose to ignore Brahe and Kepler in his *Dialogo*? Because he wanted to "go down in history as the man who established [the truth of] heliocentricity," writes Christie.[30]

In Galileo's *Dialogo*, two fictitious scholars, Simplicio and Salviati, argue the merits of Ptolemy and Copernicus, respectively, to a layman named Sagredo.

---

*Galileo conveniently neglected to tell the Pontiff about the injunction of 1616 against Copernicanism. This omission would later enrage the Pope.

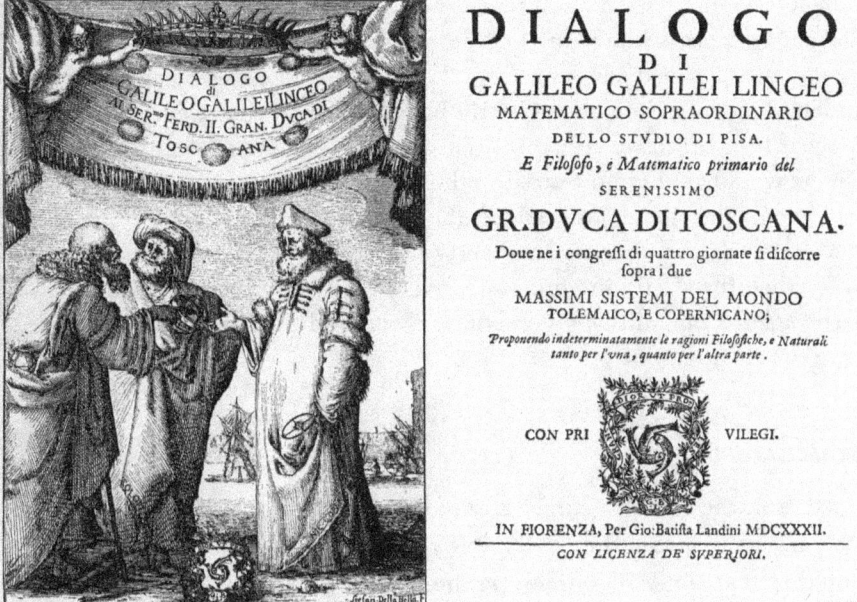

Figure 23.1. Frontispiece and title page of Galileo's *Dialogue* of 1632. Depicted in the figure from left to right are Aristotle, Ptolemy, and Copernicus. The latter "holds a model of a planet orbiting the Sun."[29]

Simplicio, which sounds like the Italian word for *simpleton*, was portrayed as a "pompous Aristotelian philosopher."[31] Salviati the Copernican was on the other hand of "sublime intellect" — one who supported his arguments with the telescopic evidence of his "friend" Galileo. Included were Galileo's observations of the Moon, the moons of Jupiter, phases of Venus, and sunspots. The Tuscan scientist also presented his erroneous theory that ocean tides are caused by Earth's motion.[32]

As noted, the Pope had told Galileo to include his words on how "God could bring about the same effects in ways unimagined by man" in *Dialogo*. Galileo complied at the end of the book. He has Simplicio say: "I once heard from a most eminent and learned person, and before which one must fall silent . . ." and proceeds to give the Pope's theological argument.[33]

Galileo puts the words of the Pope in the mouth of Simplicio, his least credible character. He then places statements by Salviati throughout the text which undermine the Pope's argument. In effect, through Salviati,

Galileo says that "God always acts in the simplest way possible — and *not* through miracles when natural causes would suffice."[34]

Galileo also has Salviati say ". . . for the few [scientific propositions] understood by the human intellect, I believe our knowledge equals the divine in regards to objective certainty." This equating of human intellect with the Divine would get him in trouble with religious authorities, as we shall see.[35]

Galileo judiciously added the words to his title page: "Proposing inconclusively the philosophical and physical reasons as much for one side as for the other." His attempt at neutrality failed. Interspersed with sarcastic barbs against Simplicio's arguments, Salviati and the theory of Copernicus were the clear victors.[36]

Rotating Earth

Since from at least the time of Gerbert d'Aurillac in the tenth century, scholars knew that if the Earth truly rotates daily, the surface of our planet would spin at some 1000 miles per hour (1600 km/hr) relative to its center. If so, they argued, wouldn't we be tossed off the face of the Earth? Wouldn't we feel the wind as a mighty torrent as we charged through our planet's atmosphere at this great speed? As we have seen, these and other arguments against a moving Earth go all the way back to at least Aristotle.[37]

Galileo addresses these concerns in *Dialogo* with his famous ship at sea example. He asks readers to consider a ship in *uniform motion* — moving at a constant speed and in a constant direction — on calm seas. He points out that observers inside the hold of this ship would feel no motion. Why? Because they and all objects inside the hold travel at the *same speed and direction* as the ship.

A modern example: We feel no motion inside an airplane — as long as it is moving at constant speed and in a constant direction in smooth air. When we pour water from our can of juice or soda, it does not fly off but flows directly into our cup below. If we toss an object straight up into the air, it does not fly to the rear but lands in our lap. Why? Again, because we and all these objects are traveling at the same speed and direction as the airplane.

The movement of the Earth is also uniform (to a first approximation). Just like on the ship or airplane, we and all things on Earth move through space at the same speed and direction as the Earth. Thus we sense no motion.[38]

Recall that Jean Buridan gave a similar "feel no wind" argument in the fourteenth century. His student, Nicole Oresme (c. 1325–1382) made a comparable argument: on a moving Earth, an arrow shot straight up from a bow would come straight down.[39] It is likely Galileo knew the views of these French scholars, yet made no reference to them in *Dialogo*.

Troubles Ahead

Galileo was warned to expect trouble from his book. Jesuit reviewer Melchior Inchofer was particularly offended; "[Galileo] declares war on everybody and regards as mental dwarfs all who are not Copernican. It is clear enough what he has in mind."[40]

This was a particularly bad time for Pope Urban. He was under severe criticism for not supporting Catholic Spain in the Thirty Years War against German Protestants. Every day his regime fell further into debt financing the troops in the War of Mantuan Succession in northern Italy. And he had to prepare what defenses he could against the Plague, which was making a horrific return to Italy.[41]

"The arrival of copies of the book in Rome caused an uproar." writes physicist Hans Ohanian. "The Pope flew into a rage." His Holiness accused Galileo of betraying him and the censors of "incompetence and collusion."[42] The vainglorious Urban VIII was particularly stung by inflammatory remarks that "Galileo had played him for a fool."[43]

In August of 1632, the Holy Office suspended publication of *Dialogo*. Sales were halted and unsold books confiscated. In September, a Special Commission found the book had indeed violated the injunction of 1616 against Copernicanism.[44]

Galileo had dedicated *Dialogo* to his new Medici patron Grand Duke Ferdinand II, son of deceased Cosimo. The new grand duke appealed to Urban VIII to pardon the Tuscan scientist. The Pontiff was in no mood to compromise. He threatened the duke that if he continued to support Galileo, "his own life in Tuscany could become very uncomfortable."[45]

In October of 1632, Galileo was summoned to face the Supreme Sacred Congregation of the Roman and Universal Inquisition in Rome. Galileo kept putting off his departure, citing "ill health and travel difficulties due to the Plague." Only when the Inquisition threatened to drag him to Rome in chains did he leave Tuscany.[46]

The "the most celebrated scientist in Europe" arrived in Rome on February 13, 1633. After a wait of two months, he was brought before the Commissary General, Cardinal Vincenzo Maculano de Firenzuola and formally charged.[47]

The Inquisition

Galileo's encounter with the Holy Office has contributed to . . . an irrefutable cleavage between faith and reason which persists to this day.[48]

— Theoretical physicist Josef M. Jauch

Note to reader: The following is a brief synopsis of the "trial" of Galileo as generally understood by historical scholars. It is based on papal archives and contemporaneous letters of key figures. The story lacks certain critical details, as we shall see.

Galileo was interrogated in three sessions. Commissary General Maculano supervised while his assistant Fr. Carlo Sincero, Prosecutor of the Tribunal, did the actual questioning. The ten Cardinals of the Inquisition serves as judges in the case. They did not attend the sessions. Testimony was sent to them to be read later and discussed at weekly Inquisition meetings.[49]

Those facing Inquisition charges were typically imprisoned during the length of their trial.[50] Not Galileo. For most of the time he was permitted to reside at the Tuscan embassy in Rome. This exception was made "out of respect for the Medici Grand Duke," as well as Galileo's advanced age and frail health.

Prior to interrogation, Tuscan ambassador Francesco Niccolini warned Galileo "to be submissive and assume whatever attitude the inquisitors seemed to want of him," Dava Sobel writes.[51]

First Interrogation (April 12, 1633) — In his opening testimony, Galileo is shown a copy of *Dialogo*. He acknowledges he is its author. He is then questioned at length on the injunction of 1616 against Copernicanism and his meeting with now deceased Cardinal Bellarmino. Asked about what was decided at the meeting, Galileo states:

[As to the] "opinion of the sun's stability and earth's motion, it was decided by the Holy Congregation of the Index that the opinion, taken absolutely, is repugnant to Holy Scripture and is to be admitted only hypothetically, in the way Copernicus takes it."[52]

Galileo then produces the 1616 letter from Bellarmino which had stated he was effectively free of suspicion. Pressed about the injunction of 1616, he says:

". . . it may be that I was given an injunction not to hold or defend said opinion, but I do not recall it since it is something of many years ago."

Pressed further, he hedges:

"I do not claim not to have in any way violated that injunction." He further states: "I do not remember that the injunction was that I could not hold or defend, and maybe even that I could not teach."

"I do not recall . . . that there was the phrase 'in any way whatsoever,' but maybe there was . . . I do not retain them in my memory I think because they are not contained in the said certificate."

He is then asked why he didn't inform censor Niccolò Riccardi about the 1616 injunction when he sought approval to publish *Dialogo*. He states:

"I did not judge it necessary . . . since with [*Dialogo*] I had neither held nor defended the opinion of the earth's motion and sun's stability; on the contrary . . . I show . . . that Copernicus's reasons are invalid and inconclusive."

Such is Galileo's fear of the Inquisition.

Four days later, on April 17 a three-member Special Commission unanimously agree that *Dialogo* violates the 1616 injunction against Copernicanism. Based on the commission's findings and Galileo's declining health, Maculano writes to Cardinal Francesco Barberini. He is the pope's nephew, Secretary of State, and supporter of Galileo.[53] Maculano's letter of April 22* states:

> *Last night Galileo was afflicted with pains which assaulted him, and he cried out again this morning. I have visited him twice, and he has received more medicine. This makes me think that his case should be expedited very quickly . . . in light of the grave condition of this man . . . the case could immediately be brought to a prompt settlement, which I expect is your feeling in obedience to the Pope.*

"In effect, Galileo would plead guilty to some as yet unspecified minor offense for writing *Dialogo* in return for a lighter sentence," writes Richard Blackwell, emeritus professor of philosophy at St. Louis University.[54]

Upon learning that what we might now call a "plea deal" has been approved by the Holy Office, Maculano writes a second letter to Cardinal Barberini on April 28. At the time, the Cardinal was at the papal vacation retreat at Castel Goldolfo with his uncle, Pope Urban.[55]

The letter reads in part:

> *The case is brought to a such a point that it may be settled without difficulty. The Tribunal will maintain its reputation; the culprit can be treated with benignity; and, whatever the final outcome, he will know the favor done to him.*

---

*Maculano's letter of April 22 was discovered in 1998. Source: Scotti, p. 258

Second Interrogation (April 30) — Presumedly now under the "plea deal," Galileo, having reread his book, gives a squishy confession: "I freely confess that it [*Dialogo*] appeared...to be written in such a way that a reader... would have had reason to form the opinion that the arguments for the false side [Copernicanism], which I intended to confute, were... capable of convincing because of their strength... two arguments, one based on sunspots and the other on the tides, are presented... as being strong and powerful, more than would seem proper for someone who deemed them to be inconclusive and refutable."[56]

He goes on to say: "My error then was ... one of vain ambition, pure ignorance, and inadvertence."[57]

Upon leaving the session, Galileo offers "to add one or two more days of discussion to *Dialogo*" to further refute the Copernican point of view. His offer was ignored.[58]

Third Interrogation (May 10) — Galileo presents his formal defense against the charges, a pro forma step for an Inquisition at the time. He includes "the original copy of Bellarmino's 1616 letter and a brief written defense of his actions."[59]

Then things turn ugly. As required, the Holy Office submits a Summary Report of the trial to the cardinals and pope for judgment.[60] The harsh and highly negative report throws the proverbial book at Galileo.[61]

Along with the three depositions of Galileo's testimony, the Holy Office report contains Frs. Tommaso Cassini and Niccolò Lorini's complaints against Galileo in 1615; passages against Copernicanism from the Bellarmino's letter and injunction of 1616; Galileo's letter to Fr. Benedetto Castelli expressing his views on science and the Bible; excepts from his *Letter on Sunspots* favoring heliocentricity, and other material derogatory to Galileo's case.[62]

The Summary Report ends with:

> [Galileo] *begged to be excused for having been silent about the injunction issued to him, since he did not remember the words "to teach in any way whatever... He said all this not to be excused from the error, but so that it be attributed to vain ambition rather than to malice and deception.*

The report omits any mention of a plea bargain. It is an undated, unsigned secret "internal document of the Holy Office." Galileo never sees it.[63]

At some point, Urban returns to Rome from his papal retreat and "reinserts himself in the deposition of Galileo's case." On June 16, he "presides over the weekly meeting of cardinal inquisitors."[64] Here the Pope lays out his harsh order. *Dialogo* is to be condemned. Galileo is to be summoned once again before the Inquisition to determine his "true purpose," under torture if necessary.[65]

On the 21st of June, the Tuscan scientist is once again interrogated before Maculano about his intentions — this time under threat of torture. If he "answers in a satisfactory manner," he is to be sentenced as commanded by the Pope.[66]

Defeated and cowed, Galileo states in part:

"... after the above-mentioned decision, assured by the prudence of the authorities, ... I held, as I still hold, as very true undoubted Ptolemy's opinion, namely the stability of the earth and the motion of the sun."[67]

"In regard to writing of Dialogo already published. . . for my part I conclude that I do not hold and, after determination of the authorities, I have not held the condemned opinion."

Questioned again, he repeats that he does not hold "the opinion of Copernicus and did not hold it" after being ordered by the injunction to abandon it. He then states:

"For the rest, I am in your hands: do as you please."

Formal Sentencing (June 22)

Galileo Galilei is led before the Cardinals of the Inquisition to hear the results of their deliberations. "Arrayed in their blood-red robes," as Dava Sobel put it, seven cardinals find him guilty of suspected heresy.[68]

By tradition, the Inquisition gives the accused the opportunity to recant — this to save one's soul and avoid the eternal tortures of Hell. Garbed in the white robes of a penitent and kneeling before the Holy Tribunal, the seventy-year-old scientist "reads aloud the Abjuration statement that had been prepared in advance for him." It says in part[69]:

... desiring to remove from the minds of Your Eminences and every faithful Christian this vehement suspicion, rightly conceived against me, with a sincere heart

and unfeigned faith I abjure, curse, and detest the above-mentioned errors and heresies . . . I swear that in the future I will never again say or assert, orally or in writing, anything which might cause a similar suspicion about me . . . if I should come to know any heretic or anyone suspected of heresy, I will denounce him to this Holy Office, or to the Inquisitor or Ordinary of the place where I happen to be.

Legend has it that while kneeling before the Inquisition, Galileo muttered under his breath *Eppur si muove* or "Yet it moves!" — meaning the Earth really does move. Historians of science tells us it is a fable. Even so, I love the story.[70]

As commanded by Pope Urban, Galileo is sentenced to "formal imprisonment at the pleasure of the Inquisition."[71] This is commuted on the following day to house arrest for the remainder of his life. *Dialogo* is placed on "the Index of Prohibited Books* where it remains for nearly two hundred years."[72] In a later declaration, publication of all of Galileo's works — including any he might write in the future — are forbidden.[73]

The Pope adds a stipulation to his decree that "it be widely distributed" and "read publicly to university professors and in parish churches throughout Catholic Europe."[74] He even has "posters and flyers printed to make the public aware of it."[75] This action by the Pontiff is unprecedented.

Galileo would express "nothing but contempt for his judges for the remainder of his days."[76]

Questions and Mysteries

Why was Commissary General Maculano's gentler "plea deal" sabotaged? All we have for original sources are the two April letters from Maculano to Cardinal Barberini. Sobel suggests that it was perhaps the Cardinal who had "persuaded the Holy Congregation to let Maculano deal extrajudicially with Galileo."[77]

What about the Pope? He and the cardinal were at the retreat when the cardinal received the letters. Did the cardinal tell the Pontiff of Maculano's plan? Did the Pope approve? We don't know.

---

* "The Lutheran Protestant Church also rejected the heliocentric hypothesis but never formally banned it." Source: Christie, "The emergence of modern astronomy — a complex mosaic: Partr XXXII."

As to the hostile Summary Report, it was unsigned and undated, as noted. We do not know exactly who wrote it or when it was sent. Was it written by the anti-Galileo faction in the Holy Office? Under the direction of certain Cardinals of the Inquisition? Under the Pope's direction?

Why did the Pope turn on Galileo in the June 16th meeting of the Inquisition? Why did only seven of the ten Cardinals of the Inquisition attend the sentencing and sign the sentencing document? It seems there was dissent amongst the cardinals over the treatment of Galileo, even in the face of the Pope's wrath.

After all these years, we still have more questions than answers.[78]

Aftermath

Pope Urban's public humiliation of Galileo had a stifling effect "on science, especially in Italy," writes Josef Jauch. "It is probably the chief reason why the center of gravity of scientific activity [in Europe] moved north-west."[79]

Ironically, the Pope's actions made Galileo more famous than ever. For much of the public, the Tuscan astronomer came to represent a martyr to science against oppression by religious authority.

It is said that the banned *Dialogo* became even more popular due to its notoriety. With the power of the printing press and the black market, it spread across Europe. Many on the continent spoke out "against the injustice of his condemnation." It later became "common in Italy, England, France, and Germany to advocate science, experimentation, and theories despite the opposition of religious leaders," write Brody and Brody. (Christie disagrees. See endnote.)[80]

Galileo received the great fame he so deeply craved — at a price. He was devastated by the ordeal. "The poor man has come back more dead than alive," observed Tuscan ambassador Niccolini.[81]

Still, the great scientist was not done — not by a long shot.

The Final Years

[Galileo's Two New Sciences], even more than his support of Copernicus, was to be the genesis of modern physics.

— Stephen Hawking[82]

In 1638, still under house arrest, the now seventy-year-old Galileo completed his final book: *Discorsi e dimostrazioni matematiche intorno a due nuove scienze (Discourses and Mathematical Demonstrations Relating to Two New Sciences.)* The two new sciences were the study of the motion of bodies and strength of materials.

This compendium of Galileo's life's work in physics was smuggled to a publisher in Holland. It embodied Galileo's vision of physics as mathematically-based science supported by empirical evidence.[83]

"Literarily brilliant" and ground-breaking scientifically, *Discourses* covered trailblazing work in impetus, moments, and centers of gravity. Also included were Galileo's pendulum and inclined planes experiments — and his famous time-squared bodies in free fall equations.*[84]

Galileo's work would help form the basis for Newton's laws of mechanics and law of universal gravitation some 50 years later. They would also reach across three centuries of separation to guide the thinking of Albert Einstein. Galileo's dictum that all speeds are relative would become the first postulate of special relativity of 1905. Galileo's principle that all objects fall at the same rate regardless of mass (neglecting air effects) would be key to Einstein's development of general relativity of 1915.

Called the Renaissance era's "scientific counterpoint to Michelangelo's art," dubbed the father of modern science by Albert Einstein, Galileo Galilei remains a giant among giants.[85]

His fear that future scientific evidence for heliocentricity would come to embarrass the Catholic church came true. It was not until 1835 that all books in support of a Sun-centered solar system were finally dropped from the Index of Banned Books.[86] In 1992, Pope John Paul II issued a declaration which acknowledged that "errors had been made" by the Church in the handling of the Galileo affair.[87] In March 2008, the Vatican announced a plan to honor Galileo with a statue inside its walls. It was suspended a month later.[88]

The Torch is Passed

... the greatest light of our times.

— Ferdinand II De' Medici on Galileo.

---

*"The mean speed formula, the basis of the mathematics of free fall, was known to the Oxford Calculatores and the Paris Physicists in the fourteenth century. The laws of free fall were known to Giambattista Benedetti in the sixteenth century." Source: Christie, "Galileo's the 12th most influential person in Western History — Really?" thonyc,wordpress.com.

Now totally blind, Galileo Galilei died on January 8, 1642 — just short of his seventy-eighth birthday.[89] A remorseful Ferdinand II had planned to have him buried in Santa Croce, the great Franciscan Basilica in Florence, next to the tomb of his father Vincenzo — and to erect a marble mausoleum in the Tuscan scientist's honor. Pope Urban VIII over-ruled him once again.[90]

Galileo's body was kept in a room behind the sacristy in Santa Croce for nearly a century. Under the reign of Florentine pope Clement VII, his remains were moved in 1737 with great ceremony to a grand mausoleum "in the body of the church in front of the tomb of Michelangelo." It remains there to this day.[91]

As if by providence, a boy was born the same year as Galileo's death in a small farmhouse in the English countryside who would build on the advances of Galileo to become the greatest scientist of all. His seminal works would generate the most profound challenge yet to religious orthodoxy.

We tell his story in the next chapter.

PART V
Enlightenment

Chapter 24

THE GREAT ONE

I do not know what I may appear to the world, but to myself I seem to have been only like a boy playing on the seashore, and diverting myself in now and then finding a smoother pebble or a prettier shell than ordinary, whilst the great ocean of truth lay all undiscovered before me.

Isaac Newton.[1]
1642–1727

Despite their momentous achievements, the cosmic works of Copernicus, Brahe, Kepler, and Galileo were incomplete. Copernicus thought planets orbited the Sun in perfect circles. Brahe continued to argue that the Earth was stationary. Kepler considered gravity a magnetic force. Galileo believed in circular planetary orbits, thought Earth's rotation caused ocean tides, and proposed that comets were "anomalous fluctuations in the air."[2]

It would take a new scientist to both explain the successes of his predecessors and resolve their errors. Possessing both extraordinary mathematical skill and the keenest of physical insights, English physicist Isaac Newton synthesized the heliocentric universe of Copernicus, the planetary measurements of Tycho Brahe, the elliptical orbits of Kepler, and the dynamics of Galileo into a Universal Law of Gravitation.

His masterwork would provide for the first time a theoretical basis for why all celestial *and* Earth-bound objects behave as they do in the presence of gravity.

Newton's World

Sr Isaac had the happiness of being born in a land of liberty & in an age where he could speak his mind — not afraid of The Inquisition as Galileo was . . .

— John Conduitt, Newton's assistant[3]

In the early seventeenth century (1600's), the center of world science began to shift from Germany and Italy to Atlantic Europe. Three natural philosophers in particular broke with Aristotelian tradition to advance science into the modern age: Francis Bacon, Robert Boyle, and René Descartes.

English philosopher and scientist Francis Bacon (1561–1626), a devout Protestant, believed that "God had fashioned a rational, orderly universe."[4] In 1620, he established the "scientific method," which called for verification of a scientific hypothesis through observation and experiment.[5] Bacon's empirical process was a direct rejection of the medieval emphasis on metaphysical explanations for physical phenomena. This resulted in a shift towards a more utilitarian approach to science, especially in England.[6]

In his masterpiece *The Skeptical Chymist* (1661), Anglo-Irish chemist Robert Boyle (1627–1691) rejected the ancient Greek theory that all matter is composed of earth, air, fire and water. He proposed, rather, that everything "was composed of minute particles of single universal matter."[7]

> *Cogito ergo sum*
> *I think, therefore I am.*
>
> — René Descartes

French philosopher extraordinaire René Descartes (1596–1650) was among the first Europeans to propose the concept of inertia, along with Galileo. His work on motion and forces "helped lay the foundations of modern dynamics theory." In his *La dioptrique* (*Optics*) of 1637, the Jesuit-educated Descartes derived the law of refraction we now call Snell's Law. A celebrated mathematician, Descartes created the Cartesian co-ordinate system and pioneered analytical geometry — the use of algebraic equations to describe geometric forms.[8]

The seventeenth century also saw a political revolution in England. It would come to have a profound effect on the West and Isaac Newton.

Royal Power Limited

Squabbling between Parliament and the monarchy broke out into the English Civil War of 1642. It ended nine years later with the victory of Parliament's Oliver Cromwell and his Roundhead army over royalist forces. After the execution of King Charles I, Cromwell appointed himself "Lord Protector."[9]

Parliament at the time was dominated by prosperous Puritan industrialists and merchants. The aim of this fundamentalist sect of Calvinism was to promote education and hard work in service to God. It also sought to eliminate all ostentation and frivolity, such as the celebration of Christmas with its pagan roots. Puritans also strived to "purify" England of all Roman Catholic influences.[10]

Power struggles between Parliament and the monarchy continued for nearly four decades. They culminating in the non-violent "Glorious Revolution" of 1689. At the time, Protestant William of Orange had chased Catholic King James II out of Britain. To be crowned King, William as well his wife Mary, had to agree to Parliament's new Declaration of Rights.[11]

Founded on the thinking of John Locke, the Declaration of Rights limited the power of the monarchy well beyond that of the Magna Carta of 1215. It established free elections, instituted freedom of speech (for Parliament), and rejected cruel and unusual punishment for all citizens. It also prohibited a Roman Catholic to sit on the throne of England.

Into this era of scientific progress and growing freedom (for Protestants) came a Puritan natural philosopher named Isaac Newton.

Troubled Genius

Nature and nature's law lay hidden in the night;
God said, let Newton be! and all was light.

— Alexander Pope, carved above the fireplace
in the room of Newton's birth at Woolsthorpe.[12]

It is Christmas day, 1642 — four months after the beginning of the English Civil War. Cries of pain sound out in a tiny farmhouse nestled in the bucolic village of Woolsthorpe-by-Colsterworth some 94 miles (150 km) north of London. Nineteen-year-old Hannah Ayscough Newton gives birth to her first child prematurely. She had attended the funeral of her young husband, an illiterate yeoman, nearly three months earlier.[13]

From a family of "gentlemen," Hannah had married below her station. Now her three-pound baby fights for breadth. "It will be a miracle if he lives until we get back," her midwives say as they rush for the doctor. Of sturdy stock, the child survives.[14]

Hannah names the boy Isaac, after his late father. She later says he was so tiny "they could put him into a quart pot."[15] The small stone farmhouse in which he is born still stands. (See Fig. 24.1.)[16]

The Newton Farmhouse

Figure 24.1. Isaac Newton's Birthplace. Woolsthorpe-by-Colsterworth, Lincolnshire, England.

With little means of support, the young widow abandons Isaac three years later.

Hannah departs for the nearby village of North Witham to wed wealthy 63-year-old rector Barnabas Smith. She leaves her son at their 100-acre farm in Woolsthorpe in the care of her mother, Margery. Why did Hannah leave her only child behind? To, as biographer Gail Christianson points out, establish young Isaac's position "as the future lord of the manor and fend off future claims to the farm by other Newton's."[17]

Too young to understand his mother's motives, Isaac watches her leave with bewilderment. He would come to be a petulant, untrusting adult, a loner with seething anger just below the surface. At age 19, he would confess a list of 58 "sins." One of them was "threatening my [step] father and mother Smith to burne them and the house over them," as he put it.[18]

The School Years

Rector Smith passed away in 1653. Hannah, now well off, returned to Woolsthorpe with ten-year-old Isaac's half-brother Benjamin and two half-sisters Mary and Hannah in tow.[19]

At 12 years old, Isaac was enrolled at King's (grammar) School in Grantham seven miles (eleven kilometers) north of Woolsthorpe. Here he was taught Latin, Greek, and the Bible, with a small amount of instruction in arithmetic. Initially the young Newton showed "little promise in academic work ... school reports described him as idle and inattentive."[20] After winning a fight with an older student, Isaac was determined to cement his superiority. He soon rose to the top of the class.[21]

The young Newton "often visited windmills and built scale models of them. He carefully tracked the Sun's movements and constructed sundials and a water clock,"[22] William Stukeley, one of Newton's later associates and first biographer recalled:

> *Every one who knew Sr Isaac, or have heard of him, recount ... when a boy, his strange inventions and extraordinary inclination for mechanic ... he always busyed himself in making knickknacks and models of wood in many kinds: for which purpose he had got little saws, hatchets, hammers and a whole shop of tools, which he used with great dexterity.*[23]

When he was around age sixteen, Hannah took her eldest son out of school to manage her now substantial estate. The absent-minded youth showed neither talent nor interest in the job.[24] He seldom completed the tasks at hand. When "his mother ordered him into the fields to look after the sheep, the corn, or any rural employment," Stukeley wrote. "his chief delight was to sit under a tree, with a book in his hands, or to busy himself with his knife in cutting wood for models . . . or go to a running stream, and make little mill wheels to put into the water . . . without even remembering dinnertime."[25]

Trinity College, Cambridge

The young dreamer was rescued by his uncle Henry Stokes, headmaster of King's School. On his uncle's advice, Isaac was returned to school in Grantham in 1660 at age seventeen. The following year, Stokes convinced Hanna to have Isaac attend Trinity College, Cambridge, some 60 miles ((95 kilometers) to the southeast.[26] Cambridge University today is "renowned for scientific research and other scholarship," Ohanian writes. In Newton's Day it was far behind Oxford and had "little reputation for anything."[27]

How the young country bumpkin from Woolsthorpe must have felt when he first came upon Trinity College. Founded "by Henry VIII a century and a quarter before," writes Christianson, it was considered "the stateliest College in Christendom." It featured a towering entrance, Great Gate and massive exterior. Tudor Gothic facades enclosed the grand court. Beyond was the stately Chapel, Master's Lodge, and "magnificent dining hall with its hammer-and-beam ceiling."[28]

Unsupported by his mother's wealth, Isaac paid his way through college for the first three years. He waited tables, "fetched beer, bread and firewood, polished boots, cleaned rooms, and emptied chamber pots" for wealthier students and fellows (faculty).[29] In his fourth year, he "attained his majority," (reached age twenty-one). Only then did he receive substantial sums of money from Hannah.[30]

Newton's initial goal in college was to obtain a law degree. Grantham had provided him meager preparation in mathematics and science.[31] Isaac Barrow, first Lucasian Professor of Mathematics at Cambridge saw promise in the young student and encouraged Isaac to study mathematics.

Newton received large doses of "classical" authors. This included Euclid, Ptolemy, and particularly Aristotle, who's teaching still dominated instruction at Cambridge. In his third year, Newton found himself attracted to the astronomy of Copernicus and Galileo, as well as Kepler's *Optics*.[32]

Newton's academic performance was undistinguished. Bored by standard course offerings, he found inspiration in the modern thinkers of his time. Studying on his own, Newton mastered the works of Bacon, Boyle, Descartes, Gassendi, Hobbes and other contemporary natural philosophers.[33]

Having been born on Christmas day, Newton felt his was a "divinely ordained mission," as Christianson put it. This served to justify his demanding demeanor and fiery temperament.[34]

Terrible Savant

Newton was of the most fearful, cautious, and suspicious temper that I ever knew.

— William Whiston, Newton's assistant at Cambridge.[35]

Isaac Newton was said to be obsessive, temperamental, irritable, and "overly sensitive to the slightest criticism." Like Copernicus, he had a great reluctance to publish. He craved recognition, yet possessed a deeply held fear of disapproval. In most cases he withheld publication of his brilliant work, often revealing his landmark discoveries only when threatened to lose precedence.[36]

He was secretive to the point of paranoia. He sought revenge for all slights, real or imagined. The brilliant Isaac could not bear intellectual mediocrity or apathy in others.[37]

In terms of religious belief, Newton was deeply devout, even more so than Kepler. Newton's private papers on religion were greater in size than all his scientific writings combined. He believed science was a holy pursuit which demonstrated the "presence of the Creator" in the natural world. He hated oaths and drove to "strip religion of all but its most fundamental rites and doctrines." This was founded in his Puritan upbringing — as was his fanatic work ethic.[38]

Newton had no wife, no family, no pastimes. He devoted nearly every waking hour to his work. The English savant often spent "weeks and months in intense concentration to the point of physical exhaustion and collapse."[39]

He had few friends. He acquired most of them later in his life when he was famous. They were generally younger men who idolized the Great

One and had learned to treat the touchy savant with extreme patience, high praise, and absolute loyalty — a "ballet choreographed on eggshells," as Christianson put it.[40]

The Plague Years

Isaac received his bachelor's degree in January of 1665. That summer, the dreaded Bubonic Plague made its return to Cambridge. It shut the school down for two years. Returning to his boyhood home at Woolsthorpe, twenty-two-year-old Newton began a quest which would revolutionize the fields of mathematics, physics, astronomy, and optics.[41]

> *All this was in the two plague years of 1665 and 1666, for in those days I was in the prime of age and invention, and minded mathematics and philosophy more than at any time since.*
>
> — Isaac Newton[42]

The story goes that in the relative quiet of his boyhood home Isaac Newton conceived his "method of fluxions" — the basic rules of differential and integral calculus. He also established the foundations for his laws of motion and law of universal gravitation. And he developed his theory of colors. He did not publish any of this work.

Newton himself told this story late in his life. It is amazing. Except it is not quite true.

We know from Newton's own notes that he spent some six years, not two on these efforts. His quest began back in April of 1664 while still at Trinity studying for scholarship exams. It continued through the plague years and beyond to 1669 — the same year he was appointed to the Lucasian Chair of Mathematics at Trinity.[43]

Newton's claim that he developed "the direct method of fluxions" and "the inverse method of fluxions," as he later put it is basically true.[44] Though, again, he began this effort in 1664 at Trinity. Newton wrote up this work in an unpublished paper of October 1666.[45]

His achievements in gravity were far from what he later claimed. He did show through calculations that that the same force which governs how the proverbial apple falls to the ground on Earth also holds the Moon in its orbit around our home planet. It was quite clever. Nonetheless, his work was inaccurate and incomplete — far from his later comprehensive theory of universal gravitation.*[46]

Why did Newton exaggerate so? (That is a kind word for it.) Because, as Christie suggests, he wanted to establish precedence. As we shall see, German polymath Gottfried Leibniz had also developed a form of calculus. English natural philosopher Robert Hooke had made key inroads in universal gravitation. Both Hooke and Dutch polymath Christiaan Huygens had done the same in Optics. Newton wanted to be recognized as "first inventor" in all three areas.[47]

Let There Be Light

Light it self is a Heterogeneous mixture of differently refrangible Rays.

— Isaac Newton[48]

As noted, Newton contended he had conducted experiments which verified his famous color theory while sequestered at Woolsthorpe. It is difficult to determine exactly when this actually took place. "The best guestimate," Christie writes, "is that this programme of experiments took place over the period 1660 to 1670."[49]

Aristotle and every scientist until Newton mistakenly believed that white light was "fundamental" — a basic entity not made of other ingredients. Newton's experiments with sunlight and prisms revealed that white light is made up of individual colors through a process called refraction. Looking at the Sun, he nearly blinded himself.

What is refraction? Take a spoon and put it into a glass half filled with water. The part of the spoon in the water appears to bend. Of course, it is not really bent — the light rays from the spoon are bent by passing through the water. Newton discovered this bending or refraction *changes* with color, e.g., blue rays bend more sharply than red rays.

---

*Newton's method was to gather, collate, and study all works he could find on a subject. He would then expand upon them. In mathematics, he explored the pre-calculus writings of "Archimedes, Bonaventura Cavalieri, Kepler, and John Wallis." In physics, his studies included "Descartes, Christiaan Huygens, and Galileo." In optics, "Euclid, Ptolemy, Ibn al-Haytham, Kepler, Snel, Descartes, and Hooke." Source: Christie, "Annus Mythologicus," thonyc.wordpress.com.

"In his 1666 essay 'Of Colours,' Newton told of placing a prism near the hole in a shutter of his darkened chambers," writes Christianson. "He then placed a second prism a few yards away from the first and carefully observed the result."[50]

> *Ye purely Red rays refracted by ye second Prisme made <u>noe other colours</u> but Red & ye purely blew ones noe other colours but blew ones. (My underline.)*
>
> — Isaac Newton.[51]

By putting the individual colors through a second prism, Newton showed that individual colors could not be divided further — they were fundamental. With his "color theory," the English savant was the first to explain why "the Colours of the *Rainbow* appear in falling drops of *Rain*," as Newton put it.[52] Like a prism, water droplets in the air refract white sunlight into a rainbow of colors.

To his delight, Newton's prism experiments showed that his "illustrious French predecessor," René Descartes was wrong. The latter had proposed colors are produced when light passes through an invisible aether — a medium of tiny spherical particles which lies between an object and an observer. Newton on the other hand had demonstrated that "color is an inherent property of light" itself.[53]

Newton sent a letter outlining his theory of Light and Color to the prestigious Royal Society of London on February 6, 1672. This was an organization founded twelve years earlier to foster the advancement of science and mathematics. Its motto was "*Nullis in verba*." It loosely translates to "Take nobody's word for it." Galileo would have loved it.[54]

> *. . . the proper Method for inquiring after the properties of things is to deduce them from Experiments.*
>
> — Isaac Newton[55]

Newton emphasized the experimental nature of his findings: "What I tell . . . is *not a Hypothesis* but the most rigid consequence . . . evinced by ye meditation of *experiments* concluding directly & without any suspicion of doubt." (My italics.)[56]

His brilliant discovery on the behavior of light was criticized in England and even more so on the Continent. Like Galileo, Newton was frustrated by those who doubted what he had seen with his own eyes.

Newton had not "provided details on his process" nor the specific glass technology or prism shapes required. Others found it difficult to replicate his findings. It was not until 1721 — nearly 50 years later — that his optical experiments were reproduced independently and its findings confirmed.[57]

Newton Discovered

Newton's "first major public scientific achievement was the invention, design, and construction" of a *reflecting* telescope. Prior working telescopes had been constructed with lenses. Newton's contained only mirrors — which he had ground and polished himself. (See Fig. 24.2.)[58*]

Again, Newton had discovered that different colors of light are bent or refracted at different angles, e.g., blue light is bent at a greater angle than

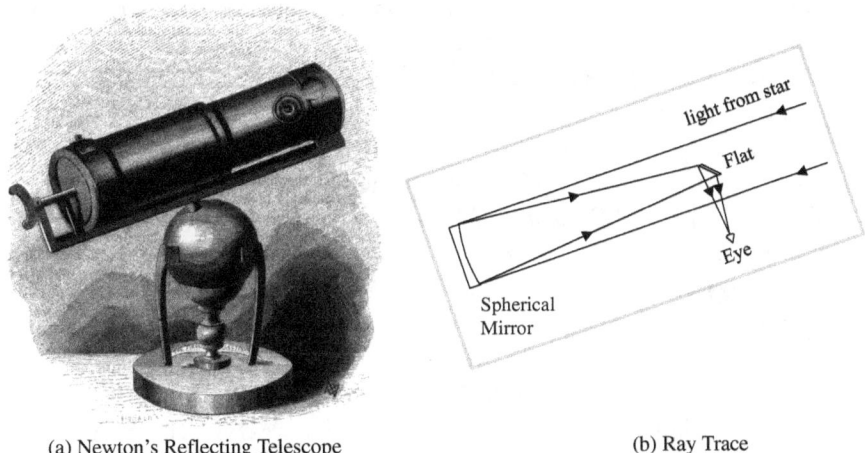

(a) Newton's Reflecting Telescope   (b) Ray Trace

Figure 24.2. Newton's Reflecting Telescope of 1668. (a) Telescope: Six-inch aperture, 40X magnification. (b) Ray Trace: Virtually parallel light from distant star enters the telescope. Concave spherical primary mirror collects light and reflects it back to focal point. Diagonal flat secondary mirror intercepts light and reflects it to eyepiece on the side of telescope.*

---

* A *spherical* mirror does not bring the incoming parallel light from stellar objects to a sharp focus. A much more difficult to grind *parabolic* mirror does. James Hadley managed to produce such a mirror in 1722. Reflecting telescopes were now able to compete with refractors. Source: Christianson, p. 163.

red. Thus, a single lens does not bring all the different colors of white light to the same focus point — an effect called *chromatic aberration*.[59]

Unlike lenses, mirrors reflect all colors at the same angle. So, they do not produce chromatic aberration. Newton's primary motive for building his reflective telescope was to demonstrate this phenomenon and validate this principle from his theory of colors.[60]

In addition, early refractive telescopes had two lenses — an objective lens at the front of the tube to focus the light and an eye piece at the other end to "spread the light across the retina." They tended to be very long to reduce the bending angles and thus minimize chromatic aberration.[61]

With his reflecting telescope, Newton did not have to worry about this effect. For the same size diameter, reflecting telescopes were shorter in length than refractors. Newton constructed an instrument with a magnification of 150X that was only six inches (15 cm) long.[62]

Newton sent his reflecting telescope to the Royal Society for examination in 1671. The "small but powerful telescope caused a minor sensation." Several fellows of the Society carried the instrument to the Palace of Whitehall for a special demonstration to King Charles II. Newton was delighted — though his precious anonymity was lost.[63]

Newton was made a fellow of the Royal Society the following year. He also consented to have his letter on light and color published in the Royal Society's journal *Philosophical Transactions* for February 19, 1671–1672. It was arguably his first scientific paper. An article on Newton's reflecting telescope appeared five weeks later. These works would become the foundation of his later masterful treatise on experimental physics called *Opticks*.[64]

Not everyone accepted Newton's radical ideas on light. His proposal that light consists of small particles was criticized by Royal Society's curator of experiments, Robert Hooke. Dutch physicist Christiaan Huygens also objected. Why? Because Newton had based his conclusions on experiments alone. This criticism reinforced Newton's reluctance to publish.[65]

Hooke threw salt on the wound in 1675 by claiming that "Newton had stolen some of his optical results." The charges were without merit. Newton reacted by turning away from the Royal Society, where Hooke was still one of its leaders.[66]

Newton continued an intermittent "vicious semi-public quarrel with Hooke" for three decades. He stubbornly waited until after Hooke's death in 1703 to publish a complete account of his optical research, the now famous *Opticks*. He published it a year later in 1704.[67]

Newton's seminal treatise "overflowed with detailed accounts of the phenomena of reflection and refraction, the spectral decomposition of white light, and the colors of the rainbow," writes Christianson. He also explained the construction of the reflecting telescope, "the color circle (the first in the history of color theory), and experiments on what would later be called 'interference effects' in conjunction with Newton's rings."[68]

Appended to the first edition of *Opticks* in 1704 were two mathematical treatises by Newton — his first works on calculus to be published.[69]

Calculus

The development of calculus gave Newton a critical mathematical instrument with which to develop his physics. It would serve as a major advantage over his peers.

Newton realized that the derivative is the inverse of the integral. Put simply, the derivative is the "*exact* slope at any one individual point on a curve." This mathematical device enabled him to determine the instantaneous velocity of an accelerating object at any point in time — such as a body falling in a gravitational field. The integral gives the *exact* area under a curve. For example, in "the graph of velocity against time, the area under the curve represents the distance travelled." (See Fig. 24.3.)[70]

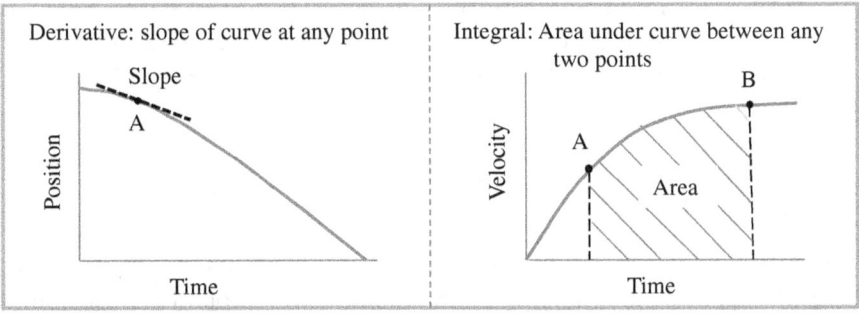

Figure 24.3. An Object Falling in Air and Calculus. (1) This is a plot of the object's position (altitude) versus time. The curve straightens out when object reaches terminal velocity. The derivative gives the exact slope of the curve at point A. Its value is the instantaneous velocity of the object at that point in time. (2) This is a plot of object's velocity versus time. The integral gives the exact area under the curve between points A and B. Its value is the distance the object has travelled between the two points.

Calculus enabled Isaac Newton's greatest achievement in physics — one which would revolutionize our world view.

The Road to a Theory of Gravity

Newton's "law of gravity combined with his laws of motion would explain both celestial and earthly motions" for the first time in history, writes Paul Heckert, professor emeritus of physics and astronomy at Western Carolina University.[71] How exactly did the idea of "universal gravitation" first come to Newton?

According to his own now legendary story, one fine day in May of 1666, 23-year-old Isaac was in his childhood home at Woolsthorpe. He gazed out a window which overlooked mother Hannah's garden. When he saw an apple fall from a tree, a thought came to him.

Like Kepler, Hooke, Huygens, and others, Newton contemplated the idea that "celestial bodies have a power which attracts other bodies — and thereby influences their motion."[72] Perhaps, he thought, the force which holds the Moon in its orbit around the Earth is the *same* force which causes an apple to fall to the surface of the Earth. If the Moon were at rest, perhaps it too would fall to the Earth.[73] Like Kepler, he referred to this force as *gravitas*, after the Latin word for "heaviness" or "weight."[74]

Although controversial, a number of historians suspect the legendary apple story is at best a simplification. "Evidence suggests that Newton's memory was not accurate . . ." historian Robert Hatch, emeritus professor at the University of Florida, tells us, "The concept of universal gravitation did not spring full-blown from Newton's head in 1666 but was nearly 20 years in gestation."[75]

Beginnings

It appears that Newton put together his laws of gravity in three stages. In the 1660's, he assumed incorrectly that the Moon travels around the Earth in a *circular* orbit. He was not yet able to deal with Kepler's elliptical orbits.[76] He also imagined that the Earth's attractive force counter-balanced the Moon's centrifugal force. He did not yet realize that the Moon's gravity also attracts the Earth, as do all masses.[77]

He tested this relationship with observations of the Moon's motion. His calculations using circular orbits were "pretty near" to the observed motion, but not accurate enough to satisfy him.[78] He abandoned the gravity problem for the time being.

Hooke as Catalyst

Newton's interest in gravitation was renewed by his nemesis Robert Hooke. In an exchange of letters beginning in 1679, Hooke proposed that the orbit of a planet or moon is the result of two things;

(1) the object's tangential motion and
(2) an "attractive motion towards a central body."

Hooke was referring to his 1674 paper, which also said:

> ... *all Coelestial Bodies ... have an attractive or gravitating power towards their own Centers ... they do also attract all other Coelestial Bodies that are in the sphere of their activities ... these attractive powers are so much more powerful ... by how much nearer the body wrought upon is to their own Centers ... I have not yet experimentally verified ...*
>
> — Robert Hooke, *Attempt to Prove the Motion of the Earth from Observations* (1674, republished 1679).[79]

It seems Hooke had come very close to a universal theory of gravitation — at least conceptually. But it isn't enough to come up with concepts in physics. You have to produce equations which tell exactly how the concepts work quantitatively — equations which Hooke lacked.[80]

Intrigued by Hooke's insights, Newton quietly pursued his own gravitational theory once again. Here he brought his prodigious mathematical powers to bear.[81]

Hooke, along with Christopher Wren, and Newton's friend, English astronomer Edmund Halley had struggled in vain to solve the problem of planetary motion. They knew Kepler had proposed that planetary orbits are elliptical. They had also surmised that the gravity between two bodies weakens with the *square* of the distance between them, the so-called inverse-square law.[*] But how to reconcile the two mathematically? In the summer of 1684, a frustrated Halley decided to pay a visit to Isaac Newton.[82]

"What type of curve does a planet describe in its orbit around the Sun, assuming an inverse square law of attraction?" Halley asked Newton.

---

[*] Halley, Hooke and Wren, architect of the new St. Paul's Cathedral in London, met in the alleged "Coffee House" meeting of January 1684. Here Hooke claimed that he had derived the inverse-square rule and the laws of planetary motion. He was never able to produce the mathematics to back up his assertion. Sources: Turnbull et al, paraphrase of 1686 report by Halley, pp. 431–448.; Christie, "The man who invented gravity." thonyc.wordpress.com.

"An ellipse," Newton answered.

When Halley asked the English genius how he knew this, Newton said he had derived it mathematically four years earlier. Newton couldn't find the proof among his papers. So, he rederived it and sent Halley an improved and amplified version of the proof three months later — his 9-page manuscript, *De motu corporum in gyrum* (*Of the motion of bodies in an orbit*).[83]

The Inverse Square Law and Kepler's Elliptical Orbits

Newton proposed, as Hooke et al had suspected, that gravity weakens with the square of the distance — the aforesaid inverse square law. Why inverse? Because you *divide* by the distance squared to calculate the drop-off in gravity.

For example, if the *distance* between the Earth and the Moon were *twice* as great as it is now, the attraction or *force* of gravity between them would be *four* times weaker. Two squared equals two times two equals four. Similarly, if the Earth and Moon were *three* times further apart, the gravity between them would be three times three or *nine* times weaker, and so on.

Newton had sought to reconcile Kepler's law of elliptical orbits with his inverse square law. It would not be easy. Others had tried. Only Newton possessed "the mathematical ability to do so."[84]

> *Newton's proof of the connection between elliptical orbits and inverse-square forces ranks among the 'top ten' calculations in the history of science.*
>
> — J. Prentis, Physics Professor, University of Michigan[85]

Using a geometric form of his second law of motion — force equals mass times acceleration or $F = ma$ — and the calculus that he had invented — Newton came up with a mathematical relationship between the *shape* of a planet's orbit and *how fast* the force of gravity drops off with distance.[86]

Newton determined that different orbital shapes give a different drop-off in gravity. For the case of an *elliptical* orbit where the Sun is located at one focus of the ellipse, the force of gravity drops off as the *square* of the distance.[87]

Thus, Newton tells us, since the force of gravity falls off as the distance squared, planets circumnavigate the Sun in elliptical orbits. In effect, from the inverse-square rule, Newton had derived Kepler's first law of planetary motion. (He derived Kepler's second and third laws as well.)[88]

With the inverse-square law of gravity, the Great One had produced a physical explanation for Kepler's Laws — one that applies for "any celestial object orbiting another object."[89]

Newton's Magnum Opus

After seeing the breath and scope of Newton's *De motu* in November of 1684, the excited Halley encouraged the now forty-two-year-old to pull together his notes and "write a full treatment of his new physics." He warned Newton that "others may scoop him."[90]

Inspired by Halley's enthusiasm and shaken by his threat, the Great One began working full-time on the problem. Here he would apply his fanatic work ethic and prodigious genius to produce a three-volume masterpiece. Stephen Hawking called it, "the most important single work ever published in the physical sciences."[91]

Of all Newton's achievements, none had greater impact on science, culture, and humankind's worldview. His celebrated creation is presented in the next chapter.

Chapter 25

THE HEAVENS AND THE EARTH

What Des-Cartes did was a good step. You have added much several ways . . . If I have seen a little further it is by standing on the sholders of Giants.

— Isaac Newton, Letter to Robert Hooke, 1675[1]

All in all, it took the poor farmer's son some 28 months to change the world. A storm of ideas, all that he had thought about since his college years flooded his mind. Bringing to bear all his mental powers as never before, "he worked himself at a fever pitch." He examined and reexamined each of his deductions again and again.[2]

His secretary Humphrey Newton (no relation) described his employer's efforts during this period[3]:

> *So intent so serious upon his studies y^t he eat very sparingly, nay ofttimes he has forgot to eat at all . . . He rarely went to Bed, till 2 or 3 (am) of y^f Clock, sometimes not till 5 or 6 (am) lying about 4 or 5 hours . . . 'I never saw him take any Recreation or Pastime, either in Riding out to take y^f Air, Walking, Bowling, or any other Exercise whatever . . . he seldom left his chambers . . .*
>
> *[He] very rarely went to Dine in y^f Hall unless upon some Publick Days, & then, if He had not been minded, would go very carelessly, wth shoes down at Heels, Stockins unty'd, Surplice [vestment] on & his Head scarcely combed . . .*

The results of Isaac Newton's fierce labors were at once intricate and comprehensive, profound and revolutionary. He produced groundbreaking physical laws which explained the workings of the cosmos with a broadness of scope and richness of theory never before seen. Grounded on Earthly experiments and heavenly observations, they would tie together sundry behaviors into a single set of physical laws for the first time in human history.

On April 28, 1686, some 18 months after Halley's initial request, Newton forwarded the first book of his masterwork to the Royal Society. Two more books would follow in March and April of 1687, respectively.[4] When the Royal Society backed out, Halley himself financed and managed the publication of Newton's *tour de force*.[5]

The Great One titled his life's greatest achievement *Philosophiæ Naturalis Principia Mathematica (Mathematical Principles of Natural Philosophy)*. Commonly called the *Principia* for short,[6] it would be published on July 5, 1687.

Parliament would enact the Declaration of Rights two years later. Despite its limitations, the Declaration would represent a watershed event in the advancement of human rights. Despite its loose-ends, the *Principia* would represent a watershed event in the advancement of science.[7]

Let's take a brief journey through Newton's masterpiece.

The *Principia*

In the *Principia*, Isaac Newton turned his 9-page *De Moto Corporum in Gyrum (On the motion of bodies in an orbit)* of 1684 into a 500-page, three-book exposition on motion and gravity.[8]

In a rare act for him, Newton thanked Halley in its Preface:

In the publication of this work the most acute and universally learned Mr. Edmund Halley not only assisted me in correcting the errors of the press and preparing the geometric figures, but it was through his solicitations that it came to be published . . .[9]

In return, Halley famously wrote of his hero in a prefix:

Nearer to the gods no mortal may approach.[10]

Unlike Galileo, Newton published the *Principia* in classical Latin. Abstruse in style, it was an extremely difficult read, even for experts. Ever sensitive to the slightest criticism, Newton later boasted he had deliberately made the book obscure "to avoid being bated by little smatterers in mathematics."[11]

The *Principia* of 1687 would be followed by a second edition twenty-six years later in 1713, and a third edition thirteen years after that in 1726.[12]

When Robert Hooke learned of Newton's gravitational theory, he justifiably claimed "his letters of 1679–1680 earned him a role in Newton's discovery," writes Christianson.[13] As noted, Hooke, Halley, and Wren had also proposed the inverse square law of gravity — without the mathematics to connect it to elliptical orbits.[14]

Hooke demanded Newton acknowledge his contribution. The petulant Newton "threatened to withhold his masterpiece from publication." Haley tried to smooth things over. Newton agreed to publish — with one condition. Every mention of Hooke's name must be deleted from the manuscript.[15]

. . . in this experimental philosophy, propositions are deduced from phenomena and are made general by induction.

— Isaac Newton.[16]

The Great One presented a number of mathematical approaches in the *Principia*. This included a geometric form of his calculus based on "ratios of vanishing small geometric quantities." He would then rigorously test mathematical predictions against experimental results.[17]

Newton conducted an "impressive array of experiments" to ground his theoretical work, Tufts University philosophy of science professor George Smith tells us. They included "pendulum-decay experiments; different size bobs in air, water, and mercury; and vertical-fall experiments in water and air. The latter included dropping objects from the top of the dome of the newly completed St. Paul's Cathedral."[18]

Newton divided the *Principia* into three books: Book 1 is on mechanics and his three laws of motion. Book 2 is on the "motions of bodies in surroundings that offer resistance, such as air or water." Book 3 is on his celebrated law of universal gravitation.[19]

Let's take a brief look at each book.

<u>Book 1</u>: *De motu corporum* (*On the motion of bodies*)

> ... *Rational Mechanics will be the sciences of motion resulting from any forces whatsoever . . . accurately proposed and demonstrated . . .*
>
> — Isaac Newton. Preface, *Principia*[20]

Isaac Newton presented his famous Laws of Motion in his first book. They were based primarily on the work of Galileo and Descartes. Newton's three laws were deceptively simple — and brilliant. They would become the foundational laws of the science of mechanics, how forces affect physical bodies, for the next two hundred years.

Newton's three Laws of Motion are presented below in modern terms.[21]

(1) *The Law of Inertia* — an object at rest or in *uniform* motion* stays at rest or in uniform motion until acted upon by an outside force.
(2) *The Force Law* — Force equals mass times acceleration ($F = ma$).
(3) *The Force Pairs Law* — For every action, there is an equal and opposite reaction.

The Law of Inertia — Newton's first law has been called "possibly the most important concept on the formation of modern physics."[22] Aristotle and Kepler held that a body in uniform motion requires the constant application of force to keep it moving. Newton argued correctly that a body in uniform

---

* Again, in physics, "uniform motion" is defined as motion at a constant speed and in a constant direction. The term "velocity" includes both speed and direction. Hence uniform motion is also constant velocity. The term "acceleration" is defined is a change in speed (faster or slower) and/or a change in direction; i.e., change in velocity.

motion stays at that same speed and in the same direction indefinitely, with no need for an outside force.[23]

Think of a car hydroplaning on a wet or icy road. No force is needed to keep it moving. Even if you shut off the engine, it will still glide along. Since the surface of the road is nearly frictionless, to the driver's dismay, the car will continue at (nearly) the same speed and in the same direction until an outside force stops the car (e.g., collision with a light pole).

The idea of inertia is an ancient one. Chinese philosopher Mozi (ca. 400 BC) said: "The cessation of motion is due to the opposing force.[24] If there is no opposing force ... the motion will never stop." The great eleventh century Islamic scientist Ibn al-Haytham had a similar idea: "An object will move perpetually unless an (outside) force causes it to stop or change direction."[25] Newton got his Law of Inertia from Descartes, who got it from Dutch scholar Isaac Beeckman (1588–1637). See endnote for details.[26]

The Force Law — Newton second law tells us "how much force is required to accelerate an object." Here *force* is proportional to a body's *mass* times its *acceleration*. Newton defines "mass" as the quantity of matter in a body, *à la* Kepler. The greater the mass of a body, the more force it takes to accelerate it — that is to change its speed and/or direction.[27]

This is why, for example, you need a more powerful engine, i.e., more force, to accelerate a massive dump truck than a much less massive sports car.

The Force Pairs Law — Unlike his first two laws of motion, Newton's third law is wholly original with him. It is also called the *Law of Action and Reaction*. More formally, "when one body exerts a force on a second body, the second one exerts a force on the first equal in magnitude and opposite in direction."[28] Examples: When you fire a bullet out of a rifle, the rifle recoils in the opposite direction — each with equal and opposite force.[29] Blow up a balloon with air and let it go. The balloon pushes (expels) the air backwards. The air in turn pushed the balloon forward with equal and opposite force.[30]

<u>Book 2</u>: *De motu corporum* (*On the motion of bodies*) second part.

The second book of the *Principia* deals with resisting mediums, such as air and fluids. Why would Newton dedicate an entire book to such a relatively minor subject?

> *The hypothesis of vortices is beset with many difficulties.*
>
> — Isaac Newton[31]

The ever-vindictive Newton wrote Book 2 "largely to refute René Descartes' vortex theory of planetary motion." The French philosopher's construct competed with Newton's new theory of gravity. Though incorrect, it had gained "somewhat wide acceptance at the time."[32]

Newton would not have it. He spent some 80 percent of Book II in repudiation of Descartes' theory.[33] For a brief explanation of the theory, please see endnote.[34]

Newton's pettiness also extended to the title of his great work. In 1686, a year before its initial publication, he changed its title to <u>Philosophiae Naturalis Principia Mathematica</u> (my underline). This was an "obvious poke in the eye" at Descartes's *Principia Philosophia*, in which he had presented the vortex theory some forty-two years earlier. The Great One even had the first and third words of the title page to his first edition printed in larger type; as *Philosophiae Naturalis Principia Mathematica*. (See Fig. 25.1.)[35]

No action was too small for Newton in mocking a competitor — even the now deceased Descartes. Book II contains what we now know are fundamental errors in Newton's physics of fluids. It would come to have the least impact of all his books on science.[36] Karma?

In his third book, Newton presented his magnum opus: his theory of universal gravitation.[37]

<u>Book 3</u>: *De mundi systemate (On the system of the world)*[38]

In the final book of the *Principia*, Newton proposed that "a mysterious grand force permeates the cosmos" — an invisible force which he called gravity, *à la* Kepler.[39]

Figure 25.1. Newton's own first edition (1686) copy of Principia. Within are handwritten corrections for the second edition. Wren Library, Trinity College, Cambridge.

There is a power of gravity tending to all bodies proportional to the several quantities of matter [masses] which they contain."[40]

Newton is telling us that every object attracts every other object in the universe. The more massive an object, the more its gravitational attraction.

Your body right now and all the other objects in your surroundings are pulling towards each other due to their respective gravity. You don't notice this attraction because the masses involved are way too small to produce an appreciable effect. For massive objects like the Earth, the effect is dramatic. As you read this, it is the gravitational force generated by the mass of the Earth which holds you and all the other objects to the ground.

What produces Earth's net gravitational attraction? In a key insight, Newton realized that gravity applies to all bodies with mass, macroscopic and microscopic. Thus, it is the accumulation of mass from all microscopic parts of the Earth which produces its overall gravity.[41]

Newton founded his theory of gravity on his Laws of Motion and an ingenious analysis of what he termed "centripetal force." He defined this new conception as "a force by which bodies are drawn or impelled, or in any way tend, *towards*[*] a point as to a centre."

Whirl around a ball on a string in a circle over your head. The tension on the string represents centripetal force. This force is directed towards the center of the circular motion — your hand. This force keeps the ball from flying off tangent to the circle.[42]

Newton first established how a body moves under centripetal force. He then extended it to a gravitational force with an inverse square drop-off. Then he considered a spherical body of spherically-symmetric density. In a "brilliant argument," he showed that the gravitational attraction of such a body could be treated as though *all its mass* were concentrated at the *center* of the sphere. This "center-of-mass" insight greatly simplified modeling the gravitational effects of celestial bodies mathematically.[43]

Newton demonstrates in the *Principia* that the force of gravity which governs the fall of objects to the ground here on Earth is the *same* force which governs the orbits of celestial objects in the heavens.

The Cannonball, the Apple and the Moon

Millions saw the apple fall, but Newton asked why.

— Bernard Baruch[44]

---

[*]Christian Huygens coined the term "centrifugal force ""in his 1659 *De Vi Centrifuga*. This is an (apparent) force related to inertia which acts *outward* on a body moving around a center point.

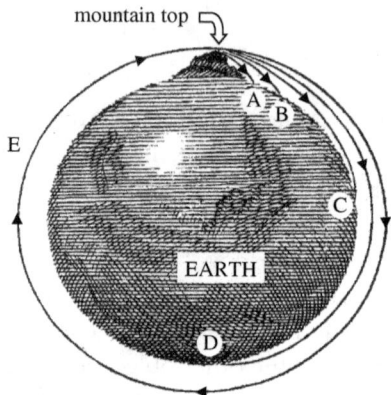

Figure 25.2. Newton's Orbiting Cannonball. From mountain top, you fire the cannonball horizontally (parallel to Earth's surface). (A): Ball drops to surface due to gravity. (B), (C), and (D): The greater the ball's speed, the further it goes. (E): If you fire the cannonball at sufficient speed, it will orbit Earth! (Not to scale. Mountain height greatly exaggerated. Air effects ignored).

Newton asks us to imagine a cannon placed on top of a very high mountain here on Earth. The cannon is set up to shoot a cannonball horizontally, i.e., initially parallel to Earth's surface. Once the cannonball explodes out of the cannon, it immediately begins to fall towards the surface of the Earth. Newton shows this in the *Principia* with his famous drawing. (See Fig. 25.2.)

Newton knew the ball has *two* components of motion:

(1) a *horizontal velocity*. It wants to continue in uniform motion *in a straight line* due to his law of inertia, and
(2) a *downward acceleration*. It wants to fall due to Earth's gravity.

A combination of the two motions determines the cannonball's *net* motion. The cannonball starts out moving tangent to the Earth's surface, falls in an arc, and soon lands on the surface of the Earth.

Now Newton gets clever. He asks: what happens if I put more and more powder into the cannon? The cannon shoots the ball out at greater and greater speeds. Thus, the cannonball travels farther before it hits the ground.

What if the cannonball travels such a great distance that we have to take into account the *curvature of the Earth*? In this case, as the ball travels in its arc towards the ground, the ground curves away from it. (Again see Figure 25.2.)

What if the cannonball travels so fast that *it falls at the same rate as the ground is falling away*? Newton realized that cannonball will never reach the Earth's surface. It will orbit the Earth.[45]

Now here is Newton's brilliant leap (with help from Hooke): The Moon also orbits the Earth. It too possesses two motions:

(1) straight line motion tangential to the Earth, and
(2) accelerating motion (falling) towards the Earth. The combination of these two motions results in its net motion orbiting the Earth.

In other words, the Moon is constantly falling towards the Earth, but its motion tangential to the Earth is so great that it orbits the Earth rather than plunging into it.[46] If the Moon had no tangential velocity, it would simply fall to the Earth, just like an apple falling from a tree.[47]

This is a most profound thought. It tells us that the *same* force which makes the apple fall to the Earth also holds the Moon in its orbit around the Earth. And what is true for the Moon is also true for other stellar objects. The same physical law applies here on Earth and in the heavens.

From his third law of motion, Newton deduced that "the Moon pulls on the Earth with the same force as the Earth pulls on the Moon."* Then, from the size of the Moon's orbit, the time it takes the Moon to complete one revolution, and the rate of fall of the cannonball (or an apple) here on Earth, he confirmed that gravity indeed *weakens as the square of the distance*.[48]

Please see Appendix C for mathematical details.

The Law of Universal Gravitation

> ... *all matter attracts all other matter with a force proportional to the product of the masses and inversely proportional to the square of the distance between them.*[49]

In summary, Newton tells us the force of gravity between two bodies is determined by two things: their masses and the distance between them. More precisely, the force of gravity is proportional to the product of their masses divided by the square of the distance between their centers of gravity.[50]

---

*The Moon is about 82 times less massive than the Earth, so this "same force" of gravity affects the Moon much more than it does the Earth.

For example, the gravitational force between the Earth and Moon is proportional to the mass of the Earth times the mass of the Moon. And this force weakens with the square of the distance between the centers of gravity of the two bodies.

The general equation for the force of gravity, F, is:

$$F = G\, m_1 m_2 / r^2$$

Where:

G is the gravitational constant.* Its "value depends on the units used for mass and distance."[51]

m_1 and m_2 are the respective masses of the two bodies, and
r is the distance between the centers of mass of the two bodies.

This simple formula is humankind's first *equation of the universe*. It describes the "gravitational pull of all bodies" in the cosmos, and "governs everything from the drop of a stone to the motions of the planets," writes historian and mathematician Amir Alexander.[52]

The scope and applicability of Newton's law of universal gravitation was unprecedented. It explained the behavior of a number of natural phenomena previously thought to be unrelated or unknown. Today we take this great leap of science for granted. In the seventeenth century, it was nothing short of astonishing.

The Newtonian Universe

This section is chiefly based on George Smith's comprehensive treatise on the Principia.[53]

Isaac Newton wrote in the *Principia*:
"I derive from the celestial phenomena the forces of gravity to which bodies tend to the sun and the several planets. Then from these forces, by other propositions, which are also mathematical, I deduce the motions of the planets, the comets, the moon, and the sea . . ."[54]

---

*The gravitational constant, G, is determined experimentally. Its present measured value in standard metric units is 6.674×10^{-11} m^3·kg^{-1}·s^{-2}. The actual masses of the Earth, Moon, Sun, and planets were unknown at the time of Newton, so he could not establish the value of G. He used *relative* masses in his analyses.

Newton's theory of universal gravitation explained a myriad of physical phenomena on Earth and in the heavens. Let's take a look.

Motion of Planets around the Sun

> ... *the common centre of gravity of the Earth, the Sun and all the Planets is to be esteem'd the Centre of the World.*
>
> — Isaac Newton, *Principia*

From astronomical data, Newton estimated the *mass ratios* of the Sun to Jupiter and the Sun to Saturn. These are the most massive solar system bodies by far. (Again, their absolute masses were unknown at the time.) From this he determined their common center of gravity — effectively the center of gravity of our solar system— is just slightly off from the center of the Sun. Voila![55]

Newton's computations once and for all extinguished the geocentric cosmology of ancient Greek science and medieval Christian theology — as well as Brahe's geo-helio compromise. Copernicus, Kepler, and Galileo were right. The Earth and all other planets do orbit the Sun.[56]

Newton's analysis of astronomical data also confirmed the inverse square law of gravitation — to an "accuracy that was high by the standards of Newton's time," Smith tell us.[57]

Elliptical Planetary Orbits?

> ... *planets neither move exactly in ellipses nor revolve twice in the same orbit. Each time a planet revolves it traces a fresh orbit, as happens also with the motion of the Moon, and each orbit depends upon the <u>combined motions of all the planets</u>, not to mention their <u>actions upon each other</u> ... the simple orbit that is the mean [average] between all vagaries will be the ellipse ...* (My underlines.)
>
> — Isaac Newton[58]

The brilliant Newton saw through the confusion of observational data. He deduced from his theory of gravity that the gravitational attraction of each body in the solar system affects all others. In other words, "every planet is attracted not only by the Sun but also (much more weakly) by other planets."[59]

Picture all the planets going around the Sun, and the various Moons going around some of the planets. Due to their gravity, they are all trying to pull one another towards each other. The results in a complex set of dynamic perturbations — disturbances which change over time — which constantly influence the orbit of each planet and moon.

Newton deduced mathematically that if a planet, say Mars, were the only planet going around the Sun and the Sun were motionless, the planet would indeed trace out an exact ellipse.[60] This explained for the first time why it had been so difficult to predict the future positions of a planet. They attract one another gravitationally — especially when one passes near another. This produces "very small irregular motions" in each planet which "cause its ellipse to deviate somewhat," as Newton put it. These multiple perturbations add up over time, which render long-term predictions difficult to make.[61]

Equatorial Bulge of Earth

Isaac Newton along with Christiaan Huygens proposed that our planet is not perfectly round, but in fact slightly "swollen at the equator and squashed at its poles." This oblate spheroid shape, Newton tells us, is due to Earth's daily rotation around its axis. Modern measurements verify this equatorial bulge: "The distance from the center of the Earth to the equator (at sea level) is roughly 13 miles (21 kilometers) greater than at the poles."[62]

Precession of Earth's Axis

Astronomers knew since ancient times that the point around which the stars appear to rotate in the sky changes very slowly over time. In the northern hemisphere, this point is presently very close to the star Polaris, the "north star" in the constellation Ursa Minor. This change is due to the precession of Earth's spin axis over thousands of years.

Newton presented the first physical explanation for this phenomenon in the *Principia*. As noted above, he had deduced that Earth is not quite a sphere. And that the gravity of the Sun and Moon cause our slightly oblate planet to slowly wobble on its axis like a spinning top — what astronomers call the "precession of the equinoxes."* The modern value to complete a precession cycle is 25,772 years. (See Fig. 25.3.)[63]

---

*The equinox is the two times in the year (Spring and Fall) when the ecliptic and the celestial equator intersect. The ecliptic is the plane of Earth's orbit around the Sun. The celestial

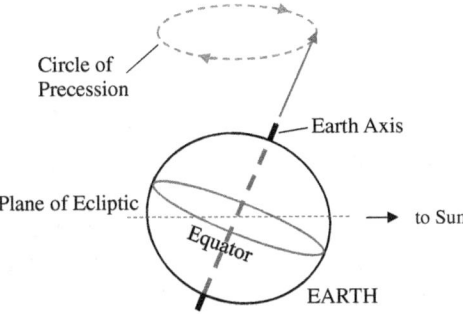

Figure 25.3. Precession of Earth's Spin Axis. The Earth rotates once a day as it moves in its orbit around the Sun. The axis of this rotation is tilted about 23.44°. This axis very slowly precesses over time like a spinning top. As noted, due to its diurnal rotation, Earth is an oblate spheroid. It bulges slightly at the equator and is flattened slightly at its poles. As a result, the gravitational pull of the Sun on the planet is not uniform. This causes the Earth to wobble or precess as it rotates. It takes 25,772 years to complete one full precession cycle (circle of precession).

Tides of Earth's Seas

At least as far back as Ptolemy, astronomers had associated the timing of Earth's tides with the location of the Moon in the sky. Ninth century Persian astronomer Abu Mashar evoked similar thinking. William Gilbert of magnetism fame stated: "the Moon produces the movements of the waters and the tides of the sea . . ."

Johannes Kepler had speculated in *Astronomia Nova* that a mass-dependent mutual attraction explained the Moon's influence on the Earth's tides. He pointed out that sailors had long known that ocean tides are higher when the Earth, Moon and Sun are in syzygy. (Syzygy is when three or more celestial bodies are in a straight line.)[64]

Newton correctly identified the physical *cause* of ocean tides for the first time in the *Principia*. They are due to the Moon and to a lesser degree the Sun's *gravitational pull* on the Earth. His analysis accounted "approximately for marine tides including . . . spring and neap tides" across the globe, George Smith tells us. Though correct conceptually, Newton's explorations were merely quantitative.[65]

(footnote continued) equator is the great circle (with Earth at its center) perpendicular to Earth's axis of rotation. Source: *Encyclopedia Britannica*, Jun 5, 2017, https://www.britannica.com/science/precession-of-the-equinoxes.

Lunar Motion

> *I find this Theory (of the Moon) so very intricate . . .*
>
> — Isaac Newton, letter to John Flamsteed, February, 1695.[66]

Despite its unprecedented accuracy and scope, Newton's theory of gravity left a number of loose ends. Most glaring was his attempt to model the Moon's orbit based on the inverse-square rule.

To precisely describe the behavior of the Moon required solving for the combined dynamic gravitational perturbations of the Earth, Moon, and Sun — a classic "three-body" problem. The mathematics of his theory of gravity worked well for the simplified case of two bodies, but was extremely difficult to solve for three* or more.[67]

From his theory, the Great One was able to mathematically derive three lunar "inequalities" which were in good agreement with observations. They were the recession of the Moon's axis of rotation, the "fluctuating inclination of its orbit," and its departure from Kepler's area rule.[68]

A fourth known inequality proved more difficult — the *precession* of the Moon's *orbit*. Astronomers had observed that the Moon's slightly elliptical orbit precesses as it goes around the Earth In other words, the ellipse itself slowly rotates over time. (See Fig. 25.4.) Observational measurements indicated that the Moon's orbit made a complete revolution about once every **9 years**. (The modern measurement is 8.85 years).[69]

In the first edition of the *Principia*, Newton proposed that the precession of the Moon's elliptical orbit is due to gravitational disturbances from the Sun. He attempted to calculate this lunar precession from his equations. If his theory and mathematics were correct, the value should have agreed with what astronomers had observed.

His computations were "excessively complicated and clogged with approximations," as he put it. With much difficulty, he arrived at a value

---

*Nineteenth century German mathematician Heinrich Bruns determined that there is no "general analytical solution to the three-body problem given by simple algebraic expressions and integrals." In modern times, analytical methods coupled with computer numerical integration are used to derive theoretical values which approach the precision of observations. Sources: Cartwright, Jon. "Physicists Discover a Whopping 13 New Solutions to Three-Body Problem." *AAAS Science*, sciencemag.org. March 8, 2013. https://www.sciencemag.org/news/2013/03/physicists-discover-whopping-13-new-solutions-three-body-problem.: Cook, Alan. "Success and failure in Newton's lunar theory" **Astronomy & Geophysics**, Vol. 41, Issue 6, Dec. 2000, 6.21–6.25. https://academic.oup.com/astrogeo/article/41/6/6.21/225623.

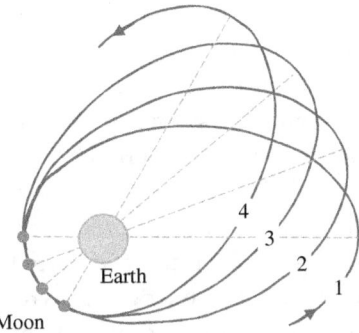

Figure 25.4. Precession of Lunar Orbit. The slightly elliptical orbit of the Moon around the Earth rotates over time. As a result, the closest position of the Moon to the Earth — its perigee — also moves over time. Shown here are four lunar perigee locations (grey dots). They are associated with four orbital cycles. The dashed lines represent the different major axes of the Moon's elliptical orbit. (Figure not to scale. Elliptical orbit and Earth location at one focus greatly exaggerated.)

of **18 years** for a complete revolution of the Moon's orbit. That was *twice* what astronomers had observed. Something was wrong with Newton's mathematical model of the Moon — very wrong.[70]

The Great One became obsessed with getting his flawed lunar model right for the next edition of the *Principia*. Thus began his vicious quarrel with John Flamsteed, Britain's first Astronomer Royal.

Newton had obtained a variety of observational data from Flamsteed in the past. Now he asked the Astronomer Royal for additional data on the Moon — hoping it would resolve the problem with his lunar model.

Like Newton, the Astronomer Royal was a "puritanical, single-minded" perfectionist. He kept his observations under seal until he could confirm them beyond doubt.[71] He reluctantly turned over 50 preliminary lunar observations to Newton. The Great One pressed for more. Flamsteed then gave Newton an additional 150 data sets — some which still contained errors.[72]

Even with the new data, Newton could not resolve the lunar orbital precession issue. Over the next 15 years or so, he continued to badger Flamsteed for more and more lunar data. Flamsteed did not work for Newton. He reported to Queen Anne. The Astronomer Royal had had enough. He refused — again telling Newton his data was incomplete.[73]

Newton was furious. Upon appointment as President of the Royal Society in 1704, he had Flamsteed's name stricken from the list of Society fellows. He used his political influence to obtain a warrant from the Queen.

It effectively gave him authority to oversee and dictate all aspects of the Astronomer Royal's work.

In a fit of pique, Newton had Flamsteed's unfinished Star Catalogue published in 1712 — without the Astronomer Royal's assistance, consent, or knowledge. He appointed his friend Edmund Halley, Flamsteed's competitor, as its editor.

Flamsteed called this unauthorized publication a "villainous outrage."[74] He "managed to gather three hundred of the four hundred printings and burned them." He uncovered errors in Halley's version, corrected them and published his own version of the Star Catalogue.[75]

As for Newton, he published the second edition of the *Principia* in 1713, his Theory of the Moon still incomplete. In it he deleted fifteen mentions of Flamsteed's name.[76]

Isaac Newton "never found a way of deriving the precession of the lunar orbit from his theory of gravity," George Smith writes. "Consequently, he never succeeded with a complete, gravity-derived account" of the Moon's orbit.[77]

Isaac Newton ends Book 3 of the *Principia* with a ground-breaking exploration of comets. His comprehensive analysis takes up about a third of the book.

The Paths of Comets

In the *Principia*, Newton showed (with much difficulty) that comets, like planets, go around the Sun. And they do so in obeyance with his inverse square law of gravitation. This was a striking achievement, considering that, unlike the planets, all he had to work with was a "small number of imprecise, one-shot observations," Smith points out.[78]

A comet had been observed streaking towards the Sun in November of 1680. The following month, a comet was seen traveling away from the Sun. John Flamsteed proposed the radical notion that the two sightings were in fact *the same comet*. Newton agreed. His gravitational analysis confirmed that some cometary orbits can indeed button-hook around the Sun.

Newton then made an even more radical proposal in the *Principia*: A comet in an extreme elliptical orbit may leave our sky and return many years later. Through a painstaking historical search, Edmund Halley determined that the comets of 1531, 1607, and 1682 were in fact the same comet.

Using the mathematics of Newton's theory of gravity, Halley successfully predicted the comet's return in 1758. This occurred after his demise and that of Newton. Halley's Comet, as it is now called, was last seen in 1986 (See Fig. 25.5). It will appear again in 2061.[79]

Figure 25.5.   Return of Halley's Comet in 1986. *Taken by NASA.*[81]

Newton wrote in the Principia:

. . . it is manifest that the heavens are lacking in resistance. For the comets, following paths that are oblique and sometimes contrary to the course of the planets, move in all directions very freely and preserve their motions for a very long time even when these are contrary to the course of planets.

— Isaac Newton, *the Principia.*[80]

The Great One's analysis also showed that, like planets, comets move freely in space without resistance. This was Newtons' most convincing argument against Descartes' invisible fluid vortices alleged to permeate interplanetary space. It effectively destroyed Descartes' vortex theory.

The notion of comets was a particularly troubling one for Isaac Newton. How could a benevolent God create a thing that could destroy life on Earth? Newton argued in the *Principia* that comets do not present a danger to Earth, that "the planes of their [the comets] motion are not in the plane of the ecliptic . . . [thus the Earth is] out of the way of the comet's tails."[82]

Of course, Newton was wrong. The planes of some cometary orbits are at small angles with respect to the ecliptic plane. Though extremely rare, comets and asteroids do strike the Earth. Collisions can be devastating.*[83]

---

*For example, a massive asteroid collision with Earth 66 million years ago is said to have led to the extinction of non-bird dinosaurs. Asteroids are rocky objects. Comets are made of ice and dust. Both orbit the Sun. Sources: Osterloff, Emily. "How an asteroid ended the age of the dinosaurs" nhm.ac.uk. https://www.nhm.ac.uk/discover/how-an-asteroid-caused-extinction-of-dinosaurs.html. Retrieved Jun 16, 2020.; "NASA Science" Jun 3, 2020. https://spaceplace.nasa.gov/asteroid-or-meteor/en/.

Caveat

Hoping again to avoid criticism, the hyper-sensitive Newton limited his claims for his theory of gravity. As there was insufficiently accurate observational data on stellar objects before Brahe, he said his theory was "a provisionally established conclusion" — one that is relevant only from the time of Brahe to his present writings. And that only to a high approximation.[84]

Reaction

With the publication of *Principia*, Isaac Newton received the fame he had secretly longed for. Still, the reception of his masterwork was not without criticism.

The German magazine *Acta Eruditorium* called Newton "the foremost mathematician of the time."[85] Scottish mathematician David Gregory wrote to Newton: "you justlie deserve the admiration of the best Geometers and Naturalists, in this and all succeeding ages."[86]

The French scientific periodical *Journal des Savants (Paris, 1688)* attacked Newton's *Physique* (physics) for being "purely hypothetical" — even though *Principia* contained extensive observational data on the motions of heavenly bodies.[87] Even so, the anonymous French reviewer called Newton's (three laws of) mechanics "the most perfect . . . one can imagine."[88]

Though he was convinced of the inverse-square drop-off of gravity, eminent Dutch physicist Christiaan Huygens called Newton's principle of universal mutual attraction a "manifest absurdity."[89]

The greatest scholarly objection to Newton's theory of gravity was its lack of a mechanism for so-called "action at a distance." How does the Sun, for instance, reach out some 93 million miles (150 million km) of empty space and hold the Earth in its orbit? Newton's non-answer:

> *I have not as yet been able to deduce from phenomena the reason for these properties of gravity, and I do not feign hypotheses . . .*
>
> — Isaac Newton[90]

Because Newton could not identify how gravity propagates through space, supporters of Descartes' vortex theory complained that the *"Principia* was a work of mathematics, not physics." They were in effect accusing Newton's theory of being like Ptolemy's model for the motion of planets — merely a computational device.[91]

German polymath Gottfried Leibniz and his supporters went even further. They declared Newton's action at a distance as "tantamount to the Aristotelian belief in occult qualities." — as though objects "possess a distinct spiritual presence."[92] This objection is indicative of the growing separation between science and religion.

It would take some 50 years before Newton's theory of gravity began to be generally accepted outside England.[93]

Newton's brilliant *Principia* set the standard for all future scientific treatises. Newtonian physics would mark the death knell of 2000-year-old Aristotelian physics, as well as the metaphysics of Descartes. Newton's masterwork would become the crown jewel of the Scientific Revolution.

Religious Issues

I am a friend of Plato, I am a friend of Aristotle, but truth is my greater friend.

— Isaac Newton, 1663.[94]

Newton's greatest revelation was that gravity is *universal*. Recall Aristotle and Church dogma held that there are two sets of natural laws — one for the impure Earth and another for the pure heavens. Newton showed that Earth and the heavens are governed by the *same* laws of physics.

Galileo would have been thrilled to learn that Newton had verified what he had suspected — the physical laws which govern how objects fall from the Leaning Tower of Pisa, how a ball rolls down an inclined plane, and how a cannonball's arc follows a parabolic trajectory here on Earth are *the same laws* which govern the orbits of all the planets, the orbits of the moons of the Earth, Jupiter, and Saturn, and the paths of comets in the heavens.

In a sense Newton liberated Western thinking from the Biblical curse of Eve's original sin. The world was no longer divided into the pure, incorruptible heavens and the transitory, impure, sinful Earth. In terms of physical laws, we humans live on a planet that is no different than the rest of the natural universe.

Impact

Newton's gravitational theory explains Galileo's observations on falling objects here on Earth. It provides "ingenious proofs" of all three of Kepler's laws of planetary motion. It ties the two disparate physical-mathematical schemes of Galileo and Kepler under a single overarching construct. And most important, it has proven superbly accurate in predicting a vast multitude of terrestrial and solar system gravitational phenomena.[95]

Newton's physics would "guide science for over the next 200 years."[96] His equations of classical mechanics are still used in the 21st century for applications where speeds are a small fraction of the speed of light, and where masses are not stupendously large. (Otherwise, we need the relativity equations of Albert Einstein.)[97]

Newton's construct and his methods has also changed the pursuit of science and a number of other disciplines.

A New Scientific Approach

> *Nature does nothing without purpose or uselessly.*
>
> — Aristotle[98]

Newton's Law of Universal Gravitation is essentially a mathematical construct. It identifies no purpose or reason behind gravity as the governing force of planetary motion.

Newton's method for establishing a physical theory set the example for modern empirical science: equations which accurately represent the behavior of physical phenomena verified by observation and measurement. No purpose need be identified.

The universe would come to be described by scientists not according to Scripture but according to science and its so-called laws of nature. The language of the cosmos would not be Hebrew or Latin or Arabic. It would be mathematics.

The Industrial Revolution

"Isaac Newton's laws of motion and his law of gravity — and the corresponding belief in the power of science — would be integral to the great Industrial Revolution. It is not a coincidence that it began in Great Britain.

The rise of manufacturing by machine and factories for the mass-production of goods gave rise to unprecedented income and population growth — a transformation which continues across the world to this day.[99]

Endings

In his lifetime, Isaac Newton received many honors. He was appointed to the Lucasian Chair at Cambridge University at age twenty-six, a position recently held by Stephen Hawking. In his later years, Newton served as a member of parliament (MP) from 1689 to 1690 for the University of Cambridge — where he never said a single word. This is the same parliament which passed the Declaration of Rights.[100]

From 1696 to 1727, the Great One served as Warden then Master of the Mint. He ruled with an iron fist and arrested over 100 counterfeiters. He served as the dictatorial president of the Royal Society from 1703 till his death in 1727. Knighted by Queen Anne in 1705, Sir Isaac was the first scientist accorded this honor.

(For more on Newton's life outside science, especially his religious writings, please see Appendix D.)

The troubled, domineering, brilliant, obsessive, "tyrannical, autocratic, suspicious, neurotic, and tortured" Isaac Newton died unmarried in London on March 20, 1727. The three-pound preemie from Woolsthorpe-by-Colsterworth has lived for eighty-four years.[101]

He is buried at Westminster Abbey — near the white and grey marble monument in his honor:

> *Hic depositum est, quod mortale fuit Isaaci Newtoni.*
> *Here lies that which was mortal of Isaac Newton.*
>
> — Inscription on Newton's grave[102]

Post-Script: Heliocentricity Confirmed

Newton had come into the world in 1642. By that time, Galileo's s work on the motion of Earthly bodies and Kepler's laws on the motion of planets had led most natural philosophers to accept that Earth indeed moves — even though there was no experimental confirmation.[103]

The proof would come in 1727, the same year Isaac Newton died. English Astronomer Royal and priest James Bradley accidentally discovered what came to be known as *stellar aberration*. His telescopic observations over a twelve-month period showed that Earth's motion around the Sun very slightly alters the apparent angle and location of stars. This constituted the first direct empirical evidence that Earth does indeed orbit the Sun![104]

The Next Chapter

Isaac Newton remains arguably humankind's greatest scientist, and one of its three greatest mathematicians along with Archimedes and Carl Frederick Gauss. His scientific discoveries and the mechanical universe they engendered would come to have a profound effect on our worldview.

Deep cultural changes would exacerbate the separation between science and religion, a turn to agnosticism and atheism for some, and an increasingly secular culture in the West.

This is the subject of the next chapter.

Chapter 26

REASON

We are lost. What are we to believe? They tell us the wisdom of the ancients is false, that Scripture is fiction, that religious authorities have been telling us lies. They say that science is the path to enlightenment. Yet we look to it for truth and find it lacking. It is cold, indifferent. We are but specks in a vast, uncaring cosmos. Where is spirituality? Where is compassion? Where is love? Where is our need to believe in a higher power? Where is God?

Shortly after the death of Newton in 1727, scientists began in earnest to address the "loose ends" outlined in the *Principia*. French mathematician, astronomer, and geophysicist Alexis Claude Clairaut was involved in resolving three key issues.[1]

Oblate Earth — In the 1730s, French scientists conducted astronomical and geodetic measurements to determine the roundness of our planet. They organized expeditions to Lapland led by Pierre Louis Maupertuis, and to equatorial South America led by Charles Marie de La Condamine and Pierre Bouguer. Alexis Clairaut was among the expedition group to Lapland.

Their collective measurements showed that "Earth is indeed flattened at the poles just as Newton had claimed." This was the first empirical evidence that our planet does spin on its axis.[2]

Lunar Anomaly — Recall that Newton's calculation of the precession of the Moon's orbit was off by a factor of two. Astronomical observations indicated a period of about 9 years for a full cycle, while Newton's calculations gave twice that value.

In 1748, Clairaut determined that the problem was with Newton's approximations regarding the Sun's gravitational perturbations of the Moon. Utilizing Newton's inverse square law, Leibniz' version of calculus, and more careful higher order approximations, Clairaut calculated a value of 9 years — in agreement with observation.[3]

Clairaut's model was the "first truly successful descriptive account of the motion of the Moon in the history of astronomy," George Smith writes.[4]

Halley's Comet — Edmund Halley had predicted the return of his namesake comet as sometime between late 1758 and early 1759 — based on Newton's law of gravity and observational data. Clairaut, along with astronomer Jérôme Lalande and Nicole Lepauté sought to make a more accurate prediction.

The trio took into account perturbations on the comet as it passed by massive planets such as Jupiter — both on its way in and back. They used the "most ambitious program of numerical integration" to date. Their prediction agreed with Halley's comet's actual return to within a month.[5]

Scientists and lay people alike saw the verification of an oblate Earth, resolution of the lunar anomaly, and the accurately predicted appearance of Halley's comet as triumphs of Newton's law of universal gravitation.[6]

These and other validations of Newton's theory of gravity had both religious and philosophical implications.

A Clockwork Universe

This most beautiful System of the Sun, Planets and Comets, could only proceed from the counsel and dominion of an intelligent and powerful being. And if the fixed Stars are the centers of other like systems, these being form'd by the like wise counsel, must be all subject to the dominion of One* . . .

— Isaac Newton[7]

Isaac Newton saw his theory of universal gravitation as the handiwork of God. To him, the order and regular movements of the planets was no accident. The fact that all planets orbit the Sun in the same direction and along the same plane was further evidence of God's hand.** The cause for all this order was "not blind & fortuitous," as Newton put it, but an act of One "very well skilled in Mechanicks & Geometry."[8]

Newton's *Principia* of 1687, and *Opticks* of 1704 for that matter, had revealed a universe which behaves in a mathematically predictable manner. The power and universality of the "laws of nature" reached to the very heavens — a realm heretofore thought to belong to religion alone.[9]

Newton's use of reason to develop these majestic natural laws moved a number of intellectuals and eventually many of the public to rank rational thinking on par with religious beliefs — or even superior to them.

The deeply religious Newton would have found this abhorrent. The Puritan farm boy had unwillingly and unwittingly helped spark a new intellectual and philosophical movement — one which would emphasize reason over faith.[10]

The Enlightenment (c. 1685–1800)

Some historians say it began in England with the non-violent Glorious Revolution of 1689 and Parliament's assertion of power over the monarchy with its ground-breaking Declaration of Rights. Others contend it was Isaac

---

* Here in "other like systems" Newton suggests the existence of exoplanets. His use if the word "One" for God is a not-so-subtle reference to his rejection of the Trinity.

** We now know that about 4.5 billion years ago the pull of gravity formed the Sun, planets and rest of the solar system from a spinning cloud of dust and gases rotating in the same direction and the same plane. Source: Redd, Nola Taylor. "How Did the Solar System Form?" space.com. Jan 31, 2017. https://www.space.com/35526-solar-system-formation.html. Retrieved Mar 12, 2019.

Newton's *Principia* which set off the movement. Still others argue that no single event marked its birth.

Whatever its origin, the period from about the late 1600s to the end of the 1700s was called the Age of Reason by its advocates. In the 1800s, the era would come to be known as the Enlightenment — a kind of insult term as it implies that people in earlier eras were unenlightened, specifically the Church and its followers.[11]

What had originated in England soon spread to Scotland, France, and Germany.[12] The era's themes included the use of reason over faith; science founded in experiment; knowledge gained through direct sense-experience; and rigorous skepticism. The movement rejected religious authority, as well as religious dogma, miracles, and magic. It called for individual freedom, equality, and government by consent of the governed — lofty goals not generally realized (in part) until well after the era ended.[13]

Not that religious belief and practices ended. Far from it. Religion still held great power and influence among a number of intellectuals and of course the rest of the population.[14]

Still, a new intellectual freedom was in the air. The written word, the gift of the Sumerians, once again aided by paper and the printing press would spread the ideals, principles, and arguments of the Enlightenment across Europe.[15]

A brief summary of key Enlightenment figures and their influence is given below:

<u>Enlightenment Thinkers</u>

> *... being all equal and independent, no one ought to harm another in his life, health, liberty, or possessions.*
>
> — John Locke[16]

John Locke
(1632–1704)

Englishman John Locke has been called one of the "most influential political philosophers" in modern history. In 1689 — two years after Newton had put out his first edition of the *Principia* — Locke published his seminal work *Two Treatises on Government*.[17]

That same year, Locke met Isaac Newton in London. The two believers in God and science would become close friends — a particular rarity for Newton.[18]

Building on the Glorious Revolution, Locke espoused a radical new vision of liberal government and human rights. He proclaimed that men are "by nature, all free, equal, and independent." They are in "a state of perfect freedom to order their own actions . . . within the bounds of the law of nature." He then declared *reason* as that law of nature.[19]

Locke called for government by the consent of the governed, where the "legislature or supreme power of any common-wealth [is] . . . directed to . . . the peace, safety, and public good of the people." Locke also argued for freedom of religion and religious tolerance — up to a point. He did not support full British citizenship for non-Protestants.[20]

Lockean political philosophy would come to have a profound impact on other nations. The American "experiment" in representative government is a prime example.

> . . . *all men are created equal, that they are endowed by their Creator with certain unalienable Rights, that among these are Life, Liberty and the pursuit of Happiness.*
>
> — Thomas Jefferson *et al.*, *Declaration of Independence* (1776).[21]

The famous beginning of the Declaration of Independence is Thomas Jefferson's "application of John Locke's ideas." The first paragraph speaks of "the Laws of Nature and of Nature's God" — language right out of the Enlightenment. The Constitution of the United States also mirrors Enlightenment principles."[22]

Locke had declared that slavery was "morally reprehensible." This message escaped Jefferson, a slave-owner himself — as were a number of founding fathers including Washington and Madison. How they could sign a document stating "all men are created equal" and refuse to end this most despicable practice is beyond comprehension. We humans can be both good and evil — at times at the same time.

Religion was also a frequent subject of Enlightenment writers.

Hume — Scottish skeptic David Hume argued that religion was an "irrational tradition," miracles a violation of the laws of nature, and angels and demons as "relics of past superstition." Influenced by Newton, Hume dubbed his masterful *A Treatise of Human Nature* (1739) with the subtitle: "An Attempt to Introduce the Experimental Method of Reasoning into Moral Subjects."[23]

The Enlightenment spread to France in the mid-1700s under the absolute monarchy of King Louis XV. French Enlightenment writers tended to be more radical and at times move conservative than their English counterparts.[24]

Diderot — French philosopher Denis Diderot was head of "a group of atheist philosophers." He published the first volume of his mammoth 35–volume *Encyclopédie, ou dictionnaire raisonné des sciences, des arts et des métiers (Encyclopedia, or a Systematic Dictionary of the Sciences, Arts, and Crafts)* in 1751. With the help of 150 scientists and philosophers, Diderot attempted to compile as much known "knowledge as possible." Featuring Enlightenment polemics, his works helped spread "liberal ideas throughout Europe — along with searing attacks against Christianity."[25]

Rousseau — Jean-Jacques Rousseau, an atheist in Diderot's group, published *The Social Contract* in 1762. Echoing John Locke, the French philosopher maintained that "the authority of state comes from the combined will of its citizens."[26]

Montesquieu — French Baron Charles-Louis de Secondat De Montesquieu promoted the separation of governmental powers "among executive, legislative, judicial branches." The American constitution would adopt this principle. Ironically, the French revolution would ignore it.[27]

Voltaire — Under the pen name Voltaire, French provocateur François-Marie Arouet, wrote "revolutionary satires, plays, and poems" which attacked religion and the traditional values of 18th century French society. The agitator ridiculed "religion, theologians, government, armies, philosophies, and philosophers," journalist and author Ian Davidson tells us.[28]

Voltaire's works were widely banned. He was twice imprisoned and spent many years in exile for mocking royal officials with his anti-establishment writings.[29] In ancient Athens, he would have been ordered to drink hemlock. In Galileo's era, he would have been burned at the stake by the Church.[30]

Yet Voltaire loathed democracy. He argued for an absolute monarch — a "philosopher-king" who would rule through reason and justice. He was also against atheism. He worried that the absence of fear of God to punish evil would undermine society and moral order. (Though much discussed, atheism actually had few proponents.)[31]

Voltaire's scandalous satire, the novella *Candide* (1759) was seen at the time as "religious blasphemy, political sedition, and intellectual hostility." It is now regarded as "one of the seminal literary works in history."[32]

Smith — Adam Smith was David Hume's greatest pupil. Smith argued, as did his teacher, that humans are "driven by emotion" which could "only be partially overcome . . . by careful exercise of reason." He famously advocated for a free market economy.

Analogous to Newton's invisible force of gravity, he contended that an "invisible hand" — the market force of supply and demand — automatically guides a free market to reach economic equilibrium.[33] His famed book, *An Inquiry into the Nature and Causes of the Wealth of Nations* (1776) has been called "the first work of modern economics."[34]

Kant — German humanist philosopher Immanuel Kant took a more conservative stance. He sought to reconcile religious belief with rationalism, political authority with individual liberty. In *The Critique of Pure Reason* (1781), Kant "defended belief in God and was deeply concerned with morality, ethics, and law."[35]

Kant's attempt to stem the rebellion against religious orthodoxy was too late. The freedom to publicly express anti-Christian and even atheist views had been established.[36]

Wollstonecraft — Near the end of the 18th century, pioneering English feminist Mary Wollstonecraft also argued for a society based on reason — one where "women as well as men should be treated as rationale human beings." In *A Vindication of the Rights of Woman* (1792), she wrote:

> *I wish to persuade women to endeavor to acquire strength, both of mind and body . . . almost all the civilized women of the present century are anxious only to inspire love, when they ought to have the nobler aim of getting respect for their abilities and virtues.*[37]

Born out of the Enlightenment, a new religious philosophy would take hold amongst a number of the British intelligentsia — one which attempted to reconcile reason and faith, science and religion.

Deism

Deists believed in God — and held that He had set the universe in motion and left it alone to work according to the laws of nature. They saw belief in Jesus as divine as irrational. They considered "religious texts like the Bible as helpful moral guides but not revelations from God."[38]

Deists argued that "reason rather than revelation should form the basis of religions." Deism "embraced all creeds" and hoped its universal language would bring them together under the umbrella of religious tolerance.[39].

Voltaire, Jean-Jacques Rousseau, and Immanuel Kant, as well as American revolutionaries Benjamin Franklin, Thomas Jefferson, Thomas Paine, and Ethan Allen are said to have held deist beliefs.

Deism would soon die out — perhaps because it had no charismatic prophet to inspire belief, no deep appeal to the transcendental or spiritual. Still its tenets of religious tolerance and rational thinking remain with us to this day.[40]

The End of the Enlightenment

The Enlightenment inspired a series of revolutions from the late 1700's to the early 1800's — in the United States, France, and Latin America.[41]

It was the French revolution (1787–1799) with its fall into class warfare, chaos and slaughter which effectively ended the movement — its idealistic hopes for the supremacy of reason in the conduct of human affairs crushed by blood and guillotine.[42]

Enlightenment ideals and principles would live on in Western culture and beyond. The world's democracies today owe their existence in no small part to this philosophical and political movement.

Biblical Criticism

Out of the Enlightenment, the evaluation of biblical texts versus actual history would come "to the forefront of academic work in the 19th century (1800's)." Scholars used Enlightenment themes of rational thinking and science to establish the likelihood of biblical stories and events. Cultural and political causes were examined. Divine intervention onto human affairs, miracles, as well as angels and devils were rejected.[43]

In a number of ways, this book is a product of the Enlightenment.

Newton's Influence

Rooted in Newton's friend John Locke's philosophy, intellectuals would attempt to apply Newtonian perspectives outside of natural world and seek universal laws for government, society, and the economy.

Karl Marx would later argue that he had found the true "economic law of motion of modern society." Sigmund Freud would write about "physical processes as quantitatively determined states of specific material particles . . ." The inevitability of Newton's Laws in controlling physical objects would lead B. F. Skinner to propose that free will is an illusion.[44]

Again, Newton was also key to the Industrial Revolution. It began in 1760's Britain and ended in 1840. A second one extended from the mid-

1800s to the early 1900's. traveling to Continental Europe, North America, Japan, and eventually the entire planet.⁴⁵

Newton had established that *"discernable laws* govern the physical world." This worldview along with his laws of motion and Robert Boyle's work led to the "machines that launched the Industrial Revolution." Chemistry went beyond alchemy. Agriculture became based on science rather than "folklore."⁴⁶ Critical advances in biology and medicine occurred as well.⁴⁷

New Discoveries in the Heavens

In the late eighteenth century and into the nineteenth, two mathematical astronomers would make remarkable predictions regarding our solar system. The first would raise deep religious questions. The second would provide the greatest verification yet of Newtonian physics.

We discuss these accomplishments in the next chapter.

Chapter 27

TWO GIANTS

Mathematics is the music of reason.

— James Joseph Sylvester,
19th century English mathematician[1]

It seems Newton's "For every action there is an equal and opposite reaction" applies at times to historical eras. The Enlightenment and its insistence on reason and logic would lead to an intellectual counter-movement known as Romanticism. Extending from the late 1700s to around the mid-1800s, it advocated emotion over reason, intuition over logic. Followers of the movement promoted the subjective, the irrational, the supernatural, and the occult.[2]

Romanticism was also a reaction to the "population growth, urban sprawl, and industrialization" produced by the Industrial Revolution. The movement glorified nature over urbanization. It also supported the French revolution's ideals of liberty, equality, and fraternity.[3]

Straddling the Enlightenment and Romanticism eras was the "chilly rationalist" Pierre-Simon Laplace, a mathematical genius and one of the greatest scientists of all time. His body of work would have a profound impact on science, mathematics, and astronomy.[4]

The French Newton

All the effects of Nature are only the mathematical consequences of a small number of immutable laws.

— Pierre-Simon Laplace[5]

Pierre-Simon Laplace
(1749–1827)

Pierre-Simon Laplace was born in *Beaumount-en-Auge*, a bucolic village in Normandy on March 23, 1749. His father Pierre was a prosperous cider trader, his mother Marie-Anne Sochon "from a fairly wealthy farming family."[6]

Similar to Newton, Pierre-Simon was aloof, petty, vain, and selfish. Unlike Newton, he was not religious. Though born a Catholic, he was variously described by those who knew him as a Deist, an agnostic, or an atheist.[7]

In 1765, Laplace's father sent Pierre-Simon to the University of Caen to study theology to prepare him for a career in the Church. There he discovered his love of mathematics. Though he did not graduate, the nineteen-year-old's mathematical ability so impressed esteemed French mathematician Jean Le Rond d'Alembert that he arranged for Pierre-Simon to be appointed professor of mathematics at the *École Militaire*.[8]

In his career, Laplace was deeply influenced by French mathematician Adrien-Marie Legendre and others. He gave them little credit in his writings — a not uncommon practice at the time."[9]

The Origin of the Solar System

In 1796, Laplace published his non-mathematical *Exposition du système du monde (Exposition on the System of the World)* or simply *Exposition*. It is considered a "masterpieces of French literature." Here he presented his "nebular hypothesis."

"Only sketchily worked out," it proposed that the solar system originated from the gravitational contraction and cooling of a "large, flattened, slowly rotating cloud of incandescent gas" — in basic agreement with our modern understanding.[10]

In his extraordinary lifetime, the French savant went on to make seminal contributions in:

(1) mathematics, from probability theory to differential equations to Laplace transforms;
(2) physics, from classical physics to mechanics to the kinetic theory of molecular motion;
(3) physical and mathematical astronomy, including the idea of what we now call black holes.
For a more comprehensive list of his remarkable achievements, please see endnote.[11]

Laplace put his expertise in all three disciples to work on what was to become arguably his greatest accomplishment.

The Calculus of the Heavens

> ... *offer a complete solution of the great mechanical problem presented by the Solar System, and bring theory to coincide so closely with observation that empirical equations [those based on based on observations] should no longer find a place in astronomical tables.*
>
> — Pierre-Simon Laplace[12]

Beginning in 1773, Laplace labored for over twenty-five years on a comprehensive model of the solar system. His goal was no less than a mathematical model so accurate that observational corrections would no longer be necessary. To facilitate this lofty objective, he transformed Newton's Law of Universal Gravitation into the language of differential calculus.[13]

Recall that in the *Principia*, Newton had used a cumbersome geometric calculus to model the Solar System. Laplace used mathematical calculus-based techniques to account for higher order effects of planetary interactions that Newton had missed. The results: A more accurate understanding of the motions of known planets and moons in our solar system.[14]

Recall that there is no exact solution to the "three body" problem, such as the dynamics of the Earth, Sun, and Moon. Laplace faced an even bigger problem. He considered the combined motions and inter-gravitational effects of all the planets and moons. Here he made brilliant use of differential equations, approximations, interpolations and numerical iterations to establish the accuracy of his models.[15]

From 1799 to 1825, Laplace published his work in a five-volume magnum opus, the densely mathematical *Traité de mécanique céleste (Treatise of Celestial Mechanics)*.[16]

"The first two volumes, published in 1799, contain methods for calculating the motions of the planets, determining their figures, and resolving tidal problems," British mathematician W. W. Rouse Ball wrote in 1908. Here Laplace provides the "first major theoretical advance" in modeling tidal behavior since Newton's rather weak attempt. "Known to tidal scientists as the Laplace tidal equations, the equations remain the basis of tidal computation to this day."[17]

"The third and fourth volumes, published in 1802 and 1805, contain applications of these methods, and several astronomical tables. The fifth volume, published in 1825, is mainly historical, but it gives as appendices the results of Laplace's latest research."[18]

A challenge to read like the *Principia*, Laplace's *Celestial Mechanics* contained "1500 pages of dense analysis and calculation." In an analysis spanning centuries of orbital motion, Laplace's equations of celestial motion "successfully accounted for all *observed deviations* of the planets from their *theoretical orbits*," according to British mathematician and science historian Gerald James Whitrow.[19] His equations modeled the future positions of the known planets and their known moons to a predictive accuracy never before achieved.[20]

Celestial Mechanics and its masterful mathematical model of the solar system made Laplace a celebrity."[21]

One of the most impressive results of Laplace's efforts was his analysis of the *stability* of the solar system. Of all his work, it represented his greatest impact on the conflict between religion and science.

No Need for God?

> *... the Planets move one and the same way in Orbs concentrick, some inconsiderable Irregularities excepted which may have risen from the mutual Actions of Comets and Planets upon one another, and which will be apt to increase, till this System wants a Reformation.*[22]
>
> — Isaac Newton, Opticks.

Recall from Newton that the gravitational attraction of each planet perturbs the orbits of all other planets. Though a small effect, Newton feared such disturbances would eventually destabilize orbits and possibly result in future planetary collisions. He believed that a merciful God would perform periodic adjustments to planetary orbits to avoid such collisions and maintain stability.[23]

In a brilliant analysis, Laplace declared that the solar system is inherently stable. He showed mathematically that the *average* motion of each planet does not vary over time. More precisely, he demonstrated that "eccentricities and inclinations of planetary orbits to each other always remain small, constant, and self-correcting," write J. J. O'Connor and E F Robertson, noted authors of the MacTutor History of Mathematics Archive.[24]

> *... the conditions of the arrangement of the planets and their satellites are precisely those which ensure its stability.*
>
> — Laplace, Exposition[25]

Do you see it? Yes, each planet perturbs all the others. But the net results are "small oscillatory corrections to unperturbed orbits." The overall effects on planetary disturbances are "conservative and periodic, not cumulative and disruptive. There are no runaway behaviors."[26]

Laplace's demonstration of the stability of the solar system* has been called "the most important advance in physical astronomy since Newton."

---

*Modern computer simulations indicate that the solar system will remain stable for at least the next hundred million years. Beyond that there is an estimated 99% probability it will remain so until the Sun runs out of nuclear fuel. At that point some 5 billion years from now, the Sun is projected to turn into a red giant, "swallow up the inner planets and incinerate the outer ones." Source: Tremaine, Scott. "Is the Solar System Stable?" IAS 2011. https://www.ias.edu/ideas/2011/tremaine-solar-system. Retrieved April 16, 2020.

It won him associate membership in the French Academy of Sciences that same year."[27]

Laplace and Napoleon

Laplace was appointed examiner of the French Royal Artillery Corps in 1784. In this capacity, he inspected and "passed the 16-year-old Napoleon Bonaparte" in 1785. He later served briefly as Minister of the Interior under Emperor Napoleon.

Laplace allegedly attended an audience with Napoleon in c. 1802. Astronomer William Heschel was also said to be in attendance (more on him later). Napoleon asked Laplace why he hadn't mentioned God in his famous treatise on the stability of the solar system. "*Je n'avais pas besoin de cette hypothèse-là*," or "I had no need for that hypothesis," the great French savant is said to have answered.[28]

If the essence of the conversation is correct, Laplace's answer was not that there is no God. It was rather that the solar system does not need God to maintain its stability. According to his analysis, it is inherently stable.

For some, God had become at best a distant power who does not interfere with everyday physical processes, or perhaps an unnecessary higher power who does not exist.

Determinism

> *An intellect which at a certain moment would know all forces that set nature in motion, and all positions of all items of which nature is composed . . . would embrace in a single formula the movements of the greatest bodies of the universe and those of the tiniest atoms . . .*
>
> — Pierre-Simon Laplace[29]

Laplace published *Essai philosophique sur les probabilités* (*A Philosophical Essay on Probabilities*) in 1814. In the introduction, "he extended an idea of Gottfried Leibniz" — the notion of physical determinism.[30]

Imagine you somehow knew the exact positions, forces, and motions of all objects in the entire universe. You could then, in principle, predict their exact behavior into the infinite future and deduce what their behavior had been in the infinite past.

This implies that man, using the laws of nature could possess "God-like powers."[31]

Emulating the ancient Greek stoics, Laplace saw the "appearance of chance as the result of human ignorance." He said as early as 1783 and later popularized in his *Essai* of 1814: "The word *chance* only expresses our ignorance of the causes of the phenomena that we observe . . ."[32]

This notion of determinism and the potential power of human knowledge in science further questioned the need for God. Determinism would become a fundamental tenet of classical physics until the early 20th century, when it would be challenged by the Uncertainty Principle of Quantum Mechanics, as we shall see.

<u>We Chase Phantoms</u>

Near the end of his life Laplace is said to have told "Swiss astronomer Jean-Frédéric-Théodore Maurice that 'Christianity is quite a beautiful thing.'" Laplace applauded its civilizing influence. Nonetheless the great French thinker "remained a skeptic" to the end of his life.[33]

When Laplace lay dying, French mathematician and physicist Siméon Denis Poisson complimented him on his "brilliant discoveries." The French savant replied, "Ah! we chase after phantoms [*chimères*]."[34]

Pierre-Simon Laplace passed away on March 5, 1827 a few weeks shy of his 78th birthday. His physician, François Magendie removed his brain and held on to it for many years. It was eventually exhibited in a mobile "anatomical museum in Britain."[35]

Six years after the death of Laplace, English polymath William Whewell "coined the word 'scientist' at an 1833 meeting of the British Association for the Advancement of Science." The term "natural philosopher" would go the way of the dodo bird. It confirmed that science was no longer a as a subset of philosophy.[36]

As a scientist, it is one thing to explain the physics behind a known phenomenon or observation. It is quite another to use a scientific theory to predict the existence of something never before seen.

The Prediction

Pythagoras of ancient Greece proposed the Sun, Moon, and planets emit a "unique hum" based on their respective orbital revolutions. The sounds were "imperceptible to the human ear." Johannes Kepler postulated that "these harmonies could be heard by the soul — music in the imitation of God."[37]

Our story begins at the height of the Enlightenment back in 1781. German-born musician, music teacher, and composer William Hershel spends his nights in his garden in Bath, England searching for binary stars with his homemade telescope. The amateur astronomer's nearly 7-foot (2 m) long Newtonian reflector is amongst the largest and finest of its day. (See Fig. 27.1.)[38]

On the night of March 13, 1781, he spots a curious diffuse object, a "greenish disc of light" too large to be a star. He observes the strange object again and again for the next month. It is moving against the background of "fixed" stars. He reports it as a comet[40]:

. . the comet appeared perfectly sharp upon the edges and extremely well-defined without the least appearance of any beard or tail.

— William Herschel[41]

Figure 27.1. Herschel's Homemade Telescope. Grinding and polishing the speculum-metal mirrors himself, Hershel's Newtonian reflector had a 6.1-inch (15.5 cm) diameter, a focal length of six- and-a-half feet (almost two meters) and a magnification of 227X. He would go on to construct "two massive reflecting telescopes, a twenty-foot (six-meter) and a forty-foot (twelve-meter) instrument."[39]

A comet? A comet that bright had to be close to the Sun. Yet it had no tail. And it moved so slowly.[42]

Professional astronomers soon trained their telescopes on the object. Three in particular studied its motion: Pierre-Simon Laplace, Finish astronomer and mathematician Anders Lexell, and comet hunter Jean Baptiste De Saron. By May, the trio came to the "same conclusion independently." Hershel's so-called "comet" moves in a nearly circular orbit around the Sun.[43]

Hershel had stumbled upon a discovery of the ages — a seventh *planet* orbiting our home star. This was the first planet to be identified since ancient Babylonians counted "wandering stars" some three thousand years earlier.[44]

German astronomer Johann Bode, director of the Berlin Observatory, proposed naming the new planet Uranus (Fig. 27.2) — after the Greek god of the sky and father of Saturn, the former outermost planet. It stuck.[45]

Figure 27.2. The "Ice Giant" Planet Uranus. From NASA's Hubble Space Telescope in 2005. The seventh planet in our solar system is made of ice, gases, and liquid metal. It is covered in methane clouds. It is called an "ice giant" because its "rocky, ice core is proportionally larger than its amount of gas." The planet's strange tilt of 97.7° "causes its axis to point nearly directly at the Sun." It is four times wider and over fourteen and a half times more massive than Earth. It has 27 known moons and 13 known rings.[49]

With its discovery, Herschel had unwittingly "doubled the diameter of the known solar system."[46]

Herschel's electrifying discovery would make him world famous. England's King George III of American Revolution infamy appointed him as Court Astronomer. With an annual pension from the king, Herschel would give up music and become a professional observational astronomer and telescope maker — the greatest of his era.[47]

His sister Caroline "served as his assistant until Herschel's death. She was the first woman to discover a comet," science writer Nola Taylor Redd tells us, "ultimately finding eight. She also discovered several deep-sky objects and was the first woman to be given a paid scientific position and to receive an honorary membership into the Royal Society."[48]

Trouble in the Heavens

Some forty years after the discovery of Uranus, French astronomer Alexis Bouvard (1767–1843) attempted to construct astronomical tables for the new planet. Working under Pierre-Simon Laplace, the "brilliant and indefatigable calculator" had published successful astronomical tables for Saturn and Jupiter in 1808. He wanted to add Uranus to his 1821 update.[50] (Laplace was seventy-two years old in 1821. He would pass away six years later.)

European astronomers had observed Uranus night after night since its discovery in 1781. Now in 1821, there were some **40 years** of observational data. Uranus has a period of **84** earth-years. (Recall a planet's period is the time it takes to complete one orbit around the Sun.) Thus, observations covered only about half of Uranus's orbit.[51]

Astronomers had seen Uranus *before* its discovery. They had mistaken the slow-moving planet for a star. John Flamsteed saw it in 1690 in his telescope and labeled it star 34 Tauri. All-in-all, there were twenty such observations — extending from 1690 to 1771.[52]

Using pre-discovery and post-discovery observations, Bouvard attempted to compute the orbit of Uranus. His calculations included the gravitational pull on the new planet from the Sun and other planets. He found that the pre-discovery observations could be "well represented by points on an ellipse." So could post observations. But they were two different ellipses.[53]

What the …? Maybe the older observations were in error or perhaps less precise. So Bouvard decides to use only post-discovery observations to compute his tables.[54]

He publishes his Uranus tables in 1821 as planned. Four years later, observations at Vilna Observatory showed errors of 12 arc seconds longitude and 13 arc seconds latitude. Errors increased to 23 arcseconds longitude, 14 latitude the following year. Observatories at Vienna and Cambridge saw similar errors."[55] In 1845 — some 24 years since Bouvard had first published his Uranus tables — the longitude was now off by about 2 arc-minutes. This was a significant discrepancy given the observational accuracy at the time. And the errors continued to grow.[56] Something was wrong. Very wrong.

Alexis Bouvard would go on to be a famous astronomer in his own right and he was appointed director of the Paris Observatory in 1822. It would be another French mathematical astronomer who would resolve the mystery.

"At the Tip of His Pen"

Urbain Jean Joseph Le Verrier (Fig. 27.3) was born on March 11, 1811 in Saint-Lô in northwest France to a family of modest means. He was a devout Catholic. By the age of 28, he had worked as a tobacco engineer, chemist, astronomy teacher, and professor of celestial mechanics. It is for his later work as a mathematical astronomer at the Paris Observatory that he is most remembered.[57]

In personality, Le Verrier was similar to Laplace. He was termed "haughty, disdainful, and inflexible" by his brilliant assistant Camille Flammarion. When later appointed to director of the Paris Observatory, Le

Urbain-Jean-Joseph Le Verrier
(1811–1877)

Figure 27.3. French Giant in Theoretical Astronomy and Mathematics Urbain Le Verrier.

Verrier acted like an "autocrat" and treated "employees at the Observatory as his slaves," writes Thomas Levenson, Professor of Science Writing at Massachusetts Institute of Technology.[58]

At the Paris Observatory, Le Verrier would take up the mantle of the great Laplace and become "amongst the foremost mathematical astronomers" of the 19th century.[59]

Laplace's inheritance was unclaimed; and he boldly took possession of it.

— Jean Baptiste Dumas, colleague of Le Verrier[60]

Le Verrier worked on improvements to the orbits of the inner planets, and then Jupiter and Saturn. After publishing a series of papers on periodic comets, he turned his attention to the stability of the solar system.

Here he sought to test Laplace's famous declaration that the solar system is inherently stable. Using his expertise in celestial mechanics, he made a complete revision of planetary theoretical models. He then calculated variations in planetary orbits all the way from 100,000 BC to 100,000 AD.

In the end, he found that the lack of precision in planetary parameters, such as each planet's precise mass, made the determination of inherent stability uncertain. He published his findings in 1839.[61]

In 1845, François Arago, Le Verrier's friend and current director of the Paris Observatory, asked Le Verrier to look into the irregular orbit of Uranus.[62] Arago told him it "imposed a duty on every astronomer to contribute to the utmost of his power" towards solving the puzzle.[63]

Le Verrier began by conducting a rigorous examination of Bouvard's calculations for his 1821 tables of Uranus. He found no errors. Rather, it seemed some unknown force was "acting on the planet." He was determined to find that the source of that force.[64]

Astronomers had identified some five possible causes for the observed irregularities in the orbit of Uranus:

(1) Some kind of resisting medium or "ether" out by Uranus is slowing the planet's motion;
(2) a giant moon orbits the planet;
(3) A comet or some other stray object had collided with or passed near to Uranus, pushing it off course;
(4) Newton's law of gravity does not work the same at such huge distances from the Sun; or
(5) there is an undiscovered object beyond Uranus, perhaps another planet.[65]

Le Verrier quickly examined each possibility.

Influences on Uranus's orbit from a resisting medium were not compatible with observations. Neither was a massive moon circumnavigating the planet. The continuing changes to its orbit ruled out the effects of a one-time comet. He never doubted that "Newtonian gravity was correct."[66]

Le Verrier focused on number five: There was an unknown planet out there perturbing the orbit of Uranus — a hypothesis that Alexis Bouvard himself had also suggested.

The French mathematician then went on a detective search unparalleled in astronomical science. His goal: to solve a set of complex equations for the location, mass, and orbit of a new unseen planet — based on the mathematics of Newton's theory of gravity.

To do this, Le Verrier had to work the gravitational perturbation problem backwards. Usually, an astronomer looks at a known planet, its parameters defined, and determines how it is perturbed by the gravitational pull of the Sun and other known planets. Now Le Verrier had to examine the perturbations to Uranus and figure out how exactly a new unknown planet, its parameters undefined, might be causing them.

There are a number of unknowns here. They include: What is the hypothetical planet's mass, its ellipticity (eccentricity), its closest and furthest distance from the Sun, its period? What is its inclination, the angle of the plane of its orbit around the Sun? What is the orientation of the major and minor axes of its elliptical orbit, and the planet's position along that orbit at a specific time? All this had to be reconciled mathematically with observations from a spinning Earth also orbiting the Sun.

Le Verrier generated "a system of equations which described each of the components of the hypothetical planet's motion."[67] It was a classic three-body problem — this time for the Sun, Uranus, and the unknown planet. He also had to take into account lesser perturbations from the other planets, especially Saturn and Jupiter. The work was very difficult and extremely arduous.

His initial computations had thirteen unknowns — "too many for someone with even his gifts to solve in a timely manner." To simplify, he guessed that the angle of the plane of the mystery planet's orbit was close to that of the other planets. This and a few other approximations and assumptions, including its distance, reduced the number of unknowns to nine. It was still an immensely daunting mathematical problem.[68]

With no computers to do the tedious computations, all mathematical manipulations, numerical iterations, and calculations had to be done by hand. It required a "horrific amount of grunt work." In an intensely laborious effort, Le Verrier finally "arrived at predictions for the mass and location" of the hypothetical planet."[69]

On November 5, 1845, Le Verrier reported to the *Académie des Sciences* that the new planet cannot be interior to Uranus. Six months later, on June 1, 1846, he gave a preliminary estimate of its location. On August 31, 1846, he made his final announcement[70]:

A telescopic search would find a new planet "beyond the orbit of Uranus at a distance of about 38 Astronomical Units" (AU)* from the Sun.[71] He predicted that on January 1, 1847, the new planet will be located about 5 degrees east of the star δ (delta) Capricorn, a fairly bright star in the Capricorn constellation.[72]

Le Verrier also gave estimates for the new planet's mass and diameter.[73]

No one looked.

Despite his confidence and brilliant analysis, his fellow astronomers in Paris would not take precious telescope time away from their current activities. After all, no one had ever predicted the location of an unobserved object in the sky, let alone a new planet.

His pleas to his French colleagues grew more and more frantic.

Some two weeks later, on September 18 the frustrated French mathematical astronomer left his countrymen aside. He wrote to Johann Gottfried Galle, a young German astronomer at the Berlin Observatory, and asked if he would please search for the hypothesized planet.[74]

Galle received the letter five days later on the morning of September 23, 1846. He asked permission from observatory director Johann Encke to drop his assigned telescope activities and search for the hypothetical planet.[75]

Like astronomers in Paris, Encke had his doubts. Still, he reluctantly agreed with Galle that it was their "moral obligation" to at least look.

In a fateful moment, student assistant Heinrich Louis d'Arrest asked to join the hunt.[76]

Not On the Map

On that same night, the sky is clear. Galle mans the observatory telescope. Looking through the eyepiece, he conducts a systematic search in the region

---

*An Astronomical Unit (AU) of 1 is the average distance from the Sun to Earth. Modern measurements put an AU at 92.96 million miles or 149.6 million km.

where Le Verrier had told him to look. He attempts to find the object predicted by Le Verrier, a disc some 3 arc minutes wide. He is unsuccessful.[77]

D'Arrest suggests they consult new star charts prepared twenty years earlier "by Carl Bremiker for the Royal Academy of Sciences in Berlin." The charts had not yet been distributed outside Berlin. Galle rescans the area with the telescope. He reads out the positions of each star one-by-one as D'Arrest checks the Bremiker star charts.[78]

Somewhere between midnight and 1 AM, Galle calls out the sighting of a disc-shaped object of about 8th magnitude. "That is not on the map!" d'Arrest declares.

It is 3.2 arc seconds across, close to Le Verrier's prediction. Right ascension 21 h 53 m 25.84 sec.[79]

D'Arrest rechecks the co-ordinates. Galle then switches with him and "double and triple checks the star chart map." The object is within one degree of Le Verrier's "predicted location near the boundary between Capricorn and Aquarius."[80] (See Fig. 27.4.)

They summon the Observatory Director. Encke, Galle, and d'Arrest take turns watching "the new object until it sets around 2:30 in the morning."[81]

The next evening, September 24, the sky is again clear. They find the object. It has *moved* against the background of stars!

Galle writes back to Le Verrier[82]:

> "*The Planet whose position you indicated **really exists**. The same day I received your letter I found a star of the eighth magnitude that was not recorded on the excellent Carta Hora XXI (drawn by Dr Bremiker). ... The observation of the following day confirmed that it was the planet sought.*"

Astronomers Johann Gottfried Galle and Heinrich Louis d'Arrest are the first to identify a previously unknown planet invisible to the naked eye — the eighth and final one in our solar system (sorry Pluto). It is the highlight of their lives, an event never to be equaled.

Neptune

Urbain Jean Joseph Le Verrier had conjured up a new planet "at the tip of his pen," as François Arago put it. His achievement would be hailed as the crowning triumph of Newton's Law of Universal Gravitation. It was the most important astronomical discovery of the 19th century.[84]

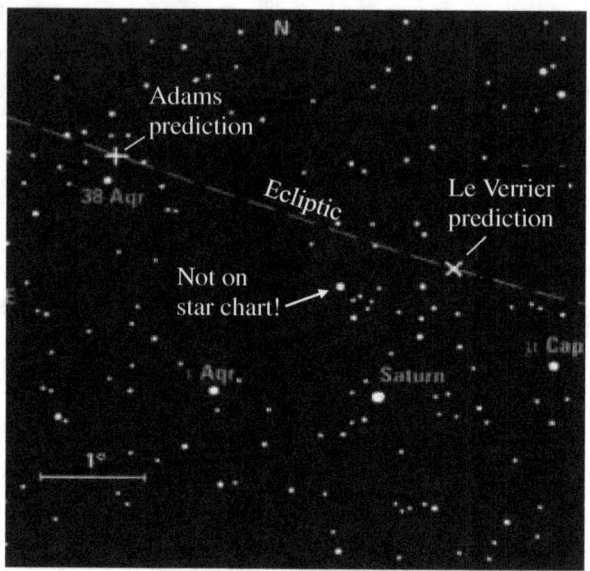

Figure 27.4. Object Found Not on Star Chart – Reconstruction. Object about a degree from Le Verrier predicted location, *as seen by Galle on the night of* September 23, 1846. Also shown is a prediction by English astronomer John Couch Adams (more on this later).[83]

The successful sighting of an eighth planet was a tribute to Le Verrier's brilliance and persistence, Galle and d'Arrest's exceptional observational skills, and the strange power of mathematics to unearth the mysteries of the physical world. As Thomas Levenson put it, "mathematics became the scaffolding of the universe — math, without God."[85]

Le Verrier advised returning to the tradition of naming the planets after the Roman gods. He proposed Neptune, Jupiter's brother and god of the sea. (See Fig. 27.5.) Uranus remains the only planet to be named after a Greek god.[86]

Le Verrier had used Bode's law to guess that Neptune would be 38 AU from the Sun. It turned out to be much closer at an average distance of about 30.1 AU. Nonetheless, Le Verrier's prediction and the Galle/d'Arrest observation had again increased the size of the known solar system — this time by about 1.5X.[87] More precisely, Neptune is 2.77 to 2.83 billion miles (4.45 to 4.55 billion km) from the Sun.

It takes 164.8 earth-years to complete one orbit. A list of planetary masses and distances from the Sun is given at the end of this chapter (Table 27.1).

Figure 27.5. Neptune — the Eighth Planet in our Solar System. From NASA's Voyager 2 spacecraft in 1989. Like Uranus, Neptune is also an "ice giant." Unlike Uranus, Neptune is not visible to the naked eye — which makes its discovery all the more remarkable. It is almost four times wider and seventeen times more massive than Earth. Neptune has 14 known moons. It also has five named rings: named Galle, Le Verrier, Lassell, Arago, and Adams.[88]

A Rival Claim

English mathematical astronomer John Couch Adams had also predicted the location of the new planet. He too used observations of Uranus and Newton's law of gravity. But, unlike Le Verrier, he questioned his own calculations.

Adams predicted a number of possible locations for the new planet which ranged over some 20 degrees of the sky. In the summer of 1846, astronomers at Cambridge Observatory under its director James Chillis spent six unsuccessful weeks searching for the new planet.[89] Their search is "infamous for having observed Neptune three times, but not recognizing it as a planet," writes astrophysicist Davor Krajnović.[90]

François Arago had called the discovery of Neptune "one of the most magnificent triumphs of theoretical astronomy, one of the glories of the Académie and one of the most beautiful distinctions of *our country*." (My italics.)[91]

It seems the British wanted part of the glory. Apparently, they kept the details of the Adams/Challis misadventure a secret. Only after Neptune's discovery by Galle was a more accurate early prediction by Adams made public. For a century, Adams was given nearly equal billing with Le Verrier as co-discoverer of Neptune.

Observatory papers unearthed in the 1950s reveal a different story.[92] Adams' best prediction for the location of Neptune on the night Galle found it was significantly further away than Le Verrier's.

Sidebar: Planet Vulcan?

In 1855, Urbain Le Verrier uncovered a slight discrepancy in Mercury's orbit. Its observed perihelion shift did not agree with Newton's law of gravity.[93] Again, since its orbit is elliptical (like all planets), Mercury is sometimes closer and sometimes farther away from the Sun. The closest point in a planet's orbit to the Sun is called its *perihelion*.[94]

The shift of Mercury's perihelion is caused by the rotation or precession of its elliptical orbit over time. Le Verrier determined that this precession is due to gravitational disturbances *from the other planets* on Mercury.

Newton's theory predicted that as a result, Mercury's perihelion — its closest point to the Sun — would shift in angle by **531** arc seconds per century. Telescopic observations showed a shift of some **574** arc seconds per century. Newton's prediction was too small by **43** arc seconds. (All modern values.) Again, an arc second is 1/60 of an arc minute. An arc minute is 1/60 of a degree. So, an arc second is 1/3600 of a degree.

To resolve the discrepancy, Le Verrier proposed the existence of a new planet orbiting near Mercury. He named it Vulcan. Decades of telescopic searching failed to find the hypothesized planet.

Albert Einstein solved the problem in 1915 with his general theory of relativity. Newton's law of gravity is incorrect. All planets have an *inherent* precession in their orbits — independent of the gravitational attraction of other planets.

In other words, if Mercury were the only planet in the solar system, its elliptical orbit would still process a little bit. This is due, Einstein tells us, to the warping of space and time by the Sun's mass/energy. Mercury's inherent precession is the most pronounced because it is closest to the Sun.

Einstein derived a value for the inherent perihelion shift of Mercury of **42.9** arc seconds per century — in excellent agreement with observations.

This precession effect is true for all planets — the farther away from the Sun, the less the inherent rotation. For example, the perihelion of Venus shifts only 8.6 arc seconds per century, Earth's a mere 3.8 arc seconds per century, and Mars less than 1.4 arc seconds.[95]

Passing

Le Verrier became director of the Paris Observatory after Arago's death in 1853. He held that position "with a short interruption (1870–73) until his death."[96]

Urbain Jean Joseph Le Verrier died on September. 23, 1877 — "thirty-one years to the date of the discovery of Neptune on September 23, 1846." As he lay dying, Le Verrier is said to have uttered the words from the Prophesy of Simon in the New Testament[97]:

Nunc dimittis servum tuum, Domine, in pace
Now let your servant depart, O Lord, in peace.

— Luke 2:29

Le Verrier was buried in the Montparnasse Cemetery in Paris. His grave is marked by a large celestial stone globe.[98]

Science and Religion

The discovery of Neptune represented a historic verification of Newton's theory of universal gravitation. It marked yet again the supremacy of science in describing the heavens — with a predictive power beyond the capacity of religion.

Table 27.1 Planetary distances and masses in our solar system. Modern values.

PLANET	MEAN DISTANCE FROM SUN			MASS
	AU	Miles	Kilometers	× Earth
Mercury	0 387	36 million	58 million	0.055
Venus	0.723	67 million	108 million	0.82
Earth	1	93 million	150 million	1
Mars	1.52	142 million	229 million	0.11
Jupiter	5.20	484 million	779 million	318
Saturn	9.58	889 million	1.43 billion	95
Uranus	19.2	1.79 billion	2.88 billion	15
Neptune	30.1	2.8 billion	4.51 billion	17
Pluto (mini-planet)	39.5	3.67 billion	5.91 billion	0.003

A mere decade later, a radical new scientific theory would be proposed. Its highly unsettling views on the development of life on our planet would challenge religious beliefs as never before. This is the subject of Part VI, the next and last part of the book.

PART VI
Life

Chapter 28

RELUCTANT REVOLUTIONARY REDUX

What is Man? Man is a noisome bacillus whom Our Heavenly Father created because he was disappointed in the monkey.

— Mark Twain[1]

He was an odd duck, a dreamer who loved to go on long solitary walks and collect insects and other things. He made friends easily, found school boring, and hated controversy. A biographer would later describe him as "a gentle, straightforward, almost childishly simple man disinclined toward disputation."²

The dreamer would come to revolutionize our understanding of the origins of every living thing on our planet. His name was Charles Robert Darwin.

The Early Years

Charles Robert Darwin
1809–1882

He was born on February 12, 1809 in Shrewsbury, England, a market town in Western England nine miles from the Welsh border. Young Charles was the fifth of six children. He was a gentleman* from a prominent British family of accomplished scientists. His grandfather, Enlightenment thinker Erasmus Darwin, was a renowned physician, physiologist, inventor, poet, and slave-trade abolitionist.³

---

* In 19th century England, a "gentleman" was considered to be a "man of good family" of some rank and a measure of affluence. Source: "Gentleman" Encyclopedia Britannica, britannica.com. Aug 29, 2018, https://www.britannica.com/topic/gentleman. Retrieved July 26, 2020.

His father Robert Waring Darwin was an accomplished physician in his own right.[4] He has been described as a "benevolent tyrant" — an imposing figure, stern yet compassionate. In a rare moment of pique, he said to Charles, "You care for nothing but shooting, dogs, and rat-catching, and you will be a disgrace to yourself and all your family."

Recalling this now famous quote, Charles Darwin later wrote: ". . . my father, who was the kindest man I ever knew. . . must have been angry and somewhat unjust when he used such words . . . [he was] in many ways a remarkable man."[5]

Charles' mother Susannah Wedgwood Darwin died in 1817. He was eight-years-old. His older sisters Marianne, Susan, and Caroline then raised Charles and his younger sister Catherine. He later recalled little of his mother. Her death was so painful to his sisters that they forbade him to speak of her.[6]

The conflict between science and religion was part of Darwin's upbringing. His mother was a deeply religious Unitarian,* his father a "free-thinker" drawn to science. Following his mother, young Charles was a believer and creationist.[7]

The same year his mother passed Charles began attending day-school in Shrewsbury "By the time I went to this day-school," he later wrote, "my taste for natural history, and more especially for collecting, was well developed. I tried to make out the names of plants, and collected all sorts of things, shells, seals, franks, coins, and minerals."[8]

The following year, he boarded at "Dr. Butler's strictly classical school" in Shrewsbury. He remained there until age fifteen, "incapable of mastering any language," as he put it. He saw its curriculum of Greek and Latin along with "a little ancient geography and history" as useless.[9]

In October 1825, "doing no good at school," as Darwin later wrote, "my father sent me at a rather earlier age than usual [age 16] to Edinburgh University with my [older] brother [Erasmus, age 21]" to study medicine. This was the same institution where his father and grandfather had studied to become doctors. Charles attended lectures on geology and zoology in his second year. Except for chemistry, he found the lectures at Edinburgh "intolerably dull."[10]

In his second year at Edinburgh, he "became well acquainted with several young men fond of natural science." Charles also befriended

---

*Unitarians were a Christian Protestant sect who rejected the Doctrine of the Trinity espoused by the Church of England. In the 1800's they were no longer obliged to keep their heretical views secret, as in the time of Newton.

Newhaven fishermen, "sometimes accompanying them when they trawled for oysters, and thus got many specimens." He presented short papers before the Plinian Society — a natural history club for students at the university — on the "ova of Flustra . . .. [and] the egg-cases of the worm-like *Pontobdella muricata.*" The former are marine animals often mistaken for seaweed; the latter marine leeches.[11]

He went on long walks with zoologist Dr. Robert Edmund Grant, a leading naturalist of the early nineteenth century. "Dr. Grant took me occasionally to the meetings of the Wernerian [Natural History] Society," Darwin later wrote, "where various papers on natural history were read, discussed, and afterwards published in the *Transactions*. I heard Audubon deliver there some interesting discourses on the habits of N. American birds . . ."[12]

On the other hand, Charles put no serious effort into learning medicine. In January 1828 his father had him switch to Christ's College, Cambridge to pursue an ordinary degree — leading to ordination as a member of the clergy. "[I] liked the thought of being a country clergyman," he later wrote, ". . . I did not then in the least doubt the strict and literal truth of every word in the Bible."[13]

His father's hopes for a scholastic turnaround were in vain. "During the three years which I spent at Cambridge," Darwin wrote, "my time was wasted, as far as the academical studies were concerned, as completely as at Edinburgh and at [Dr. Butler's] school." He managed to earn a BA without distinction in January 1831.[14]

No pursuit at Cambridge . . . gave me so much pleasure as collecting beetles.

— Charles Darwin[15]

There was one bright spot. In his last year at Cambridge, Darwin read scientist-explorer Alexander von "Humboldt's *Personal Narrative*, a stirring account of his travels to the 'New World.'" He also read John Herschel's* *A Preliminary Discourse on the Study of Natural Philosophy*. These two books "stirred up in me," he later wrote, "a burning zeal to add even the most humble contribution to the noble structure of Natural Science."[16]

A curious request would set the course of the rest of Charles Darwin's remarkable life.

---

*Astronomer, chemist, mathematician, and inventor John Frederick William Herschel (1792–1871) was the son of astronomer William Herschel of Uranus fame.

The Invitation

In 1805, four years before Charles Darwin was born, the British Royal Navy defeated the combined French and Spanish fleet off the coast of Cape Trafalgar in Spain. After this last battle of the sail, Great Britain now held mastery of the oceans and seas across the planet.[17]

In an ongoing effort to enhance its military and commercial interests, in 1831 the admiralty commissioned hydrographer, meteorologist and Naval Captain Robert FitzRoy to conduct an "accurate survey of the southern coast-lines of South America." As captain of the survey ship HMS *Beagle*, his mission would include recording chronometer readings to verify and improve longitude mapping around the world.[18]

The admiralty advised FitzRoy to seek out a "gentleman" naturalist to accompany him in this 5-year circumnavigation of the globe. The previous captain of *Beagle* had committed suicide. It was hoped that "companionship and shared interests in natural philosophy would mitigate the hardships" and loneliness of command. FitzRoy asked his friend mathematician George Peacock, Vicar of Wymeswold and fellow at Trinity College, to "suggest a Cambridge man fit and willing to accompany him."[19]

Peacock recommended the Revd. John Stevens Henslow, Professor of Botany at Cambridge. Henslow declined the offer due to stern objections from his wife. Peacock's second choice was Henslow's brother-in-law, Leonard Jenyns, Vicar of Swaffham Bulbeck. Jenyns also declined. He felt he could not in good conscience desert his parish.

Henslow and Jenyns proposed a recent Cambridge graduate named Charles Darwin to accompany FitzRoy on the *Beagle* voyage.[20] Henslow had been Darwin's botany professor at Cambridge, whose lectures the young student "liked much." Darwin described the professor as a "deeply religious and orthodox" man of "admirable moral qualities."[21]

Darwin had gone on "field excursions" with the professor and other interested students "to distant places where he lectured on the rarer plants or animals which were observed." Darwin termed these excursions "delightful."[22] Henslow also introduced Darwin to the work of the esteemed Revd. Adam Sedgwick, one of the founders of modern geology. Darwin had found his first academic love, the study of geology. Thus, Darwin would serve as naturalist and geologist on Captain FitzRoy's voyage.[23]

Henslow saw something in his student in their nearly daily long walks together and in Darwin's frequent dining as a guest with the professor and his family. Here was a young man of high character and imaginative mind — one with a passion for collecting and natural history.

From "their shared passion for natural history, Jenyns had known Darwin since he was an undergraduate at Cambridge." The two had gone on "joint excursions to collect insects in the Fens and in the woods around Bottisham Hall near Jenyns' home."[24]

Young Charles Darwin was most eager to go on this trip of a lifetime. After an interview with FitzRoy, the *Beagle* captain found the grandson of the famous Erasmus Darwin most acceptable. "I like what I see of him much," FitzRoy wrote to Captain Beaufort, Admiralty Hydrographer. Now all Charles needed was permission from his father.[25]

Robert Darwin "strongly objected," as Charles later put it, "adding the words — 'If you can find any man of common sense, who advises you to go, I will give my consent.'"[26]

Charles turned to his uncle Josiah Wedgwood II, his deceased mother's brother, for help. Then he wrote to his father[27]:

My dear Father,
I am afraid I am going to make you again very uncomfortable... I have given Uncle Jos, what I fervently trust is an accurate and full list of your objections, and he is kind enough to give his opinion on all. The list and his answers [are] enclosed...

— Charles Darwin, letter to his father, Aug. 31, 1831.[28]

The enclosure listed eight objections from Robert Darwin, including the fact that two earlier candidates for the voyage had declined. Chief among his objections was that working as a naturalist would be "disreputable to [his son's] character as a Clergymen."

Uncle Josiah responded to his brother-in-law's objections in a mild and diffident manner. He wrote of Robert's chief objection: "I should not think it would be in any degree disreputable to [Charles's] character as a Clergyman. I should on the contrary think the offer honourable to him; and the pursuit of Natural History, though certainly not professional, is very suitable to a clergyman."[29]

It worked. With Uncle Josiah's support and that of Charles's older sisters, Robert Darwin relented. Though he could not have known it at the time, his decision to let his son go on this voyage would come to be the most momentous judgment of his life. It would plant the seeds of a revolution in science on par with that of Copernicus and Newton.

Charles prepared for the trip with a heretofore unseen intensity.

... working like a tiger... at present Spanish & Geology, the former I find as intensely stupid, as the latter most interesting.

— Charles Darwin, studying in preparation for voyage, July 9, 1831.[30]

The Voyage

The voyage of the Beagle has been by far the most important event in my life and has determined my whole career

— Charles Darwin[31]

On the morning of December 27, 1831, twenty-two-year-old Charles Darwin boarded the HMS *Beagle* — a 90-foot (27.4 m) long by 24-foot (7.3 m) wide armed brig — with Captain Robert FitzRoy in command. The British Man-of-War (Fig. 28.1) "left its anchorage in the Barn Pool on the west side of

Figure 28.1. H.M.S. *Beagle* in the Straights of Magellan. 90-foot long by 24-foot wide armed brig. Drawing by R. T. Pritchett, 1890.

Plymouth Sound" in Cornwall, England with "twenty-six souls on board" and set sail across the cold dark North Atlantic Ocean.[32]

Charles Darwin came on board as an unpaid volunteer sustained by an allowance from his father. On the advice of Professor Henslow, he brought with him the first volume of Scottish geologist Charles Lyell's breakthrough tome, *Principles of Geology*, which he "studied attentively."[33]

Published in 1830, Lyell's book argued that our planet is extremely old. Natural geological forces have since changed "Earth's crust through *countless small changes*." This "uniformitarian" principle was a direct challenge to a belief then prevalent "among geologists and other Christians." They held that "unique catastrophes or supernatural events — such as Noah's flood — had shaped Earth's surface." Lyell's concept of gradual slow change would come to have a great influence on Darwin's thinking, as we shall see.[34]

FitzRoy and his crew would refer to Darwin as Philos for "philosopher." For young Charles, it would be the journey of his dreams. The five-year voyage would change the young man from an idle dreamer with no direction to a man of discipline and dedication to his life's passion.[35]

On the voyage, Darwin would prove himself congenial, diligent, self-reliant, and relentless in pursuit of natural history and geology — with no small amount of personal bravery.

A man who dares to waste one hour of time has not discovered the value of life.

— Charles Darwin, 1836.[36]

Under the protection of the British flag, Captain FitzRoy conducted surveys of shallow waters and inlets along the coasts of South America. Meanwhile, Charles Darwin journeyed inland to investigate the flora, fauna, and geology of the continent.

The itinerary of the *Beagle* voyage is shown in Fig. 28.2. Overall, Darwin spent "three years and three months on land and 18 months at sea."[37]

At sea, the novice mariner passed "unknown and dangerous coasts." Often seasick, he endured squalls, gales of cold rain, heavy fog and swells, and skies "brilliant with lightning." He witnessed the harrowing crossing of the Straights of Magellan, along with "beautiful days, calm sea, and a fine breeze."

On land, he traveled on foot, horseback, muleback and stagecoach; on the waters by river raft, canoe, sail boat, and steam boat, He engaged

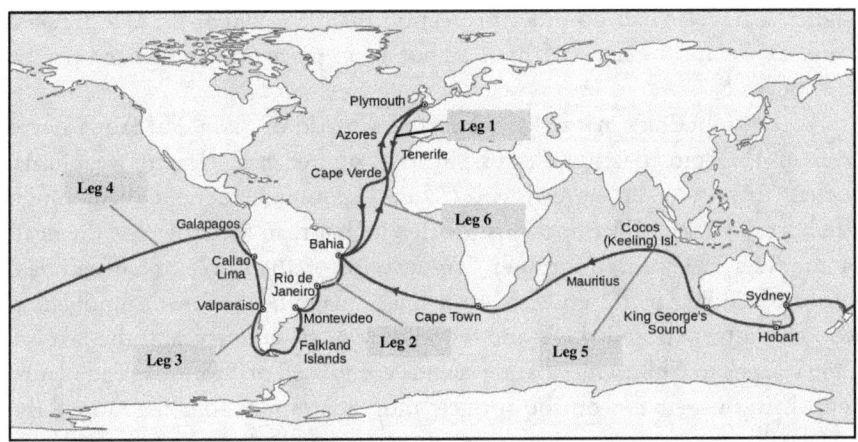

Figure 28.2. Map of Darwin's Journey on The Beagle. *Itinerary Highlights:*
LEG 1 — Leaves Falmouth, England, 27 Dec 1831. Cape Verde. St. Paul's Archipelago. Crosses equator on 17 Feb 1832. Bahia. Abrolhos Islands. Rio De Janeiro. Montevideo. Uruguay. Buenos Aires. Bahia Blanca. Back-tracking to Montevideo, Buenos Aires.

LEG 2 — Patagonia Coast to Tierra del Fuego. Hermite Island (Chile), Falkland Islands (twice). Uruguay. Argentina.

LEG 3 — Patagonia Coast again. Straights of Magellan. Port Famine, Chiloe Island. Valparaiso. Chonos Islands. Valdivia. Concepcion. St Jago (Santiago). Back-tracking in Chile. Lima, Peru.

LEG 4 — Leaves Lima, 7 Sept 1835. South Pacific. Galápagos. Tahiti. New Zealand; Sydney. Hobart, Tasmania.

LEG 5 — Indian Ocean. King George Sound, Australia. Keeling Island. Mauritius. Cape of Good Hope. South Atlantic. St Helena. Ascension. Bahia, Recife in Brazil.

LEG 6 — To North Atlantic. Cape Verde. Azores. Returns to Falmouth, 2 Oct 1836.[38]

"in cave explorations, excursions across pack ice, and treks across blazing sand. From Patagonia to Australia to the Keeling Islands of the Indian Ocean" to glorious Tahiti, the great Kauri pine forests of New Zealand, Table Mountain in Cape Town, the desolate wild valleys of St. Helena Island and bleak Ascension Island, Darwin meticulously explored and recorded the geology and life forms he encountered.[39]

Darwin hiked along marshes and lagoons, across valleys and rivulets, along beaches and land-locked plains. He took "long scrambles though the woods," followed winding creeks, crossed ravines via rickety suspension bridges, and climbed to the edges of volcanic craters. He explored regions where "a traveler had no other protection than his own arms." He trekked many days with no signs of men or horses in places never before seen by Europeans.[40]

The budding naturalist witnessed a world of flora and fauna never seen in England: passion flowers, woods of Acacias, elegant tree ferns, ancient forests, "valleys green with clover, Cardoon (Artichoke thistle) high as a horse's back, and Pampas thistle often higher than the crown of the head of the rider." He passed through "large woods of the Roble or Chilean oak, intricate winding paths, endless green hills," magnificent forests, undulating wood county, petrified shells and wood, fine grass country, and the lake of Tagua Tagua in Chile with floating islands composed of "various dead plants. With living vegetation on the surface, they float about four feet above the surface."

He was often required to hunt and kill for food, at times to ride all day with no water, at other times "exhausted with fatigue and hunger." He took ill on a number of occasions, enduring the "horrors of illness in a foreign country," as he put it, "without being able to speak one word [of Spanish] or obtain any medical aid."

At night, he slept "with thin clothes on the bare ground," under tents, on boulders, on a bog, on a sand beach, or bivouacked in the corner of a field. He stayed at dwellings from simple and uncomfortable to fancy estates. With letters of introduction, he stayed at Haciendas, Ranchos, at homes of English "gentlemen," and at large estates of wealthy landowners.

He ascended mountain peaks, including 6400-foot-high Bell Mountain with a guide and horses through treacherous snow drifts, narrowly escaping mountain snowstorms. He "crept on hands & knees under a confused mass of dead & dying tree trunks" in a failed attempt to reach the summit of San Pedro in Chiloe. He climbed to the 1600-foot summit of a "most perfectly conical hill" in Tres Montes — the path so steep he used "the trees like a ladder." In a Mule team he ascended a Mountain in the Cordilleras range some 12,000 feet high — "breathing deep & difficult" — finding fossil shells on the highest ridge.[41]

A Bountiful Collection

I have worked with every grain of energy I possess.

— Charles Darwin[42]

Darwin "collected so many samples of plants and animals that his shipmates wondered aloud whether he was out to sink the *Beagle*."[43] His specimen collection included spiders, beetles, scorpions, and a vast number of other insects; reptiles including snakes and lizards; a variety of birds; and seashells, including "specimens of the giant Chama [Giant clams]" at the Keeling Islands.

He dug into rock and slate and broke stones to excavate animal bone fossils of now extinct animals. This included fossilized remains of a "huge Llama-like animal the size of a camel" near Puerto San Julian in Patagonia.

He even collected while at sea. With the *Beagle* on its way to South America, he devised a plankton net and hung it at the aft end of the ship. "The number of animals that the net collects is very great," he wrote.[44]

Periodically, he packed and shipped his findings back to a delighted Professor Henslow in England. The professor in turn sent them to other experts for analysis and classification.

Darwin recorded his findings in a series of field notebooks. In a letter to sister Catherine in July 1834. he wrote:

> *My geological notes & descriptions of animals I treat with [great] attention . . . from knowing so little of Natural History, when I left England, I am constantly in doubt whether these will have any value . . .*[45]

Near the end of the voyage in late July at Ascension Island, he received a letter from his sisters. The news was good. "[Geologist Adam] Sedgwick had called on my father and said that I should take a place among the leading scientific men."[46]

Darwin later found out that Henslow "had read some of the letters which I wrote to him before the Philosophical Soc. of Cambridge . . . [my] collection of fossil bones, which had been sent to Henslow, also excited considerable attention amongst palæontologists."[47]

The Diary

Charles Darwin recorded his nearly five-year adventure around the world in a 400-page personal Diary. It reads as a travelogue of a bygone era — before modern cities, automobiles, superhighways, and rampant commercialization. His writing is unassuming and self-effacing, without a hint of conceit.

He notes the "remains of very numerus old Indian villages . . . clearly considerably advanced in civilization."[48] He describes the Fuegians of Tierra del Fuego; women dressed in the *saya y Manta* in Lima (Fig. 28.3); and "a young Hottentot groom" he hired "to accompany me as a guide" at Cape Town.[49]

Figure 28.3. Fuegians and Peruvians. Left: Watercolor of a native from Tierra del Fuego. Painted by Conrad Martens, ship's artist on the Beagle. Right: Women of Lima dressed in the saya y Manta, a legacy of the Moors — Muslims who fled persecution in Spain after its Catholic reconquest.[50]

He writes on indigenous tribal languages and migration — lamenting the ongoing extermination of indigenous populations under European colonialism, especially in South America. This complaint and his anti-slavery stance were rare for a nineteenth century English "gentleman."[51]

Before his voyage on the *Beagle*, Charles Darwin held the prevailing belief that all species of plants and animals, living and extinct, were "immutable" — unchanged throughout time and separately created by God. On his return to England, he pondered what he had witnessed on his great trip. He began to see a pattern in the variations of life on our planet, a connection between geology and natural science, between the "history of Earth and the history of life" as Spanish-American evolutionary biologist and philosopher Francisco José Ayala puts it.[52]

A Budding Theory

The primary source for this section is the writing of American science historian Sandra Herbert, history of science professor at the University of Maryland.[53]

On the five-year voyage, Darwin attended "Divine services" every Sunday on ship or on shore whenever possible.[54] Yet things he saw on the *Beagle* voyage began to challenge his beliefs.[55]

Near the end of the voyage, in June or July of 1836 Darwin wrote down his first recorded hint that species might not be permanent. Why, he wondered, were different varieties of birds and tortoises in the Galápagos archipelago unique to each island he had visited?

In a now famous passage in his private Ornithological Notes, he wrote:

*I have [bird] specimens from four of the larger islands . . . In each Isld. each kind is **exclusively** found . . . When I recollect, the fact that the form of the body, shape of scales & general size, the Spaniards can at once pronounce, from which Island any Tortoise may have been brought. The only fact of a similar kind of which I am aware, is the constant asserted difference — between the wolf-like Fox of East & West Falkland Islds. — If there is the slightest foundation for these remarks the zoology of Archipelagoes — will be well worth examining; for such facts [would **inserted**] undermine <u>the stability of Species</u>.* (My underline.)[56]

Undermine the stability of species! Is Charles Darwin suggesting that the species of living things may change over time?

On his return to England in October of 1836, Darwin devoted the bulk of his time to documenting his geological findings. This included his theories on the formation of coral reefs and volcanic islands. From 1837 to 1839, he published seven papers on the subject.[57]

In his spare time, he privately pursued ideas on the transmutation of species. His investigation would come to revolutionize our understanding of the formation of life on planet Earth.[58]

<u>Crucible of Creation — 1837</u>

— Had been greatly struck from about month of . . . March [1837] on character of S. American fossils — & species on Galápagos Archipelago. These facts origin (especially latter) of all my views.

— Charles Darwin[59]

Charles Darwin presented 80 mammal and 450 bird specimens to the Zoological Society of London on January 4, 1837.[60] At the end of the month, English anatomist and paleontologist Richard Owen identified a "number of Darwin's fossil mammal specimens from South America." They included the huge extinct "Llama-like" mammal Darwin had found in Patagonia. Owen christened it *Macrauchenia*.[61]

On February 17, Charles Lyell presented Owen's work in his "presidential address to the Geographical Society." The great Scottish geologist made an intriguing comment: "Extinct species were related to existing species in the same locality."[62] If each species were created independently by God, why would He have one species go extinct and then follow with a related species in the same place?[63]

In March 1837, Darwin makes his first mention of the term "transmutation of species" in his private Red Notebook. Here he considers South American mammals, such as the ostrich and Llama.

> *When we see Avestruz [Ostrich] two species . . . one is urged to look to common parent? Why should two of the most clearly allied species occur in same country?*
>
> — Charles Darwin, The Red Notebook.[64]

Darwin conjectures that the two ostrich species he observed in South America — the greater and lesser rhea — are descended from a *common* parent species. He also speculates that the extinct *Macrauchenia* identified by Owen was the *ancestor* of the current South American Llama.[65]

In around mid-July,* (date uncertain) Darwin begins his private notebooks A and B — the former on geology and the latter his first notebook exclusive to "the transmutation of species." In it, Darwin begins to write on species change with greater confidence.

He records his first ideas on "branching descent" in Notebook B — as well as his first evolutionary Tree of Life. (See Fig. 28.4.)[66]

More Progress — 1838

In February of 1838, Darwin records his first thoughts that humans and animals have a common ancestry. "Man in his arrogance thinks himself a great work worthy the interposition of a deity," he writes in his private Notebook C, "More humble, and I believe truer, to consider him created from animals."[68]

In early Spring of 1838, Darwin observes the human-like behavior of an Orangutan named Jenny at the London Zoo. "One of the keepers was teasing her — showing her an apple, refusing to hand it over. Poor Jenny 'threw herself on her back, kicked & cried, precisely like a naughty child,'

---

*A month earlier, on June 20, 1837, Queen Victoria had ascended the British throne after the death of her uncle, King William IV. The Victorian Age had begun. She would rule for over 60 years until her death in 1901. Her reign would be marked by the "expansion of Britain's industrial power and the British empire." Source: "British History: Victorian Britain" bbc.co.uk. 2014. Retrieved Jan. 8, 2020. http://www.bbc.co.uk/history/british/timeline/victorianbritain_timeline_noflash.shtml.

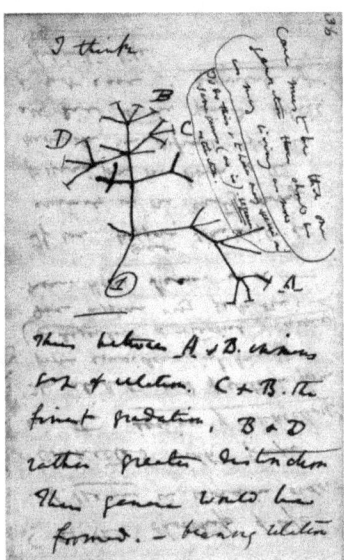

Figure 28.4. Darwin's Tree of Life. His first diagram of an evolutionary tree from his First Notebook on Transmutation of Species (1837). For interpretation of Darwin's handwriting, please see endnote.[67]

Darwin writes in a letter to his sister."[69] This deepens his conviction that man shares a common ancestry with apes.*

By June, Darwin refers to the study of species and varieties as "my prime hobby."[70]

In late September, he reads social economist Thomas Malthus's *Essay on the Principle of Population* (1798) "for entertainment." In the midst of the Industrial Revolution with its "increasingly crowded cities," Malthus argues that a geometric rise in human population outpaces the food supply. This inevitably leads to competition to survive.[71]

This gives Darwin a "new insight" for all living organisms:

"It at once struck me that under these circumstances favourable variations would tend to be preserved, and unfavourable ones to be destroyed," Darwin later wrote in his Autobiography. "The result of this would be the formation of new species. Here, then, I had at last got a theory by which to work . . ."[72]

---

*As we shall see, Darwin would not explicitly state this most controversial tenet publicly until he published *The Descent of Man* some thirty-three years later in 1871.

Delay after Delay

By December of 1838, Darwin felt he had established the core principles of his theory. What did he do next? He spent most of his time working on other things — for the next eighteen years!

He completed publication of seven geology papers (1837 to 1839). He edited and supervised the publication of *The Zoology of the Voyage of the Beagle* (1838 to 1843), a five-part description of his *Beagle* collection written by various experts. He published *Journal of Researches* (1839), later known as *The Voyage of the Beagle;* additional articles on geology (1841); a paper on mold formation (1840); and *The Structure and Distribution of Coral Reefs* (1842).

It is as though Darwin purposely avoided revealing his radical species theory to the public. It seems that, similar to Copernicus, he feared ridicule from his fellow scientists — as well as the wrath of the religious. And ever the perfectionist, he felt he needed more and more data to support his radical theory.

On a personal note, he married his cousin Emma Wedgwood, daughter of Uncle Josiah in January 1839. They would have ten children together. Three would die "in childhood and seven lived long lives."[73]

Perhaps another reason he hid his species theory — to avoid the reaction from his deeply religious wife.

Darwin was not the first to consider a theory on the mutability of species.[74]

Lamarck

In the early years of the 19th century, the great French naturalist Jean-Baptiste de Monet, chevalier de Lamarck proposed "the first broad theory of evolution." In Lamarck's paradigm, "organisms evolve through eons of time from lower to higher forms," Francisco Ayala writes, "culminating in human beings."[75]

According to Lamarck, traits *acquired* by organisms during their lifetimes are inherited by their offspring. In a classic example, giraffes stretch their necks to try and reach leaves higher and higher in trees. In the process, they develop *slightly* longer necks and forelegs in their lifetime. Their offspring then inherit these traits. This stretching continues from generation to generation. Over time, we observe the long necks and forelegs of today's giraffes. Although incorrect, Lamarck's theory opened up thinking on the mutability of species.

Darwin later recalled that while on a walk with Edmund Grant at Edinburgh, the professor "burst forth in high admiration of Lamarck and his views on evolution. I had previously read the *Zoönomia* of my grandfather, in which similar views are maintained, but without producing any effect on me."[76] Charles was well aware of the nasty reaction his grandfather Erasmus had provoked when he came out for Lamarckian evolution. This was yet another reason for his reluctance to publish.

In the 1840s — the same decade European archeologists had uncovered the first evidence of Sumerian civilization — Charles Darwin would reveal his species theory to a few close friends.

Opening Up?

In January 1842, Darwin sent a letter to Charles Lyell, now his dear friend, in which he outlined a "tentative description of his ideas" on species formation. He then drew up a thirty-five- page "Pencil Sketch" of his theory.[77]

The Essay of 1844

> *I was so anxious to avoid prejudice, that I determined not for some time to write even the briefest sketch of [my theory]. In June 1842, I first allowed myself the satisfaction of writing a very brief abstract of my theory in pencil in 35 pages; and this was enlarged during the summer of 1844 into one of 230 pages.*[78]

In 1844, Charles Darwin worked his "Sketch" into a comprehensive 230-page *Essay*. In his dissertation, he discussed a key insight regarding his species theory: the selective breeding of animals and plants by humans.

"The breeder has simply to watch for even the smallest approach to the desired end," Darwin wrote, "to select such individuals and pair them with the most suitable forms, and so continue with succeeding generations."[79]

In this way, with long patience man has selected "one breed of horses to race and another to pull . . . sheep with fleeces good for carpets and other sheep good for broadcloth . . . one dog to find game and . . . another to fetch him the game when killed."[80]

He called this selective breeding process *artificial selection*. "Why couldn't the same mechanism occur in a natural setting?" he asked himself. He would dub this process *natural selection*.

Darwin kept both his "Sketch" and *Essay* private. He left a note to Emma in his desk drawer to publish it if he should die unexpectedly.[81]

Some three years later, in January 1847 he sent a copy of his *Essay* to botanist and close friend Joseph Dalton Hooker. "I am almost convinced (quite contrary to opinion I started with) that species are not (it is like confessing a murder) immutable," Darwin had written to Hooker earlier.[82]

Still, Darwin continued to work mostly on other topics. He published *Geological Observations on the Volcanic Islands Visited during the Voyage of HMS Beagle* (1844); a second edition of *Journal of Researches* (1845); and *Geological Observations in South America* (1846).

He then began an eight-year project on the classification of barnacles — marine crustaceans which "attach themselves permanently to a variety" of underwater surfaces. Talk about wanting to avoid confrontation![83]

Darwin also suffered ill health for a long time. He was too sick to attend his father's funeral in 1848.[84] In addition, "Charles and Emma's first daughter, Anne Elizabeth died of tuberculosis" on April 23, 1851, writes Tim Berra, professor emeritus of evolution at Ohio State University. "She was ten years old. Charles never fully recovered from this tragic loss."[85]

Darwin had "the last of his barnacle monographs ready for publication" in September 1854. Now 45 years old, Charles Darwin at last turned his full attention to his species theory. He told his closest friends that he intended to "write a book on the subject."[86]

One of his first goals was to gather more evidence on domestication and artificial selection. "In 1855, he began developing a web of contacts, both in the UK and worldwide, to gather information on the origins and variation of domesticated animals — particularly poultry, ducks, rabbits, and pigeons."[87]

Darwin was now working full time on his species theory. Here he applied the same intense work ethic and thoroughness that he had demonstrated on the *Beagle* expedition.

The "Big Book"

Darwin explained his theory in detail to Charles Lyell during an April 13–16, 1856 visit by the geologist and his wife at Darwin's home at Downe. He also discussed his theory on April 26–27 at a house party at his home with friends J. D. Hooker, Thomas Huxley, and Thomas Vernon Wollaston.

In a letter of May 1, Lyell urged Charles to publish — though he had reservations on the mutability of species. Darwin responded on June 3[88]:

> *To give a fair sketch would be absolutely impossible . . . for every proposition requires such an array of facts . . .*

Darwin then added a prophetic statement:

I rather hate the idea of writing for priority, yet I certainly sh^d be vexed if any one were to publish my doctrines before me . . ."

It seems Lyell's urging had an effect. Darwin wrote in his Journal on May 14 that he had begun writing a "species sketch." By July, he had decided to produce a "full technical treatise" on his theory — a three-volume tome to be called *Natural Selection.*

He sent a letter to eminent American botanist and friend Asa Gray on July 20: "I have come to the heterodox conclusion that there are no such things as independently created species . . . I know that this will make you despise me," Darwin wrote. "I do not much underrate the many huge difficulties on this view, but yet it seems to me to explain too much, otherwise inexplicable, to be false."[89]

Darwin fell ill again the following March of 1857. Still, he continued to labor on his *Natural Selection* tome — his workdays now "ridiculously short."[90]

On September 5, Darwin sent a follow-up letter to Asa Gray — along with a 1200-word outline of his theory. It was "the briefest abstract of my notions on the **means** by which nature makes her species," he wrote. He asked Gray to "not to mention my doctrine." In other words, maintain secrecy.[91]

Scooped

As the calendar passed into 1858, all was going well. Darwin's "'big book' on *Natural Selection* was well under way. Then, on June 18, he received a letter from Alfred Russel Wallace in the Malay Archipelago. The Welsh naturalist had enclosed an attachment — a 20-page summary of Wallace's own theory on species modification.[92]

In the letter, he asked Darwin what he thought of his theory. He also asked him, if he "thought well of his essay," to forward it to Lyell for his reaction.[93]

A flummoxed Darwin dutifully did so, along with his own letter to Lyell. It stated:

"Your words have come true with a vengeance . . . if Wallace had my MS. sketch written out in 1842, he could not have made a better short abstract! . . . all my originality, whatever it may amount to, will be smashed."[94]

Wallace had come up with a theory on how new species come into existence which was nearly identical to Darwin's!

What to do? What to do? Wallace's dispatch would light a fire under Darwin as never before. With a new-found sense of urgency, he would come to produce his life's masterpiece — and engender the greatest conflict yet between science and religion.

That is subject of the next chapter.

Chapter 29

THE ORIGIN OF SPECIES

Enki made man from clay to serve the gods. YHWH made man in His own image. Jesus gave man a path to salvation. Mohammed instructed man to surrender to Allah.

Charles Darwin made man a descendent of ape-like creatures, animals without a soul, a product of random evolution.

Man the creator, man the destroyer, man the super-predator now invoking the sixth great species mass extinction — the Anthropocene.[1]

It was among the most trying times of Darwin's life. A seriously ill child. Deep anxiety over Wallace's letter. Pressed to make a decision to finally reveal his theory publicly.

What to Do About Wallace?

I have rather a more favourable opinion of the 'Vestiges' [speculative book on evolution] than you appear to have. I do not consider it a hasty generalization, but rather as an ingenious hypothesis strongly supported by some striking facts and analogies, but which remains to be proved by more facts and the additional light which more research may throw upon the problem.

— Alfred Russel Wallace, 1845.[2]

Alfred Russel Wallace
1823–1913

Alfred Russel Wallace was a thirty-five-year-old naturalist working in what is now Indonesia. Over the past few years, he had been sending domestic fowl specimens from Indonesia to England at Darwin's request, for the latter's research on domestication of species.[3]

Like Darwin, Wallace had read Lyell's *Principles of Geology*. He had also read Darwin's *Journal of Researches* (1845, second edition) "which had hinted at evolution."[4] The Welsh naturalist had published a paper in September of 1855 titled, "On the Law which has Regulated the Introduction of New Species."

In this essay, Wallace included ideas from Darwin, Lyell, and others "with his own observations." His conclusion: "every species has come into existence both in space and in time with closely allied species." Sound familiar?

Wallace's talk about "creation" as the mechanism for species change threw Darwin off.[5] In the margins of his copy of Wallace's paper, Darwin wrote: "Laws of Geograph. Distrib. nothing very new . . . Uses my simile of tree — It seems all creation with him . . ."[6]

The following month, Wallace wrote a letter to Darwin on his domestication work from Celebes (now Sulawesi) in Indonesia. It reached Darwin in May 1856. In his reply, Darwin commented on Wallace's essay. Polite and proper as usual, he wrote:

I can see that we have thought much alike & to a certain extent have come to similar conclusions. . . I am now preparing my work for publication . . . I do not suppose I shall go to press for two years . . . I have slowly adopted a distinct & tangible idea. — Whether true or false others must judge.[7]

Now in his June 18, 1858, letter, Wallace proposed a theory of evolution apparently indistinguishable from Darwin's.[*]

Darwin's Reaction

Early in the summer of 1858 Mr Wallace, who was then in the Malay archipelago, sent me an essay "On the Tendency of Varieties to depart indefinitely from the Original Type;" and this essay contained exactly the same theory as mine.

— Charles Darwin, Autobiography.[8]

Darwin was conflicted. He so wanted to be recognized as the originator of his species theory. Yet as always, he wanted to avoid controversy. His conscience was urging him to act the perfect gentleman and take the high road.

Finally, he made a decision. He told Lyell he would let Wallace have precedence on the theory. The great geologist wouldn't hear of it.

At the same time, Darwin's tenth and last child, 18-month-old Charles Waring, a Down Syndrome baby, had become critically ill with scarlet fever. The devoted father turned to his family and left the matter of precedence to his dear friends Lyell and Hooker.[9]

They concocted a compromise of sorts. Lyell and Hooker submitted a joint announcement of Darwin's and Wallace's theories to the Linnean Society of London. Included in the submission were (1) extracts of Darwin's unpublished *Essay* of 1844; (2) the outline of his theory sent in private to Asa Gray on September 5, 1857; and (3) Wallace's latest essay of 1858.[10]

The Darwin-Wallace papers were presented at the Meeting of the Linnean Society on July 1, 1858. They were the "first public announcement of the theory of evolution." What was the reaction to this historical event? Underwhelming. "Our joint productions excited very little attention," Darwin later wrote.[11]

Darwin did not attend the July 1 Linnean Society meeting. He was in mourning. His son Charles had passed away on June 28.

Wallace's response to the joint announcement would set the tone for the rest of his career as a naturalist.

---

[*] "Wallace's views differed from Darwin's in several ways," Ayala writes. "Most important, Wallace did not think natural selection sufficient to account for the origin of human beings. In his view, this required direct divine intervention." Source: Ayala, britannica.com.

The Forgotten Evolutionist

Wallace sent his reaction in a letter to Hooker in October 1858, some four months after the Linnean Society meeting. The Welch naturalist wrote that he was happy to have learned of the joint announcement. (He had not been told of it before its release.) He acknowledged Darwin as "first inventor" and recognized Darwin's much earlier and much more complete views on the subject. Wallace said that "it would have caused me much pain & regret if Mr. Darwin's excess of generosity led him to make public my paper unaccompanied by his own."[12]

Darwin wrote to Wallace in January 1859. He declared that he "had absolutely nothing whatever to do with leading Lyell and Hooker to what they thought was a fair course of action."

He added: "My abstract will make a small vol. of 400 or 500 pages ... Everyone ... thought your paper very well written & interesting. It puts my extracts (written [some] 20 years ago!), which ... were never for an instant intended for publication, in the shade."[13]

Wallace remained self-effacing and modest. He never contested priority and at least outwardly continued to give the credit to Darwin as first inventor. Unlike the Newton/Leibniz calculus affair, there was no ugliness, no squabbles between the two great men.[14]

Scientists and the public would generally come to remember Charles Darwin as the progenitor of the theory of evolution. Unlike Wallace, Darwin had been working on the theory of evolution for over two decades. He had "developed the theory in considerably more detail and provided far more evidence for it," Ayala writes.[15]

It would be Darwin who would later write a world-famous book on the subject. The humble Wallace would for the most part be forgotten. Even so, in the strictest sense it is the Darwin-Wallace theory of evolution.[16]

Let us now return to the summer of 1858. Recall, Darwin had been working on his "big book." And Wallace's essay *On the Tendency of Varieties* was now public. Lyell and Hooker urged Charles to write "a briefer account of his theory" poste-haste to establish priority.[17]

The "Small Volume"

> "In September 1858 I set to work by the strong advice of Lyell and Hooker to prepare a volume on the transmutation of species, but was often interrupted by ill-health ...

I abstracted the MS. begun on a much larger scale in 1856 [the three-volume "big book"], and completed the volume on the same reduced scale. It cost me thirteen months and ten days' hard labour."[18]

— Charles Darwin, *Autobiography*

The one-volume book would hardly be brief. Within a little over a year, Charles Darwin published a two-hundred-thousand-word tome. He titled it *On the Origin of Species by Means of Natural Selection*. It would be his life's masterpiece.

On the Origin of Species

There is grandeur in this view of life, with its several powers, having been originally breathed by the Creator into a few forms or into one; and that, whilst this planet has gone cycling on according to the fixed law of gravity, from so simple a beginning endless forms most beautiful and most wonderful have been, and are being, evolved.

— Charles Darwin, last paragraph of *the Origin of Species*[19]

The first edition of *Origin of Species* appeared on November 24, 1859 — some thirteen years after Galle and D'Arrest's discovery of the planet Neptune.[20] "It is no doubt the chief work of my life," Darwin later wrote, "It was from the first highly successful." (See Fig. 29.1.)[21]

"Twelve-hundred and fifty copies were sold on the day of publication," Darwin noted, "A second edition of 3000 copies sold soon afterwards." By the time of his autobiography in 1876, the book had been translated "into almost every European tongue," as well as Japanese. "Even an essay in Hebrew has appeared on it, [claiming] that the theory is contained in the Old Testament."[22]

I am all the more glad that you have published in this form — for the 3 vols [the "Big Book"] . . . would have choked any Naturalist of the 19th Century . . .

— J.D. Hooker, letter to Darwin of December 12, 1859.[23]

Darwin's masterwork is a "steady, not to say relentless, recounting of specifics," notes science writer and Pulitzer Prize nominee Timothy Ferris.

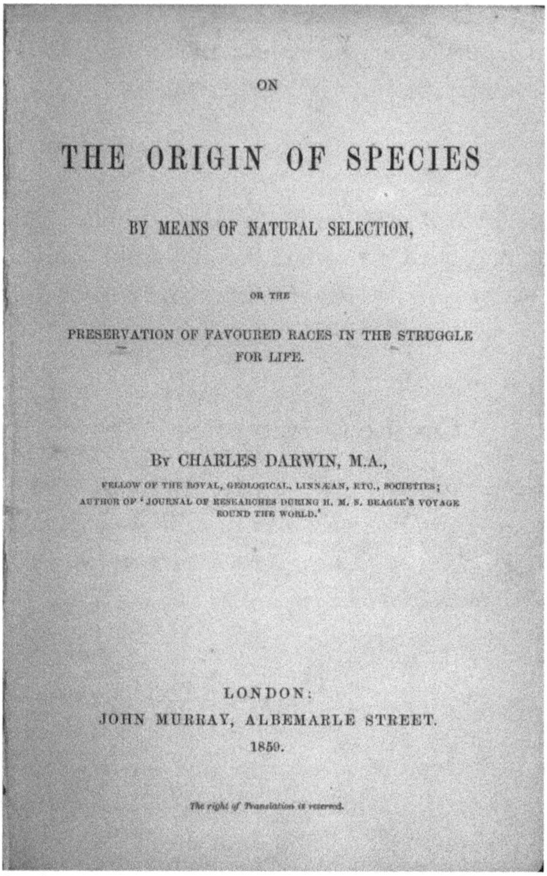

Figure 29.1. The Origin of Species by Means of Natural Selection. *Cover, First Edition, 1859.*[24]

It is "objective to the point of bloodlessness . . . a constant amassing of factual detail."[25] The book is at once exhaustive in detail and modest — so much so that its premise "struck many readers as self-evident."[26]

Darwin writes with humility and argues from reason. Where he feels evidence to support his thinking is lacking, he points it out. Aristotle would be proud.[27]

I found it a difficult read. It is over four hundred pages of dry, relentless facts supported by detail after detail (unlike the breezy and lighthearted book you are now reading.) I suspect that under this deluge of facts, a number of readers gave up before completing the book and conceded the victory to Darwin.

In *Origin of Species*, Darwin provides an exhaustive list of animals, plants, and other organisms as evidence for artificial selection. To give you some idea of his thoroughness, he discusses pigs, cows, goats, rabbits, wild rabbits, ferrets, and cats; hairless dogs, gray hounds, blood hounds, terriers, spaniels, Italian greyhounds, bull-dogs, pug dogs, Blenheim spaniels, English pointers, foxhounds, and Spanish pointers.

He writes about English carrier pigeons, short-faced tumbler pigeons, rock pigeons, Indian sub-species *C. intermedia* pigeons, and pouter-pigeon; carnivorous birds, common ducks, Strickland woodpecker, finches, game-cock, common goose, Toulouse goose, peacock, and wild Indian fowl; white sheep, ancon sheep, Leicester sheep, and merino sheep; long and short-horned cattle, Hereford cattle; racehorses, dray-horse, donkeys, dromedary, and camels; gall-producing insects and silk-worms; and more.

Of plants he includes exotic plants, cultivated plants, peach trees, common roses, wheat, orchard, cabbage, heartsease, wheat, barley, peas common gooseberry, cabbage, heartsease, pelargonium, dahlia, fuller's teasel, poppies, pear, and strawberry, and more.

And this was only in his first chapter, "Variation under Domestication."![28]

Timothy Ferris in his outstanding book, *The Coming of Age in the Milky Way*, outlines Darwin's theory in terms of three propositions: variation, overproduction, and natural selection.[29]

Let's take a look.

The Three Premises:

Variation—"Each individual member of any given species is different," Ferris writes, "Each, as we would say today, has a distinct genetic makeup."[30]

Consider a group of fruit flies or guppies or humans or any living organisms in a given location. Individuals within each group have different characteristics. Some fruit flies, for example, might have shorter wings or longer wings. Certain humans in a given group might have blue or brown eyes, be more fleet of foot, or more socially gregarious. All, even "identical" twins, are genetically unique in some way.[31]

Overproduction — A species produces "more offspring than the environment can support." Darwin's insight here came from Malthus' *An Essay on the Principle of Population*, as noted earlier.

All organisms "tend to increase in a geometric ratio whereas on an average the amount of food must remain constant," Darwin wrote, "[Therefore] there must be a severe struggle for existence . . ."[32]

"Excess production *by itself* is not a problem," University of Florida genetics professor emeritus Peter Luykx tells us, "as long as sufficient resources can be found (look at the expansion of the human population). But in most circumstances excess production of offspring will lead to competition among them."[33]

Natural Selection — In this "severe struggle for existence," favorable variations in a population "tend to be preserved, and unfavorable ones destroyed," Darwin wrote. This is "a natural means of selection."[34]

It tends "to preserve those individuals with any slight deviations of structure more favorable to the then existing conditions and tending to destroy any with deviations of an opposite nature."[35]

Darwin is telling us that differences among individuals along with environmental pressures affect the *probability* that an individual will survive long enough to pass along its characteristics to its offspring. As noted, Darwin labeled this process "natural selection," as opposed to the artificial selection — as in the breeding of dogs, pigeons, flowers, etc.[36]

From this, the great naturalist arrived at a momentous conclusion: the process of *natural selection* leads over time to the formation of new species.[37]

> *... Small differences distinguishing varieties of the same species steadily tend to increase till they equal the greater difference between species.*
>
> — Charles Darwin[38]

Here we see the influence of Charles Lyell. In his *Principles of Geology*, Lyell had argued that small geological changes "could be seen to add up over time" to produce the big changes we see today. Darwin had witnessed this again and again during his *Beagle* voyage explorations.[39]

> *I always feel as if my books came half out of Lyell's brains ...*
>
> — Charles Darwin[40]

What constitutes a new species? One which can no longer mate, or prefers not to mate, with others in a related group — or if they do mate, cannot produce fertile offspring. For example: if a female horse mates with a male donkey, they produce a mule which is typically sterile. Like the horse and donkey, the mule is a different species.[41]

Darwin summarized evolution by natural selection as follows:

> *As many more individuals of each species are born than can possibly survive; and as, consequently, there is a frequently recurring struggle for existence, it follows that any being, if it vary however slightly in any manner profitable to itself, under the complex and sometimes varying conditions of life, will have a better chance of surviving, and thus be **naturally selected**. From the strong principle of inheritance, any selected variety will tend to propagate its new and modified form [eventually into a new species].*[42]
>
> — Charles Darwin, introduction to *Origin of Species*

The process of species evolution has deep implications. Perhaps the most profound is that all life on our planet evolved from a single source.

Common Descent

"I believe that animals have descended from at most only four or five progenitors and plants from an equal or lesser number," Darwin wrote in *Origin of Species*, "Analogy would lead me one step further, namely, to the belief that all animals and plants have descended from some, one prototype . . . probably all the organic beings which have ever lived on this earth have descended from some, one primordial form, into which life was first breathed by the Creator."[43]

Darwin gave numerous cases of evidence for common descent. An example:

> *[The Rhinoceros], their short necks . . . contain the same number of vertebrae with the giraffe; their thick legs . . . [are] built on the same plan with those of the antelope, of the mouse, of the hand of the monkey, of the wing of the bat, and of the fin of the porpoise.*
>
> *[In] each of these species the second bone of their leg . . . show[s] clear traces of two bones having been soldered and united into one . . . the complicated bones of their head [have] been formed of three expanded vertebrae . . . in the jaws of each when dissected young there [are] small teeth which never come to the surface . . . In possessing these useless abortive teeth, and in other characters, these*

three rhinoceroses in their embryonic state . . . much more closely resemble other mammalia than they do when mature. And lastly, that in a still earlier period of life, their arteries . . . run and branch as in a fish, to carry the blood to gills which do not exist.[44]

In summary, "species are related to one another through a common ancestor, not created separately," Peter Luykx writes.

Darwin sent presentation copies of his book to friends and former professors, along with personal notes. To his mentor John Henslow, he wrote "I fear you will not approve of your pupil." To fossil decipherer Richard Owen, "It will seem an abomination." To Asa Gray, ". . . there are very many difficulties." He also sent a copy to Alfred Russel Wallace with the comment: "God knows what the public will think."[45]

Response to *Origin of Species* was a classic example of confirmation bias. Those who held conservative religious beliefs generally saw Darwin's book as negative. Those with liberal views saw it as confirmation of previously held opinions.

Reviews of *Origin of Species*

What a book a Devil's chaplain might write on the clumsy, wasteful, blundering low & horridly cruel works of nature!

— *Charles Darwin*[46]

The apparent cold and indifferent power of natural selection and its notion of "survival of the fittest" struck many believers for its absence of ethics and compassion in nature. To some, it seemed that evolution attacked morality itself.[47]

Unlike the orderly laws of nature discovered by Kepler, Galileo, and Newton, and the determinism of Laplace, Darwinian evolution was at its core a random process. It seemed to describe a progression of changing organisms with no designer or apparent purpose except survival and reproduction.

The Religious

University professors at the time were required to be Anglican clergy and received their funding from the Church of England. Whether from religious or financial motives, they generally came out against Darwinian evolution.

Revd. Richard Owen called Darwin's theory "a limited or inadequate view and treatment of the great problem." He labeled it an "abuse of science." Now head of the natural history department of the British Museum, he had received numerous complaints regarding Darwin's radical book.[48]

Revd. Adam Sedgwick had taken Darwin on his first geology field trip. He told Darwin: "I have read your book with more pain than pleasure." Protestant theologian Charles Hodge wrote that "the denial of design in nature is actually the denial of God."[49]

John Herschel wrote: "the principle of . . . natural selection . . . [is insufficient without] an intelligence guided by a purpose . . ."[50] Even former *Beagle* captain Robert FitzRoy chimed in. He was "very indignant with me," Darwin later wrote, "for having published so unorthodox a book (for he had become very religious)."[51]

Darwin's great friend Revd. John Henslow was somewhat kinder: "If it be faulty in its general conclusions, it is surely a stumble in the right direction." Talk about damning with faint praise.[52]

Though generally supportive, Christian biologist Asa Gray argued that "God himself is the very last, irreducible causal factor and, hence, the source of all evolutionary change."[53]

Charles Lyell called *Origin of Species* "a splendid case of close reasoning & long sustained argument throughout so many pages." Yet he struggled to "come to terms with the idea of mankind with immortal soul having originated from animals," as he put it.[54]

The Not So Religious

Botanist and religious skeptic Hewett Watson declared: "Your leading idea [of natural selection] will assuredly become recognized as an established truth in science."[55] Socialist theorist and atheist Harriet Martineau wrote:

> *What a book it is! — overthrowing (if true) revealed Religion on the one hand, & Natural (as far as Final Causes & Design are concerned) on the other.*[56]

Christian socialist country rector and novelist Charles Kingsley wrote to Darwin, "I shall prize your book."[57] Lady Aylesbury expressed indignation at Kingsley for "favouring such a heresy." He replied, "What can be more delightful to me Lady Aylesbury, than to know that your Ladyship & myself sprang from the same toad stool."[58]

The most vociferous defense of Darwin's theory came from Thomas Huxley. For this, the British press dubbed him "Darwin's Bulldog."

Huxley also broadcast that humans and apes have a common ancestor — a subject Darwin had deliberately avoided in *Origin of Species* for fear of the reaction.[59]

> *Apropos of the origin of man . . . it is as respectable to be modified monkey as modified dirt.*[*][60]
>
> — Thomas Huxley

The Descent of Man

Darwin would reveal his thoughts on human evolution to the public over a decade after the first publication of *Origin*. In *The Descent of Man and Selection in Relation to Sex* (1871), he wrote: "man is descended from a hairy quadruped, furnished with a tail and pointed ears . . ."[61]

Recall Laplace had argued that if God had created the solar system, He let it run in accordance with His Law of Gravity — without Divine interference. Darwin, still a believer at the time of *Origin's* first edition, proposed that God had created the first life form(s) and let it evolve according to His Law of Evolution by Natural Selection.[62]

Lampoons and Parodies

Ape and monkey parodies of evolution that had begun with the publication of *Origin of Species* accelerated with publication of the *Descent of Man*. In American magazine *Harper's Weekly*, a cartoon Gorilla says: "That Man [Darwin] wants to claim my Pedigree. He says he is one of my Descendants."[63] A ceramic "statuette of a monkey contemplating a human skull" circulated during the period. It would later find its way to the Kremlin desk of communist leader Vladimir Lenin.

London satire magazines *Figaro* and *Fun* joined in with their own Darwin/ape caricatures.[64] A cartoon in British weekly *Punch* declared: "apes are more intelligent than humans because they at least know when to keep silent." *The Hornet* depicted Darwin as a "Venerable Orang-outang: A contribution to Unnatural History." (See Fig. 29.2.)

---

[*] Huxley's quip on "modified dirt" is a reference to "Then the LORD God formed a man from the dust of the ground." (Genesis 2:7).

The Origin of Species 457

Figure 29.2. **"A Venerable Orang-outang."** Caricature of Charles Darwin as an ape published in *The Hornet*, a satirical magazine, on 22 March 1871. Author Unknown.[66]

Darwin's Evolving Religious Views

... disbelief crept over me at a very slow rate, but was at last complete ...

— Charles Darwin[65]

While on the *Beagle*, Charles Darwin remained "quite orthodox," as he put it. On his return to England, as he formulated his core tenets of the theory of evolution, he began to question his religious beliefs.[67]

> I had come ... to see that the Old Testament from its manifestly false history of the world" he later wrote, "... and from its attributing to God the feelings of a revengeful tyrant, was no more to be trusted than the sacred books of the Hindoos, or the beliefs of any barbarian."[68]

---

*Though quite liberal for a 19th century intellectual, Darwin still held views which offend today. They include his opinions on women, colonialism, religions other than Christianity, and cultures other than European.

> ... *the Gospels cannot be proved to have been written simultaneously with the events — they differ in many important details far too important as it seemed to me to be admitted as the usual inaccuracies of eye-witnesses... I gradually came to disbelieve in Christianity as a divine revelation.* (Though he called the morality espoused in the New Testament "beautiful.")[69]

Still, Darwin could not totally dismiss the existence of God. To wit:

> ... *the extreme difficulty or rather impossibility of conceiving this immense and wonderful universe, including man with his capacity of looking far backwards and far into futurity, as the result of blind chance or necessity...*
> *I feel compelled to look to a First Cause having an intelligent mind in some degree analogous to that of man... I deserve to be called a Theist.*[70]

Darwin seemed to hold this opinion during the writing of the first edition of *Origin of Species*. His religious views gradually became weaker:

> "... *can the mind of man,*" he asked himself, "*which has ... been developed from a mind as low as that possessed by the lowest animal, be trusted when it draws such grand conclusions? ... I for one must be content to remain an Agnostic.*"[71]

In 1873, some fourteen years after the first edition of *Origin of Species*, he wrote: "The old argument from design in Nature ... fails, now that the law of natural selection has been discovered.... There seems to be no more design in the variability of organic beings, and in the action of natural selection, than in the course which the wind blows." (See Fig. 29.3)[72]

Figure 29.3. **Charles Darwin.** Popular Science Monthly, February 1873.[74]

"I can indeed hardly see how anyone ought to wish Christianity to be true," Darwin wrote, "for the plain language of the text seems to show that the men who do not believe, and this would include my Father, Brother and almost all my best friends, will be everlastingly punished. And this is a damnable doctrine."[73]

Darwin's devoutly religious wife Emma, worried that her unbelieving husband would not go to heaven along with her. In a letter written shortly after their marriage, she wrote:

> ... *I should be most unhappy if I thought we did not belong to each other for ever.*

A note in Charles' handwriting can be found at the end of the letter:

> *"When I am dead, know that many times, I have kissed and cryed over this. C. D.*[75]

It seems to me that Emma need not have worried. Evolution is the very foundation of modern biology and medical science. Her husband's pursuit of the scientific truth, no matter the personal pain, has contributed in no small part to the saving of countless lives and the extension of human lifespans so many of us enjoy today.

If there is a benevolent God, if there is a Heaven, surely Charles Darwin, despite his existential doubts, is there amongst the greatly honored — side-by-side with his beloved Emma for eternity.

Passing

In his lifetime, Charles Darwin published six editions of *Origin of Species* and two editions of *The Descent of Man*.

His books also included *Voyage of the Beagle* (*Journal of Researches*), *Coral Reefs*, *Volcanic Islands*, *Geology of South America*, *Living and Fossil Cirripedia* (Barnacles), *Movement and Habits of Climbing Plants*, *Variation of Animals and Plants under Domestication*, *The Expression of the Emotions in Man and Animals*, *Insectivorous Plants*, *Effects of Cross Fertilization in the Vegetable Kingdom*, *Different Forms of Flowers and Plants of the Same Species*, *Power of Movement in Plants*, and *The Formation of Vegetable Mould through the Action of Worms*.[76]

The acclaimed English naturalist produced a number of these books in multiple editions. He also published his autobiography and wrote well over three-hundred articles for magazines and newspapers. Not bad for a former indolent college student.[77]

For his seminal accomplishments, Britain's Prime Minister Palmerston and Prince Albert pushed for Darwin to be named a knight. Ecclesiastic advisors to Queen Victoria quashed the notion, including Bishop of Oxford Samuel Wilberforce. Unlike Sir Isaac Newton, there would be no Sir Charles Darwin. His theory was just too controversial.[78]

The Demise

After a "three-month decline, his memory deteriorating," surrounded by Emma and their seven remaining children, Charles Darwin died of heart failure on April 19, 1882. He was seventy-three-years-old.[79]

Funeral ceremonies were held at Westminster Abbey a week later. It was attended by thousands. Included were over a hundred dignitaries — scientists, government ministers, ambassadors, and churchmen who had come to honor the great naturalist. Among the pall-bearers were J.D. Hooker, Thomas Huxley, and Alfred Russel Wallace.[80]

They choir sang from the Book of Proverbs[81]:

> *Happy is the man that findeth wisdom and getteth understanding. She is more precious than rubies, and all the things thou canst desire are not to be compared unto her... Her ways are ways of pleasantness, and all her paths are peace.*

With these words, Charles Robert Darwin, a man of pleasantness and peace, was laid to rest. He was interned in the north aisle of the nave of Westminster a few feet away from Isaac Newton.[82]

Acceptance of Darwinian Evolution

> *... that naturalists believe in the separate creation of each species was the general belief when the first edition [of Origin of Species in 1859] appeared... Now [sixth and last edition of 1872]... almost every naturalist admits the great principle of evolution.*
>
> — Charles Darwin[83]

Darwin's near complete conversion of naturalists to the general concept of evolution in a little over twelve years was remarkable. Still, *natural selection* as the driver of species change remained controversial.

A number of alternate theories became popular. They included:

(1) Theistic Evolution, where God "specifies the laws which govern the direction of evolution";
(2) Neo-Lamarckism, where "organisms inherit acquired characteristics";
(3) Orthogenesis, where organisms "have an innate tendency to evolve in a definite direction due to some internal mechanism or 'driving force,'" i.e., evolution is not random; and
(4) Saltationism, where rather than the accumulation of slow gradual changes, sudden large transformations drive evolution.[84]

Advances in genetics in the following century would turn the tide towards Darwin. Coupled with ever-mounting evidence, they would made Darwinian evolution "one of the best substantiated theories in the history of science."[85]

Yet religious objections continued to fester — reaching into the classroom itself. These are the subjects of the next chapter.

Chapter 30

THE LAW OF EVOLUTION

Nothing in biology makes sense except in the light of evolution.

— Theodosius Dobzhansky, prominent 20th century evolutionary biologist.[1]

Scientific breakthroughs since Darwin's day have transformed our understanding of how life forms have evolved. This knowledge has turned the theory of evolution into established science on par with quantum mechanics and relativity.

We begin with landmark discoveries in the field of inheritance.

The Mechanism Behind Evolution

The laws governing inheritance are for the most part unknown.

— Charles Darwin, The Origin of Species[2]

Charles Darwin was painfully aware of the missing pieces of his theory of evolution. He, and Wallace for that matter, conceived the theory of evolution without any idea how it worked at the micro level. What exactly produces "slight variations" in organisms over time? How are these changes in characteristics passed down from generation to generation?

The science of genetics, developed in the early to mid-20th century, gives us the answer.

Genetic Science

In 1866, Augustine monk Gregor Mendel (1822–1884) published his landmark theory of inheritance in Proceedings of the Natural Science Society of Brünn, Austria-Hungary (now Brno, Czech Republic). Mendel's findings were based on a series of meticulous cross-breeding experiments with pea plants "in the garden of his monastery in Brünn."[3]

In his eight-year study, Mendel artificially fertilized over 40,000 pea blossoms with hand-picked pollen over multiple generations. From this, he established the role of dominant and recessive characteristics in inheritance. His gene-based theory correctly explained for the first time how traits are handed down from one generation to the next.[4]

The unknown Mendel had published his paper in an obscure journal. It was generally ignored. So, he sent reprints to a number of leading biologists, botanists and naturalists — including Charles Darwin.[5]

Darwin was by then a world-famous naturalist who received a great deal of unsolicited mail. Perhaps this is why he never opened his copy of Mendel's paper.[6]

Darwin at the time was writing his "much revised" fourth edition of *Origin of Species* of 1866. He was also busy entertaining guests from Britain and abroad, exchanging letters with peers, corresponding with foreign publishers on translations of his books, as well as continuing his biological research.[7]

In addition, he was working on material for an upcoming book on variation. It was to include his admittedly speculative hypothesis on heredity — an incorrect theory he called Pangenesis.*[8]

While Darwin was grappling with his faulty ideas, the answer to his speculations lay unopened somewhere in his office.

Gregor Mendel passed away in 1884, knowing he had failed to convince anyone of his ground-breaking discovery. The laws of heredity would be "rediscovered" by European botanists Hugo de Vries, Carl Correns, and Erich Tschermak independently in 1900. Mendel's 1866 manuscript is now considered a founding paper of modern biology.[9]

Scientists in the 1920s and 1930s used Mendel's laws to verify the genetic basis of Darwinian evolution mathematically.[10] Experiments by Ukrainian-American geneticist Theodosius Dobzhansky further supported these theoretical findings. Dobzhansky linked natural selection and Mendelian genetics into a single model of evolutionary science in his 1937 book, *Genetics and the Origin of Species*.[11]

The next major breakthrough came in 1953 with the identification of the *structure* of the DNA molecule. And what a breakthrough it was!

DNA

Your children are not your children.
They are the sons and daughters of Life's longing for itself.

— Kahlil Gibran.[12]

Swiss chemist Friedrich Miescher had identified DNA (DeoxyriboNucleic Acid) in the late 1860s. The significance of his discovery was unappreciated at the time. In the following decades, biochemist Erwin Chargaff and others

---

* In Pangenesis theory, "each organ and tissue of an organism throws off tiny contributions of itself that are collected in the sex organs," Ayala writes. "They in turn determine the configuration of the offspring." Source: Ayala, britannica.com.

identified the "primary chemical components of the DNA molecule and how they joined with one another."[13]

The crucial breakthrough came in 1953. Supported by key "X-ray crystallography work" by British researchers Rosalind Franklin and Maurice Wilkins, American geneticist James Watson and British biophysicist Francis Crick determined "the molecular structure of DNA for the first time," Francisco Ayala tells us.[14]

DNA and the Genetic Code

DNA molecules contain a series of four organic chemical bases. They are adenine (A), guanine (G), cytosine (C), and thymine (T). These simple bases are organized in pairs in a "three-dimension double-helix" configuration. The order and frequency of the four bases set the genetic code for the cell, and ultimately the organism itself. See endnote for more details.[15]

The four bases and the DNA double helix configuration are shown in Figure 30.1. A simplified depiction of DNA replication is given in Figure 30.2.

We now know that DNA is found in all known life forms on our planet. The genes and chromosomes of every cell's nucleus contain this hereditary material. It explains how traits are encoded at the molecular level.[16]

A, T, G, and C. Four bases. Just four. Different sequences of these four bases make up the genetic code of all living things past and present.

Homeotic genes — The DNA in so-called "homeotic" genes are particularly amazing. The way the DNA is sequenced and how the genes are arranged in the chromosome give instructions on the timing and details of an organism's development.

Homeotic genes direct cells to form body parts in the embryo. They turn on and turn off certain other genes to determine the when and how of construction of our internal organs, our skeletal structure, brain, eyes, ears, nose, mouth, arms, legs, fingers, toes, and so on.

Science writer Mark Caldwell gives us a striking example:[17] "Consider your body's crowning glory, your head." The DNA in your homeotic genes "tell the bone cells in your skull to form themselves into a dome, your jaw into a trap-shaped mandible, the left side to arrange themselves in a mirror image of your right, to leave holes in just the right places for your eyes, to place your head at the top of your body." (Quote paraphrased.)[18]

DNA also programs the stages of our lives — from embryo to fetus to new-born baby to walking toddler to precocious child to risk-taking

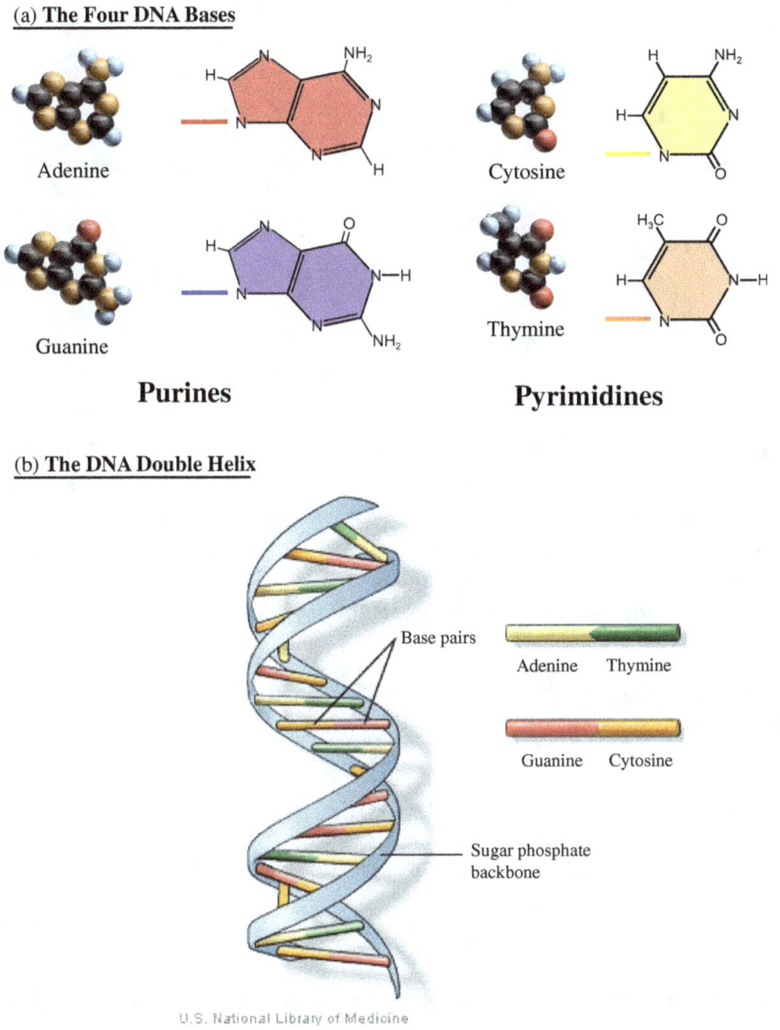

(a) **The Four DNA Bases**

Adenine

Cytosine

Guanine

Thymine

Purines　　　　**Pyrimidines**

(b) **The DNA Double Helix**

Base pairs

Adenine　Thymine

Guanine　Cytosine

Sugar phosphate backbone

U.S. National Library of Medicine

Figure 30.1. DNA (deoxyribonucleic acid). (a) The four DNA bases — a single DNA molecule contains millions of these. (b) The DNA Double Helix.

adolescence to independent young adult to mature adult to crabby senior citizen like me.[19]

The order and sequence of DNA is different depending on the organism. To accomplish all this for all life forms with the same four bases seems miraculous to me.[20]

This is all well and good. Still, it doesn't explain variation. How and why do organisms slowly *vary* in characteristics from generation to

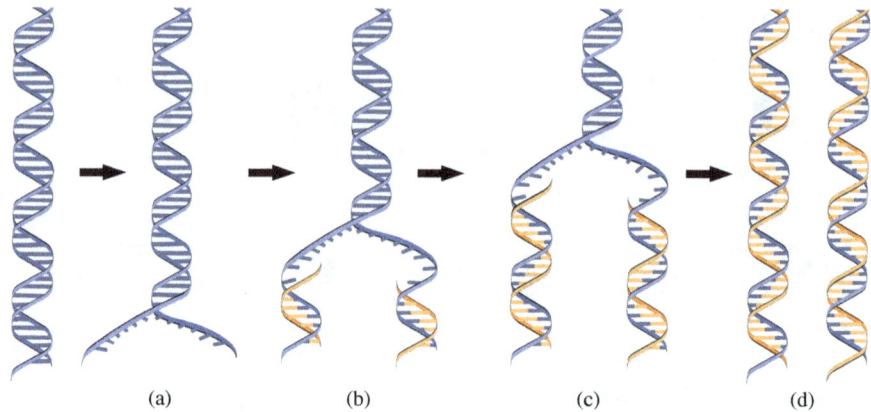

Figure 30.2. DNA replication: (a) DNA molecule unwinds. (b) Replication begins in both strands of molecule at same time. DNA bases on each strand bind to their complementary bases floating nearby in the cell nucleus — G with C and A with T. (c) Each new DNA strand pair made up of one old strand (shown in blue) and one new strand. (shown in orange). (d) Result: two new DNA molecules identical to the original one. Process occurs during cell division.[21]

generation? Darwin didn't know the mechanism for this core principle of evolution. But we do.

Mutation

Our ignorance of the laws of variation is profound.

— *Charles Darwin, The Origin of Species.*[22]

DNA molecules can and do change — in a process called mutation. It is the root cause of Darwin's "slight variations" in organisms over time. It is a key driver of evolution.[23]

Put simply, genes are segments of DNA and chromosomes are the structures that contain the genes. Mutations can affect "anywhere from a single DNA building block (base pair) to a large segment of a chromosome that includes multiple genes," writes David Finegold, Professor of Human Genetics at the University of Pittsburgh.[24]

What causes a mutation? It can occur as a DNA replication error when a cell divides to form two cells.[25] In this case, the most common mutation is a mis-pairing. A mutation can also be generated by exposure to ultraviolet light, chemical carcinogens, natural radiation such as in rocks or soil, drugs, and viruses.* In the simplest case, a single DNA base is substituted by *another* DNA base. An A can replace a G or C or T; a G can replace an A or C or T, etc.

No matter how they occur, these mutations change the genetic sequence of DNA itself.[26]

Acquired mutations — A mutation can happen in *any* cell in an organism. For humans, it could occur for example in a cell in your finger, or your knee tendon, or your stomach muscle. These are called *acquired* (or somatic) mutations and are not passed on to your offspring.[27]

Inheritable mutations — You were once a single cell. Every cell in your body today is the descendent of that single egg cell in your biological mother's womb fertilized by your biological father's sperm. Your every cell "contains essentially the same DNA."

If your father's sperm cell happens to contain a mutation, and the sperm cell fertilizes your mother's egg cell, the resultant fertilized egg cell will contain the mutation.[28]

Similarly, if your mother's unfertilized egg cell happens to contain a mutation, the resultant fertilized egg cell (fertilized by your father's sperm) will contain the mutation.

In either case, this is an *inheritable* (or germinal) mutation. As the fertilized egg cell divides again and again to form your growing embryo, the DNA mutation is replicated in each of your new cells. Your genetic code is altered. You have inherited** the mutation.[29]

How often does mutation occur? Mutation rates are usually very low. They are equivalent to "a single typographical error in Tolstoy's 1400-page *War and Peace*," Peter Luykx tells us. That is why variation and resultant diversity in organism populations is generally a very slow process — just as Darwin had predicted.[30]

Mutation also explains Darwinian *natural selection*. Organisms that happen to have heritable mutations that are harmful in a particular environment tend not to thrive and thus produce fewer offspring. Organisms with beneficial mutations — such as keener eyesight or "swifter legs" —

---

* Cosmic rays, mostly protons from the Sun, collide with molecules in our upper atmosphere and produce muons, a heavy version of electrons. A number of muons reach the surface of Earth. "About 10,000 muons pass through our bodies every minute. Some of these muons ionize molecules as they go through our flesh, occasionally leading to genetic mutations." Source: Schirber, Michael. "Death Rays from Space: How Bad Are They?" space.com. Aug. 27, 2009. "https://www.space.com/7193-death-rays-space-bad.html" Retrieved May 10, 2020.

** Some inheritable mutations are silent and result in no change to the resulting protein. Source: Walsh, Nicole. Review of draft Chapter 29. Email to Author, Oct 23, 2020.

tend to thrive and produce a greater number of offspring. Thus, over many generations, harmful mutations tend to be weeded out of a population and beneficial ones, "though initially rare, eventually become common."[31] Voila!

DNA and Us

We humans have about "20,000 genes and over 3.1 billion bases of DNA," writes Joel L. Carlin, biology professor at Gustavus Adolphus College.[32] Your DNA is 99.9% identical to my DNA and to every other human being on the planet today.[33]

> *We are more alike, my friends, than we are unalike.*
>
> — *Maya Angelou*[34]

DNA confirms Darwin's principle of common descent. We share about 98.8% of our DNA with chimpanzees, 85% with mice, 75% with chickens, and some 60% with banana trees. If that doesn't blow your mind, I don't know what will.[35]

Human DNA also confirms Darwin's theory regarding *human* descent. We are primates who "belong to the same biological group as great apes," evolutionary biologist Herman Pontzer tells us. Fossil evidence "along with studies of human and ape DNA" indicate that humans share a common ancestor with bonobos and chimpanzees.[36]

Our shared predecessor lived sometime around 6 to perhaps 8 million years ago.[37] For much of human evolution since, "more than one early human species has lived on Earth at the same time."

Humans are of the family *Hominid*, genus *Homo* and species *Homo sapiens*.[38] Others of our genus include *Homo habilis*, *Homo rudolfensis*, *Homo erectus*, and *Homo heidelbergensis*. In addition, there were archaic humans *Homo neanderthalensis* (c. 200,000 to perhaps 24,000 years ago); as well as "an early form of *Homo sapiens* called Cro-Magnon;" and the "enigmatic *Homo naledi*."[39]

Our modern human species originated in Africa some 200,000 to perhaps 300,000 years ago. All other *Homo* species are now extinct. Our species, *Homo sapiens*, is the lone survivor.[40]

So where's the proverbial missing link? Sorry, evolutionary scientists have not yet been able to establish a "complete chronological series of species leading to *Homo sapiens*." This from distinguished University of Chicago primate paleoanthropologist Russell Howard Tuttle. At least not a connection which experts can agree upon. Fossil relationships are just not that clear.[41]

We know that members of *Australopithecus* had ape-like and human-like traits. Living some 4.4 million to 1.4 million years ago, their fossils have been unearthed in southern, eastern, and north-central Africa. Paleontologists tell us they are "closely related to, if not actually ancestors of, modern human beings."[42]

Your Family Tree

If you take a DNA test, it may list the names of your second or third cousins or perhaps more. If you were somehow able to go back far enough in time, you would find that Apidima 1, Ur-Nammu, King David, Shalmaneser V, Siddhartha Gautama, Aristotle, Judas Maccabee, Cleopatra, John the Baptist, Claudius Ptolemy, Ts'ai Lun, Constantine, Hypatia, Muhammed, Ibn al-Haytham, Genghis Khan, Michelangelo, Amina Mohamud, Kepler, Galileo, Newton, Voltaire, Wollstonecraft, Tecumseh, Charles Darwin, Buffalo Calf Road Woman, Marie Curie, Subrahmanyam Chandrasekhar, Nelson Mandela, Shuji Nakamura, Marta Vieira da Silva, Adele, Howard Joel Wolowitz, Bozo the Clown, your next door neighbor, and all other humans who live and have ever lived on this planet are your 1st or 19th or 231st or perhaps 6221st cousins. Some may be way far out on your family tree but you are related to all of them — and to me. Hello cousin.

There is only one family tree.

Our current understanding of DNA and its role in genetic science is perhaps the most compelling evidence for evolution. All related scientific disciples also confirm Darwin's theory.

Modern Evidence for Evolution

To list the myriad aspects of evidence for evolution requires a book of its own.* The following is a brief summary:

In General

Today's scientific disciplines of paleontology, geology, genetics, biochemistry, comparative anatomy, physiology, philology, ecology, ethology, biogeography,

---

*May I commend you to Kenneth Miller's eminently readable book *Finding Darwin's God*. Miller presents a comprehensive and compelling compilation of modern evidence for evolution—and argues that the theory supports the existence of God.

and "especially molecular biology" all support and confirm evolution by natural selection.[43]

Evidence from these biological disciplines has "established the evolutionary origin of organisms with the kind of certainty attributable to such scientific concepts as the roundness of Earth, the motions of the planets, and the molecular composition of matter," writes Ayala.

The evidence includes:

Molecular Biology

In addition to DNA, molecular biology reveals the "working of organisms at the level of enzymes and other protein molecules," Ayala tells us. These molecules contain detailed "information about an organism's ancestry." Many hundreds of tests have confirmed without exception that all living entities "from bacteria to humans are related by descent from a common ancestor." Darwin is right again.[44]

All life forms, including humans "begin as single cells that reproduce themselves by similar division processes." All share the "ability to create complex molecules out of carbon and a few other elements." And all multicellular organisms "grow old and die."[45]

Feature Sharing

> *What can be more curious than that the hand of a man, formed for grasping, that of a mole for digging, the leg of the horse, the paddle of the porpoise, and the wing of the bat, should all be constructed on the same pattern, and should include the same bones, in the same relative positions?*
>
> — Charles Darwin, *The Origin of Species*.[46]

As Darwin noted, different species "share similar physical features." And "closely related groups tend to share features that were present in its last common ancestor."

A most striking example is the skeleton of the whale. "Its ancestors lived on the land, but they moved back to the sea." Like other mammals, it has a spinal column, rib cage, and forelimb bones with five digits like us![47] (See Figure 30.3.)

Other evidence for evolution includes:

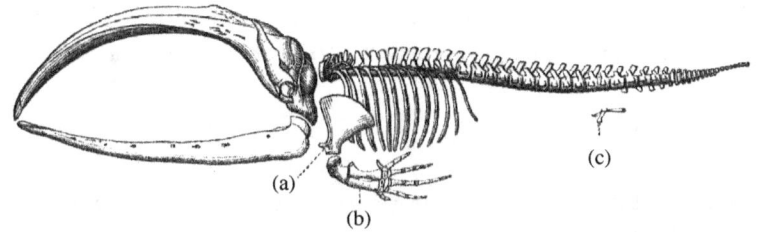

Figure 30.3. Baleen Whale Skeleton. (a) Omoplate; (b) Foreleg; (c) rest of the Hind Leg. Note the five digits on the "hand," and the spine and rib cage like other mammals, including us.

The Fossil Record

Though still incomplete, an extensive number of discoveries since Darwin's day have added significantly to the fossil record. They consistently show a connection between extinct species and present-day species over a broad range of organisms — including humans and our earlier primate ancestors, as we have seen.[48]

Resistance

Life-forms with very short lifecycles enable biologists to observe evolutionary changes over brief time periods. Examples: resistance to antibiotics in subsequent generations of bacteria, and resistance to pesticides in subsequent generations of insects.[49]

Embryos

Darwin wrote in his *Essay* of 1844: "The unity of type is wonderfully manifested by the similarity of structure during the embryonic period in the species of entire classes." For instance, "the different [varieties] of our domestic animals differ less, during their young state, than when full grown."[50]

Again, Darwin was right. For example, embryos of all vertebrates — animals with a backbone (spinal cord)* — have "gill slits and a tail during

---

* "The major groups of vertebrates include fishes, amphibians, reptiles, birds, and mammals." Source: Jollie, Malcolm T. "Vertebrate," *Encyclopedia Britannica*, britannica.com. Retrieved Jan 19, 2020. https://www.britannica.com/animal/vertebrate.

early development." This includes us humans. "Your embryonic tail is now your tailbone. Your gill slits have turned into your jaw and inner ear."[51]

Vestigial Organs

These are structural remnants of now lost functions from ancestor species. Human vestigial organs are useless or even disadvantageous features which we have inherited from our "hominid, ape or earlier ancestors."[52]

They include:

Our appendix, wisdom teeth, tailbones, goose bumps, and a group of useless ear muscles that monkeys use to pick up sound. (Some of us can use them to wiggle our ears, so not completely useless.) Sinuses and tonsils, and the Palmar grasp reflex in human infants (think earlier primates swinging in trees) may also be vestigial.[53]

Geology

Geological advances since Darwin's time have provided compelling evidence in support of the theory of evolution. They include the age of the Earth, the fossil record, continental drift, and mass extinctions. The latter two are major drivers of evolutionary change.

Please see Appendix E for more details.

Ongoing Evolution

Is evolution still happening? As in humans? Absolutely, says Laurence D. Hurst, Professor of Evolutionary Genetics at the University of Bath. Human DNA "shows evidence for recent selection for resistance of killer diseases like Lassa fever and malaria. . . [and] is still ongoing in regions where the disease remains common."[54]

"Mutations which allow humans to live at high altitudes have become more common in populations in Tibet, Ethiopia, and the Andes˚. . . . This rapid surge in frequency of a mutated gene that increases blood oxygen content gives locals a survival advantage in higher altitudes, resulting in more surviving children."[55] And, like all mutations, they are totally random. This is yet another example of natural selection.

"Although much remains obscure, and will long remain obscure," wrote Darwin in the introduction to *Origin of Species*. Below is one glaring example.[56]

The Origin of Life?

But if (and oh! what a big if!) we could conceive in some warm little pond, with all sorts of ammonia and phosphoric salts, light, heat, electricity, &c., present, that a protein compound was chemically formed ready to undergo still more complex changes . . .

— Charles Darwin, letter to J. D. Hooker, 1871.[57]

Evolution tells us that all life on this planet, past and present, stems from a single source. This utterly mind-boggling concept means that there is but one tree of life for all organisms.

Yet, just as astrophysicists do not yet know what caused the Big Bang, biologists do not know how life began on our planet. Nor do they know the exact characteristics of that original life form(s). As of this writing, it "remain completely unknown."[58]

In addition, scientists have not, as of this writing, been able to "detect the sudden 'moment' of evolution for any species," Russell Howard Tuttle tell us.[59]

Still, it is my view that the staggering totality of evidence for the theory of evolution is so compelling, so overwhelming, so detailed as well as broad in scope that it should be called the law of evolution.

Despite all the evidence, the theory of evolution is still seen as suspect in some religious circles.

Religious Conflict Today

It may surprise you to know that Darwinian evolution is generally accepted by the majority of today's Western religious authorities. Pope Pius XII acknowledged that "evolution is compatible with the Christian faith" in a 1950 encyclical. He also stated that the intervention of God "was necessary for the creation of the human soul." Similar views have been "expressed by other mainstream Christian denominations."[60] They include the United Presbyterian Church, and Lutheran World Federation — as well outside Christianity, such as the Central Conference of American Rabbis.[61]

Nonetheless, evolution remains controversial within some sects. This is especially true in the United States, where Christian fundamentalists hold a literal interpretation of Scriptures.[62]

During the 1920s, fundamentalists pushed twenty state legislatures to invoke anti-evolution statutes. Four of them — Arkansas, Mississippi,

Oklahoma, and Tennessee — passed laws which "prohibited the teaching of evolution in their public schools."[63]

The Tennessee statute would lead to the famous "Scopes Monkey Trial" — which arguably rivals Galileo's in notoriety.[64]

The July 1925 trial was held in Dayton, Tennessee. It was a media circus. Among the trial attendees was Vaudeville and Broadway performer Joe Mendi, a chimpanzee.[65]

The defendant was one John T. Scopes, a Dayton, Tennessee highschool teacher. His alleged crime: teaching evolution to his students.

Renowned criminal defense lawyer Clarence Darrow led the Scopes defense. After a stirring speech, he asked that his defendant be found guilty, as planned. The jury promptly did so. The judge fined Scopes $100 — the minimum under the statute.

Also as planned, the American Civil Liberties Union (ACLU) appealed. They hoped to challenge the constitutionality of the Tennessee law all the way to the Supreme Court, if necessary. Not to be outwitted, the Tennessee Supreme Court overturned the Scopes verdict on a technicality. Tennessee's anti-evolution law remained in force.[66]

Texas followed by banning "the theory of evolution from its textbooks."[67]

Tennessee's state legislature and governor repealed the anti-evolution law in 1967. The following year, the United States Supreme Court declared all "laws banning the teaching of evolution in public schools" unconstitutional.[68]

The separation between science and religion was now the law.

A Different Strategy

Christian fundamentalists then said, if we are not allowed to teach religion in public schools, let's relabel it as science. They introduced bills in a number of state legislatures which required so-called "creation science" to be taught alongside evolution.[69]

Creation science declares that all organisms "came into existence when God created the universe," Ayala writes. In addition, it teaches that "the world is only a few thousand years old, and the biblical Flood was an actual event where only one pair of each animal species survived" — ala Noah's Ark.[70]

Arkansas and Louisiana passed laws in the 1980's which required "the balanced treatment of evolution science and creation science" in public schools. Both laws were declared unconstitutional.[71]

In 2005, parents and teachers of Dover, Pennsylvania sued the school district for its promotion of "intelligent design" as an alternative to evolution in its classrooms. The Pennsylvania District Court agreed and "found that the district's policy impermissibly advanced religion."[72]

"Kentucky, Louisiana, Mississippi and Tennessee still have anti-evolution laws on the books," writes Ann Reid, executive director of the National Center for Science Education. According to a rigorous 2008 national survey, over "20 percent of public high school biology teachers reported pressure to downplay evolution." Most troubling, some sixty percent of teachers nationwide reported "downplaying evolution, covering it incompletely, or ignoring it all together."[73]

Believers, you need to understand that science is not your enemy. If you believe that God created the universe, then the scientific laws under which the universe operates are God's laws. The scientific evidence which confirms these laws are God's evidence.

Final Thoughts on Evolution

It seems evolution by *natural selection* is just a first step. At some point a species in the organism pool evolves to where it can influence and control the evolution of certain other species. Darwin called this second step evolution by *artificial selection*. As noted, it began thousands of years ago with we homo sapiens' selective breeding of cattle, horses, sheep, dogs, poultry, ducks, rabbits, pigeons, plants, etc.

The third step is when a species evolves to the point where it can influence and control the evolution of *its own species* — through prenatal manipulation of DNA. We are now making this somewhat troubling capability possible through a technology called genome or gene editing. It has the potential to eradicate birth defects — and to radically alter our species evolutionary path.[74]

Do you want your baby born with brown, blue, or green eyes? Violet is a popular option. Would you like black, brown, blond, or red hair? Hooked, flat, or long nose? Oval, oblong, square, round, or diamond facial shape? A stronger physique? Higher intelligence? Please see the complete list of options at mybabydna.com. Prices vary.[75]

The Last Chapter

In 1864, Charles Darwin received the Copley Medal of the Royal Society of London, its highest honor. That same year, James Clerk Maxwell (1831–1879) presented his landmark theory of electromagnetism to the Royal Society.[76]

The Scottish physicist's great work united electricity and magnetism into a single phenomenon. Out of it would come the first explanation in human history of what light is: an electromagnetic wave. His theory of electromagnetism has been called the crowning achievement of nineteenth century physics.[77]

Maxwell's breakthrough paved the way for the 20th century scientific revolution. This in turn helped usher in our modern secular worldview.[78]

This is the subject of the next and final brief chapter.

Chapter 31

THE SECULAR UNIVERSE

Science is a game we play with God to find out what his rules are.

— Cornelius Krasel[1]

We have come a long way since the days of Sumer. City-states have been replaced by nation-states. Mud brick ziggurats to the gods have given way to steel skyscrapers celebrating commerce and trade. Picture symbols in wet clay have surrendered to alphabets and characters on paper and electronic screens. The era of the moon god *Nanna* has transformed to the space age and humans on the Moon.

Priest-kings no longer rule. On much of our planet, religious authorities have no government role and little if any political power. Premarital sex, out-of-wedlock birth, pornography, and abortion are tolerated by many. Though not condoned by all, the openness by which they are practiced is indicative of the weakening of religious influence.

Still, belief in God continues and, in many places, thrives. Compassion for others remain at the root of all major world religions. Abrahamic ethical monotheism, the religions of Judaism, Christianity, Islam and others, is still alive. Over 55 percent of the world's population today identifies with one of these religions. And many still see science as a revelation of God's work.[2]

We live in an era of unprecedented prosperity and unprecedented destruction of nature. Both phenomena are due in no small part to the breathtaking progress in science and technology over the past two centuries.[3] The twentieth century in particular underwent a series of changes in science which were greater in scope than all preceding eras combined. At once strange and wonderful, these new theories further deepened the separation between science and religion.

Modern Physics

These scientific views end in awe and mystery, lost at the edge in uncertainty . . . they appear to be so deep and so impressive that the theory that it is all arranged simply as a stage for God to watch man's struggle for good and evil seems inadequate.

— Richard Feynman[4]

Initially developed in the early twentieth century, Quantum Mechanics is the most successful, most accurate, and most bizarre theory in the history of science. It describes the workings of nature at the microscopic scale. Uncertainty lies at the very heart of quantum theory — a feature of the physical world which no measurement can overcome. It tells us that nature at its most fundamental level is random. No one knows why.

Please see Appendix F for examples of quantum uncertainty — and how it applies to the randomness of mutation and variation in evolution.

Relativity

The great Albert Einstein produced two seminal theories of relativity, also in the early twentieth century. Like quantum mechanics, it challenges our fundamental understand of reality.

Special relativity describes a strange universe where time slows and space shrinks with motion.[5] The general theory of relativity reveals a cosmos where gravity slows time and stretches space.*

Countless observations, experiments, and tests over the past hundred years or so have validated virtually all predictions of quantum mechanics, special relativity, and general relativity. This tells us that the strange universe of modern physics is our universe.

The Big Bang

The Big Bang theory (Fig. 31.1) is widely accepted by physicists today as our best current explanation for the origin, evolution, and structure of the cosmos. A product of general relativity, quantum mechanics, and astronomical observations, it depicts a finite *beginning* of the universe. Like Genesis.

A number of independent measurements from vastly different observations agree with Big Bang theory predictions to a precision of 10% or better. This is most remarkable considering the measurement challenges for something which began some 13.8 billion years ago.[6]

To me, the Big Bang — a theory on the origin of the universe supported by empirical evidence — is the single greatest intellectual achievement of humankind to date.

Religious Implications

We must recognize how much the revelations of modern science have strayed from the simplistic views portrayed in ancient Scripture. None of the mind-boggling phenomena revealed by modern physics are found in the

---

*For a clear and entertaining explanation of special and general relativity for non-experts, again may I respectfully recommend *Einstein Relatively Simple: Our Universe Revealed in Everyday Language* by the distinguished and humble Ira Mark Egdall.

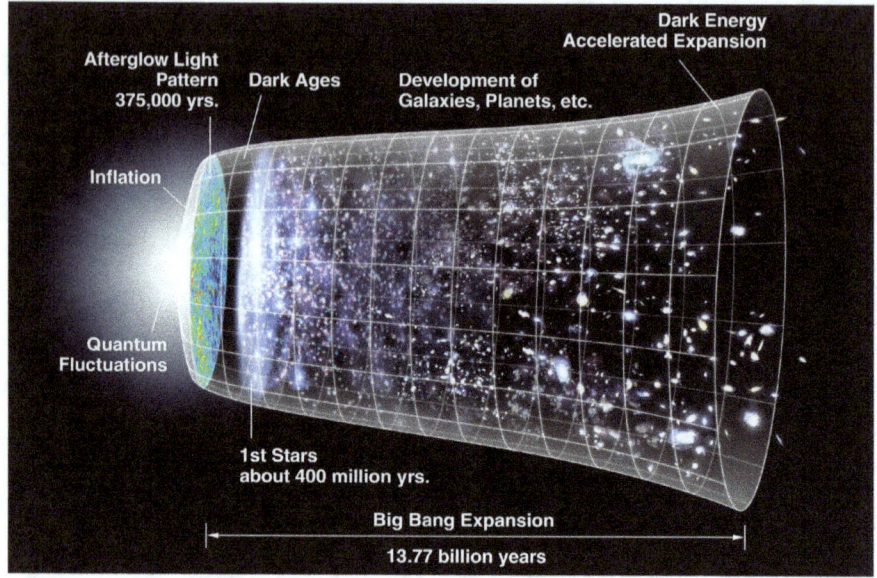

Figure 31.1. The Big Bang Theory. Our universe may have begun with some kind of quantum fluctuations. Standard Cosmology: It then expanded exponentially in a miniscule blink of time — a process called inflation. It continued to expand at a much slower rate. The Cosmic Microwave Background (Afterglow Light Pattern) appeared some 380 thousand years after the big bang. The first stars formed some 200 to 400 million years later. Observations indicate expansion is now accelerating. The unknown cause is dubbed dark energy (not to scale).[7]

Bible. They include quantum uncertainty, wave-particle duality, the relativity of time and space, black holes, the expansion of space, and much more. For many, "science has replaced religion as the source of answers to the deepest mysteries," as journalist John Hogan put it.[8]

Quantum theory confirms the inherent randomness of evolution. This challenges the notion of a purposeful God. Multiple observations and theoretical computations indicate our universe is nearly 14 billion years old. This clashes mightily with the fundamentalist Biblical interpretation that the universe began 6000 years ago.

Observations tell us there are several hundred billion stars in our Milky Way galaxy. The observable universe — the part of the universe we can see — contains some 500 billion to a trillion galaxies. Most astrophysicists believe the total universe is likely infinite in size. This further amplifies the Copernican worldview that our location in the cosmos is not special at all.[9]

Observations also tell us that exoplanets — planets which orbit stars other than our Sun — are ubiquitous. Since the first discoveries in the early 1990s, over four thousand have been found. Astrophysicists estimate there are some 100 thousand million exoplanets in our Milky Way galaxy alone. It seems just a matter of time before we find one with life. This again suggests we humans are not special.[10]

Separation

No one created the universe and no one can direct our fate.

— Stephen Hawking[11]

In summary, it wasn't one thing. It was a series of things — Copernicus' Earth as just another planet going round the Sun, Brahe's observations of a comet outside the sublunary sphere, Kepler's ellipses replacing Plato's divine circles, the telescopic discoveries of Galileo and his peers, the universal power of Newton's laws, the randomness at the heart of Darwinian evolution, the bizarre predictions of modern physics. The cumulative effect of these scientific discoveries has led in no small part to the secular world in which we now live.

In the second edition of the *Principia*, Newton wrote:

He [God] endures always and is present everywhere, and by existing always and everywhere he constitutes duration and space . . . He is omnipresent not only virtually but also substantially; for active power cannot subsist without substance.[12]

Can you imagine such a passage regarding the Divine in a modern scientific treatise? It has become standard practice in today's world for scientists to write papers with no mention of religion, whatever their personal beliefs.

It is not that scientists don't believe in God. Some do. Some are agnostics. Some atheists. It's just that the inclusion of one's religious views in scientific journals is culturally discouraged.

The high province of "objectivity" must be maintained. Science must gain its merit in the cold, analytical realm of experiment, observations, and test. It must speak only to those qualities that the universe chooses to reveal to us, and nothing more. This unspoken mandate has replaced, nay obliterated any mention of a link between the physical world and the Almighty.

The separation of science and religion is complete. The concept of God in science is dead.

Or is it?

Reunification?

The Standard Model [of Quantum Mechanics] . . . makes impressively accurately predictions of how particles interact and influence each other. But [it] can't explain the input . . . Why do the elementary particles have just the right properties to allow nuclear processes to happen, stars to light up, planets to form around stars, and on at least one planet, life to exist?

— Brian Greene[13]

Our universe . . . appears to be fantastically well <u>designed</u> for our own existence. This specialness is not something we can attribute to lucky accidents, which is far too unlikely. The apparent coincidences cry out for an explanation. (My underline.)

— Leonard Susskind[14]

There is a conversation going on amongst scientists and religious scholars. The argument from believers is compelling. The response by science is curious. The question: Is the universe fine-tuned for life?

Hidden in the core of modern physics are apparent "coincidences" necessary for our existence. They include parameters and features in the submicroscopic world of quantum mechanics as well as on the cosmic scale of the universe. The slightest variation in a number of them and the universe as we know it would not exist — including us.[15]

This supports the notion that our universe is "balanced on the knife-edge of coincidence."[16]

Astronomer Brandon Carter rationalized this with his "Anthropic Principle" in 1973. Put simply, it says our universe is "how it is because it must allow for eventual creation of us, as observers."[17] I find this a bit of circular logic.

To others, the fine-tuned universe implies intent. Purpose. Evidence for a Higher Power. Is science telling us our universe is, dare I say, designed? Or, as Stephen Hawking has proposed, do we live in just one of an immense number of parallel universes — one that happens to be just right for life?

We explore these conjectures, these mysteries, and their scientific and spiritual implications in my next book: *The Fine-Tuned Universe.*

Thank you for reading this book. I hope you enjoyed it. Now please relax and watch "The Big Bang Theory" reruns or something. I'd love to join you, but I have another book to write.

ACKNOWLEDGEMENTS

I want to thank my wife Pat for her invaluable assistance, loving support, and unwavering encouragement. She has served as first editor, reading each chapter again and again, checking for overall content and details. All this despite her ongoing health issues. You are my hero.

I'd also like to thank those who took time out from their busy lives to review drafts of this book. They are, in alphabetical order: Marsha Cohen, Paul Heckert, Peter Luykx, Darold Rorabacher, Joyce Schiffman, John Stevenson, and Nicole Walsh. Your insights and wise council have made it a better book. Special thanks to Darold Rorabacher for reviewing the entire book, twice. Once again, your critiques and recommendations were spot on. Also to Joyce Schiffman for your supportive and helpful review of the entire manuscript. I am indebted to Pastor John Stevenson for what had to have been a difficult read of the chapters on Christianity, given the agnostic bent of my writing. I also thank Nebil Husayn for your guidance on Mohammed and the Nestorians. Any remaining issues are, of course, due to my deficiencies alone.

Trying to tell the story of some 5000 years of human history, with its complex interactions and motivations was a daunting task. I faced a vast area of subject-matter, much of it well beyond my individual knowledge. I am indebted to scholars and educators who have written books, academic papers, popular expositions, web articles, and the like on the history of science, history of religion, and science itself, particularly cosmology and astronomy. I would, in particular, like to thank, those listed in the Bibliography as major sources. Special thanks to Robert A. Hatch for steering me to the blogs of Thony Christie.

I'd also like to extend my appreciation to the late Julia Rose Caruso as well as Magda Vergara at the University of Miami and Linda Maurice at Nova Southeastern University and their staffs for giving me the opportunity to give lectures on the conflict between science and religion. Presenting the works to students at their respective Osher Lifelong Learning Institutes has been invaluable and enjoyable. Student feedback and enthusiasm have encouraged me and also helped make this a better book. I want to again acknowledge the greatest library in the world: the World Wide Web. Used judiciously, it has been a great source of material literally at my fingertips.

I would finally like to thank Rochelle Kronzek, Executive Editor at World Scientific Publishing for your expert guidance and unwavering patience. Your support and collaboration have been invaluable. I am grateful once again to editor Rhaimie Wahap for your dedication and attention to detail in putting the book in its final form. I am also indebted to Phua Kok Khoo, Chairman and Editor-in-Chief of World Scientific Publishing, for agreeing with Ms. Kronzek's recommendation to once again publish a book of mine. No words can express the thrill I feel for this honor.

ABOUT THE AUTHOR

IRA MARK EGDALL

Ira Mark Egdall is the award-winning author of *Einstein Relativity Simple: Our Universe Explained in Everyday Language*. He is a retired aerospace program manager with an undergraduate degree in physics from Northeastern University. Mark now teaches lay courses in modern physics as well as the history of the science /religion conflict at Lifelong Learning Institutes at the University of Miami, Nova Southeastern University, and Florida International University. He also gives entertaining talks on modern physics and Cosmic Roots.

www.iramarkegdall.com

ALSO BY IRA MARK EGDALL

Einstein Relatively Simple brings together for the first time an exceptionally clear explanation of both special and general relativity. It is for people who always wanted to understand Einstein's ideas but never thought they could.

Told with humor, enthusiasm, and rare clarity, this award-winning book reveals how a former high school drop-out revolutionized our understanding of space and time. From $E = mc^2$ and every day time travel to black holes and the big bang, *Einstein Relatively Simple* takes us all, regardless of our scientific backgrounds, on a mind-boggling journey through the depths of Einstein's universe. Along the way, we track the Einstein through the perils and triumphs of his life — follow his thinking, his logic, and his insights — and chronicle the audacity, imagination, and sheer genius of the man recognized as the greatest scientist of the modern era.

In Part I on special relativity, we learn how time slows and space shrinks with motion, how mass and energy are equivalent. Part II on general relativity reveals a cosmos where black holes trap light and stop time, where wormholes form gravitational time machines, where space itself is continually expanding, and where over 13.8 billion years ago our universe was created in the ultimate cosmic event — the Big Bang.

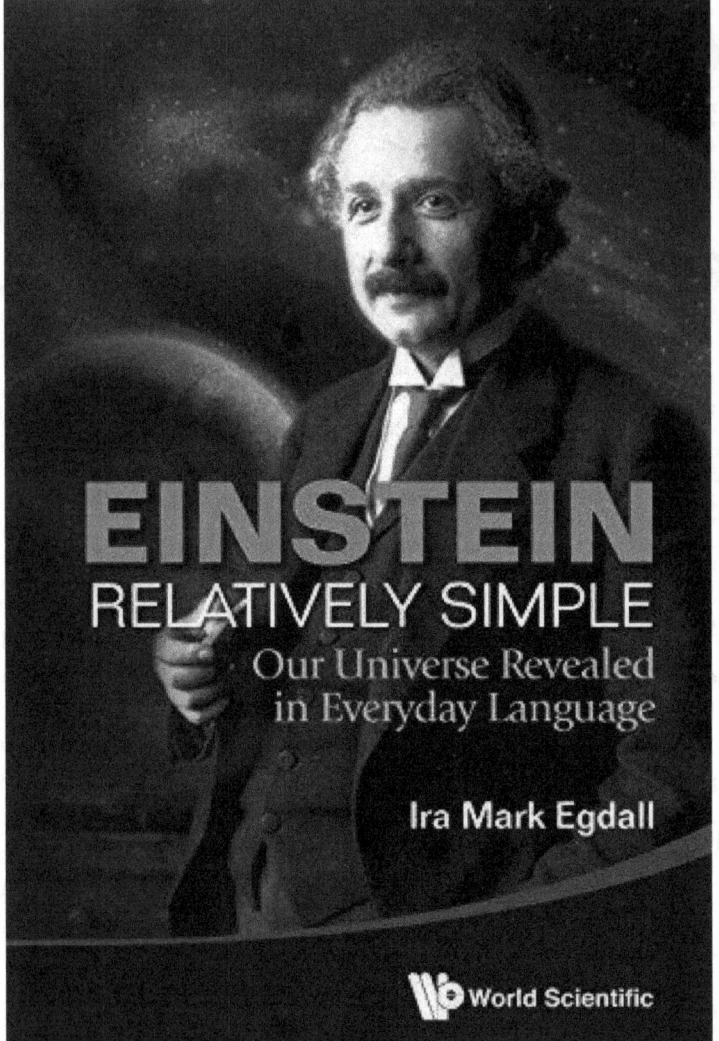

SUPPLEMENTARY MATERIAL

Supplemental material for Cosmic Roots includes:

Appendices
    Appendix A The Cosmic Model of Aristotle
    Appendix B Three Keys to the Scientific Revolution
    Appendix C The Cannonball and the Moon
    Appendix D Newton Beyond Science
    Appendix E Geology and Evolution
    Appendix F Quantum Uncertainty Examples
Endnotes
Bibliography
Figure Credits
Index

If you wish to access this material, please go to:
    iramarkegdall.com
    Books
    Cosmic Roots

Or go to:
https://www.worldscientific.com/worldscibooks/10.1142/12699#t=suppl

This green initiative saves paper, reduces fossil fuel burning in book shipments, and affords reader a reduced book price. Thank you for your participation.

www.ingramcontent.com/pod-product-compliance
Lightning Source LLC
Chambersburg PA
CBHW050523300426
44113CB00012B/1938